Boolean Differential Calculus

Synthesis Lectures on Digital Circuits and Systems

Editor
Mitchell A. Thornton, *Southern Methodist University*

The *Synthesis Lectures on Digital Circuits and Systems* series is comprised of 50- to 100-page books targeted for audience members with a wide-ranging background. The Lectures include topics that are of interest to students, professionals, and researchers in the area of design and analysis of digital circuits and systems. Each Lecture is self-contained and focuses on the background information required to understand the subject matter and practical case studies that illustrate applications. The format of a Lecture is structured such that each will be devoted to a specific topic in digital circuits and systems rather than a larger overview of several topics such as that found in a comprehensive handbook. The Lectures cover both well-established areas as well as newly developed or emerging material in digital circuits and systems design and analysis.

Boolean Differential Calculus
Bernd Steinbach and Christian Posthoff
2017

Embedded Systems Design with Texas Instruments MSP432 32-bit Processor
Dung Dang, Daniel J. Pack, and Steven F. Barrett
2016

Fundamentals of Electronics: Book 4 Oscillators and Advanced Electronics Topics
Thomas F. Schubert and Ernest M. Kim
2016

Fundamentals of Electronics: Book 3 Active Filters and Amplifier Frequency
Thomas F. Schubert and Ernest M. Kim
2016

Bad to the Bone: Crafting Electronic Systems with BeagleBone and BeagleBone Black, Second Edition
Steven F. Barrett and Jason Kridner
2015

Fundamentals of Electronics: Book 2 Amplifiers: Analysis and Design
Thomas F. Schubert and Ernest M. Kim
2015

Fundamentals of Electronics: Book 1 Electronic Devices and Circuit Applications
Thomas F. Schubert and Ernest M. Kim
2015

Applications of Zero-Suppressed Decision Diagrams
Tsutomu Sasao and Jon T. Butler
2014

Modeling Digital Switching Circuits with Linear Algebra
Mitchell A. Thornton
2014

Arduino Microcontroller Processing for Everyone! Third Edition
Steven F. Barrett
2013

Boolean Differential Equations
Bernd Steinbach and Christian Posthoff
2013

Bad to the Bone: Crafting Electronic Systems with BeagleBone and BeagleBone Black
Steven F. Barrett and Jason Kridner
2013

Introduction to Noise-Resilient Computing
S.N. Yanushkevich, S. Kasai, G. Tangim, A.H. Tran, T. Mohamed, and V.P. Shmerko
2013

Atmel AVR Microcontroller Primer: Programming and Interfacing, Second Edition
Steven F. Barrett and Daniel J. Pack
2012

Representation of Multiple-Valued Logic Functions
Radomir S. Stankovic, Jaakko T. Astola, and Claudio Moraga
2012

Arduino Microcontroller: Processing for Everyone! Second Edition
Steven F. Barrett
2012

Advanced Circuit Simulation Using Multisim Workbench
David Báez-López, Félix E. Guerrero-Castro, and Ofelia Delfina Cervantes-Villagómez
2012

Circuit Analysis with Multisim
David Báez-López and Félix E. Guerrero-Castro
2011

Microcontroller Programming and Interfacing Texas Instruments MSP430, Part I
Steven F. Barrett and Daniel J. Pack
2011

Microcontroller Programming and Interfacing Texas Instruments MSP430, Part II
Steven F. Barrett and Daniel J. Pack
2011

Pragmatic Electrical Engineering: Systems and Instruments
William Eccles
2011

Pragmatic Electrical Engineering: Fundamentals
William Eccles
2011

Introduction to Embedded Systems: Using ANSI C and the Arduino Development Environment
David J. Russell
2010

Arduino Microcontroller: Processing for Everyone! Part II
Steven F. Barrett
2010

Arduino Microcontroller Processing for Everyone! Part I
Steven F. Barrett
2010

Digital System Verification: A Combined Formal Methods and Simulation Framework
Lun Li and Mitchell A. Thornton
2010

Progress in Applications of Boolean Functions
Tsutomu Sasao and Jon T. Butler
2009

Embedded Systems Design with the Atmel AVR Microcontroller: Part II
Steven F. Barrett
2009

Embedded Systems Design with the Atmel AVR Microcontroller: Part I
Steven F. Barrett
2009

Embedded Systems Interfacing for Engineers using the Freescale HCS08 Microcontroller II: Digital and Analog Hardware Interfacing
Douglas H. Summerville
2009

Designing Asynchronous Circuits using NULL Convention Logic (NCL)
Scott C. Smith and JiaDi
2009

Embedded Systems Interfacing for Engineers using the Freescale HCS08 Microcontroller I: Assembly Language Programming
Douglas H.Summerville
2009

Developing Embedded Software using DaVinci & OMAP Technology
B.I. (Raj) Pawate
2009

Mismatch and Noise in Modern IC Processes
Andrew Marshall
2009

Asynchronous Sequential Machine Design and Analysis: A Comprehensive Development of the Design and Analysis of Clock-Independent State Machines and Systems
Richard F. Tinder
2009

An Introduction to Logic Circuit Testing
Parag K. Lala
2008

Pragmatic Power
William J. Eccles
2008

Multiple Valued Logic: Concepts and Representations
D. Michael Miller and Mitchell A. Thornton
2007

Finite State Machine Datapath Design, Optimization, and Implementation
Justin Davis and Robert Reese
2007

Atmel AVR Microcontroller Primer: Programming and Interfacing
Steven F. Barrett and Daniel J. Pack
2007

Pragmatic Logic
William J. Eccles
2007

PSpice for Filters and Transmission Lines
Paul Tobin
2007

PSpice for Digital Signal Processing
Paul Tobin
2007

PSpice for Analog Communications Engineering
Paul Tobin
2007

PSpice for Digital Communications Engineering
Paul Tobin
2007

PSpice for Circuit Theory and Electronic Devices
Paul Tobin
2007

Pragmatic Circuits: DC and Time Domain
William J. Eccles
2006

Pragmatic Circuits: Frequency Domain
William J. Eccles
2006

Pragmatic Circuits: Signals and Filters
William J. Eccles
2006

High-Speed Digital System Design
Justin Davis
2006

Introduction to Logic Synthesis using Verilog HDL
Robert B.Reese and Mitchell A.Thornton
2006

Microcontrollers Fundamentals for Engineers and Scientists
Steven F. Barrett and Daniel J. Pack
2006

Boolean Differential Calculus

Bernd Steinbach and Christian Posthoff

ISBN: 978-3-031-79891-7 paperback
ISBN:978-3-031-79892-4 ebook

DOI 10.1007/978-3-031-79892-4

A Publication in the Springer series
SYNTHESIS LECTURES ON DIGITAL CIRCUITS AND SYSTEMS

Lecture #52
Series Editor: Mitchell A. Thornton, *Southern Methodist University*
Series ISSN
Print 1932-3166 Electronic 1932-3174

Boolean Differential Calculus

Bernd Steinbach
Freiberg University of Mining and Technology, Germany

Christian Posthoff
The University of West Indies, Trinidad & Tobago

SYNTHESIS LECTURES ON DIGITAL CIRCUITS AND SYSTEMS #52

ABSTRACT

The Boolean Differential Calculus (BDC) is a very powerful theory that extends the basic concepts of Boolean Algebras significantly.

Its applications are based on Boolean spaces $\mathbb{B}$ and $\mathbb{B}^n$, Boolean operations, and basic structures such as Boolean Algebras and Boolean Rings, Boolean functions, Boolean equations, Boolean inequalities, incompletely specified Boolean functions, and Boolean lattices of Boolean functions. These basics, sometimes also called switching theory, are widely used in many modern information processing applications.

The BDC extends the known concepts and allows the consideration of *changes* of function values. Such changes can be explored for pairs of function values as well as for whole subspaces. The BDC defines a small number of *derivative* and *differential* operations. Many existing theorems are very welcome and allow new insights due to possible transformations of problems. The available operations of the BDC have been efficiently implemented in several software packages.

The common use of the basic concepts and the BDC opens a very wide field of applications. The roots of the BDC go back to the practical problem of testing digital circuits. The BDC deals with changes of signals which are very important in applications of the analysis and the synthesis of digital circuits. The comprehensive evaluation and utilization of properties of Boolean functions allow, for instance, to decompose Boolean functions very efficiently; this can be applied not only in circuit design, but also in data mining. Other examples for the use of the BDC are the detection of hazards or cryptography. The knowledge of the BDC gives the scientists and engineers an extended insight into Boolean problems leading to new applications, e.g., the use of Boolean lattices of Boolean functions.

KEYWORDS

Boolean Differential Calculus, derivative operation, differential operation, Boolean Algebra, Boolean Ring, Boolean function, Boolean equation, Boolean lattice, applications, XBOOLE

Contents

Introduction

Boolean (two-valued, dual, binary) systems have a long history and tradition in science and technology. In our daily life we normally meet the decimal number system that uses ten digits which get their meaning in a number based on powers of 10. The representation by means of dual numbers only needs two digits 0 and 1. Instead of the powers of 10 now powers of 2 are taken into consideration.

Arithmetic using dual numbers has been completely presented by G. W. Leibniz (*Explication de l' Arithmétique Binaire* (Histoire de l'Academie Royale des Sciences 1703), published in Paris 1705). In 1854, George Boole published a precedent-setting book which describes a system that formalizes the *laws of correct thinking* (Boole [1854, Reprint: 2010]). He was working with the two values *true* and *false* and created a propositional calculus which today is called *Boolean Algebra*. The next famous name is Claude Shannon. He proved that Boolean Algebra and dual arithmetic could be used to simplify the arrangement of the electromechanical relays that were used then in telephone call routing switches. Next, he expanded this concept proving that it would be possible to use arrangements of relays to solve problems in Boolean Algebra. Using this property of electrical switches to implement logic is the fundamental concept that underlies all electronic digital computers. Shannon's work became the foundation of digital circuit design. The restriction to only two digits and the utilization of the ideas of Boole and Shannon are the keys that Conrad Zuse was able to realize the first computer.

Since then a gigantic building of theory and applications has been created that requires careful and comprehensive studies. Switching theory deals with Boolean structures using the elements of $\mathbb{B} = \{0, 1\}$ and covers a wide field of applications. However, there are applications which require information about the effects of changing the values of variables and functions. One of them is the test of combinational circuits. Investigations of this topic by Reed [1954], Huffman [1958], and Akers [1959] can be seen as the roots of the *Boolean Differential Calculus* (BDC).

Basics of the BDC were published by A. Thayse and his group (Davio et al. [1978], Thayse [1981]). Comprehensive studies were carried out by a research group at the Chemnitz University of Technology where the authors are coming from. Results of this research were published in German in the monograph about the Boolean Differential Calculus by Bochmann and Posthoff [1981]. Chapters in Posthoff and Steinbach [2004] and Sasao and Butler [2010] summarize the main definitions of the BDC and present some selected applications. An extensive article about the BDC and its applications (Steinbach and Posthoff [2010a]) was published in the *Journal of Computational and Theoretical Nanoscience*.

The BDC was successfully applied in the last years to solve many problems. This success is not only based on the theoretical background of the BDC, but also on the efficient implementation of the derivative operations as part of the XBOOLE library as shown by Steinbach [1992]. The book Steinbach and Posthoff [2009] contains many examples and exercises where the BDC and the XBOOLE-Monitor (freely available over the Internet) were efficiently used. This book also provides the solution for all exercises and supplements the textbook Posthoff and Steinbach [2004].

The naming of several concepts of the BDC has been based on *formal analogies* with the *differential* and *integral calculus* for real numbers and functions. However, it should not be forgotten that we are dealing with finite algebraic structures without concepts like *limits* etc. Nevertheless, these analogies can be extended to create a *Boolean Integral Calculus*. The basics of this inverse calculus were already provided in the monograph Bochmann and Posthoff [1981]. The Boolean Integral Calculus and especially methods to solve Boolean differential equations have been presented in the Ph.D. thesis of Steinbach [1981]. The book Steinbach and Posthoff [2013a], published by Morgan & Claypool, explains these methods, their practical calculations using the XBOOLE-Monitor, and demonstrates many applications.

We also can consider the *possibility* of changing the values of variables and functions by using differentials of variables $\mathrm{d}x_i$ and functions $\mathrm{d}f_j$. Using such differentials we get an excellent possibility to describe graphs and model their properties.

The BDC basically deals with changing the values of Boolean functions. Extensions to multivalued variables and functions are given in Yanushkevich [1998], and in combination with binary concepts already in Bochmann and Posthoff [1981]. Some comments according to this approach are repeated in Bochmann [2008].

Replacing a single Boolean function by a lattice of such functions is an important source for finding optimal solutions, especially in circuit design. Therefore, derivative operations of lattices become a new problem. A recent research result of Steinbach and Posthoff [2013b, 2015] and Steinbach [2013] is the extension of the BDC to lattices of Boolean functions. These extensions and promising applications are included into this presentation of the Boolean Differential Calculus.

It is our aim to present the theory of the BDC and many applications in a well understandable manner. The classification of the single derivatives as a special case of the vectorial derivatives contributes to this aim. The reader should always try to understand the concepts and do as many examples as possible. To this end we added several exercises at the end of the chapters and give the associated solutions in the final chapter. We hope that this book helps the readers to solve their relevant problems more efficiently.

Bernd Steinbach and Christian Posthoff
December 2016

CHAPTER 1

Basics of Boolean Structures

1.1 LATTICES AND FUNCTIONS

We use the set $\mathbb{B} = \{0, 1\}$ with two different elements 0 and 1 as the initial point. An order relation $\leq$ can be defined as follows:

$$0 \leq 0 , \quad 0 \leq 1 , \quad 1 \leq 1 .$$

This order relation allows the finding of the minimum and the maximum of any two elements:

$$\begin{array}{llll} \min(0,0) = 0 , & \min(0,1) = 0 , & \min(1,0) = 0 , & \min(1,1) = 1 , \\ \max(0,0) = 0 , & \max(0,1) = 1 , & \max(1,0) = 1 , & \max(1,1) = 1 . \end{array}$$

Instead of *minimum* and *maximum* the operation signs $\wedge$ and $\vee$ are used, together with the names *conjunction* and *disjunction* instead of minimum and maximum:

$$\min(x, y) = x \wedge y , \quad \max(x, y) = x \vee y .$$

This results in Table 1.1.

Table 1.1: Conjunction and disjunction

x_1	x_2	$x_1 \wedge x_2$	$x_1 \vee x_2$	$\min(x_1, x_2)$	$\max(x_1, x_2)$
0	0	0	0	0	0
0	1	0	1	0	1
1	0	0	1	0	1
1	1	1	1	1	1

It means that we can start with the order relation $\leq$ and define the conjunction and the disjunction as the minimum and the maximum of two values. It is, however, also possible to define these two operations $\wedge$ and $\vee$ by means of Table 1.1 and take the conjunction for the minimum and the disjunction for the maximum. This is justified because in each row the value $x_1 \wedge x_2$ is less than the value of $x_1 \vee x_2$. The two approaches are completely equivalent.

Now we formulate the following laws.

1. **Commutativity:**
$$x_1 \vee x_2 = x_2 \vee x_1 \; , \tag{1.1}$$
$$x_1 \wedge x_2 = x_2 \wedge x_1 \; , \tag{1.2}$$

2. **Associativity:**
$$x_1 \vee (x_2 \vee x_3) = (x_1 \vee x_2) \vee x_3 = x_1 \vee x_2 \vee x_3 \; , \tag{1.3}$$
$$x_1 \wedge (x_2 \wedge x_3) = (x_1 \wedge x_2) \wedge x_3 = x_1 \wedge x_2 \wedge x_3 \; , \tag{1.4}$$

3. **Idempotence:**
$$x \vee x = x \; , \tag{1.5}$$
$$x \wedge x = x \; , \tag{1.6}$$

4. **Absorption:**
$$x_1 \vee (x_1 \wedge x_2) = x_1 \; , \tag{1.7}$$
$$x_1 \wedge (x_1 \vee x_2) = x_1 \; . \tag{1.8}$$

The absorption laws connect the two operations $\wedge$ and $\vee$. The other laws include only one operation.

Table 1.2 shows the proof of one of these laws; this *table method* is not difficult and very typical. We enumerate all the vectors of $\mathbb{B} \times \mathbb{B}$ and calculate the left and the right side of the equation. The equality of the first and the last column show that the equation is correct.

Table 1.2: Proof of the absorption law (1.7)

x_1	x_2	$x_1 \wedge x_2$	$x_1 \vee (x_1 \wedge x_2)$
0	0	0	**0**
0	1	0	**0**
1	0	0	**1**
1	1	1	**1**

These four laws define the algebraic structure of a *lattice*.

Now we extend this concept and build vectors with n components where each component will take one of these two values.

Definition 1.1 Boolean Space. For $\mathbb{B} = \{0, 1\}$,

$$\mathbb{B}^n = \{\mathbf{x} \,|\, \mathbf{x} = (x_1, x_2, \ldots, x_{n-1}, x_n),\; x_i \in \mathbb{B},\; i = 1, \ldots, n\} \tag{1.9}$$

is the set of all binary vectors of length n, the *Boolean space* $\mathbb{B}^n$.

It can also be understood as the *cross product*

$$\mathbb{B}^n = \underbrace{\mathbb{B} \times \ldots \times \mathbb{B}}_{n \text{ times}} = \{\mathbf{x} \,|\, \mathbf{x} = (x_1, \ldots, x_n),\; x_i \in \mathbb{B},\; i = 1, \ldots, n\} \; .$$

Here are two small examples:

$$\mathbb{B}^2 = \{(00), (01), (10), (11)\} ,$$
$$\mathbb{B}^4 = \{(0000), (0001), (0010), (0011), (0100), (0101), (0110), (0111), (1000), (1001), (1010), (1011), (1100), (1101), (1110), (1111)\} .$$

Since there are n components and each component can take on two values, $\mathbb{B}^n$ contains 2^n different elements (vectors).

For $\mathbb{B}^n$ we introduce a *partial order relation* in the following way: for $\mathbf{x} = (x_1, \ldots, x_n)$ and $\mathbf{y} = (y_1, \ldots, y_n)$ the $\leq$ of $\mathbb{B}$ has to be applied for each component:

$$\mathbf{x} \leq \mathbf{y} \quad \text{if and only if} \quad x_i \leq y_i \quad \text{for} \quad i = 1, \ldots, n .$$

The effect of this definition explains the name *partial order*: for some vectors we get $\mathbf{x} \leq \mathbf{y}$, for instance $(0010) \leq (1011)$, but for other vectors it can be possible that there we have neither $\mathbf{x} \leq \mathbf{y}$ nor $\mathbf{y} \leq \mathbf{x}$, for instance $\mathbf{x} = (1001)$, $\mathbf{y} = (0110)$. These vectors are not comparable.

Nevertheless, this relation allows again the definition of the *minimum and the maximum of two vectors* by applying the conjunction and the disjunction of $\mathbb{B}$ component by component:

$$\begin{aligned} \mathbf{x} &= (10011100) , \\ \mathbf{y} &= (00110101) , \\ \mathbf{x} \wedge \mathbf{y} &= (00010100) , \\ \mathbf{x} \vee \mathbf{y} &= (10111101) . \end{aligned}$$

The properties of a lattice are again satisfied, $(\mathbb{B}^n, \wedge, \vee)$ is also a *lattice*.

As we can see by this example, neither the minimum nor the maximum must be equal to one of the given vectors. When all the vectors of $\mathbb{B}^n$ are included, then the minimum is equal to the vector $\mathbf{0} = (0, 0, \ldots, 0)$, the maximum is equal to $\mathbf{1} = (1, 1, \ldots, 1)$, and these two vectors are also elements of $\mathbb{B}^n$.

Now a subset S of $\mathbb{B}^n$ is considered: $S \subset \mathbb{B}^n$. The conjunction of all elements of S is the minimum $\min(S)$ of S, the disjunction of all elements of S is the maximum $\max(S)$ of S. The vectors $\min(S)$ and $\max(S)$ can be an element of S, but not necessarily. If the conjunction and the disjunction of any two elements of $S \subset \mathbb{B}^n$ are again elements of S then S is a *sublattice* of $\mathbb{B}^n$.

For the representation of Boolean functions a second relation will be considered, an *order relation* where all the elements are comparable. The easiest way is the use of the decimal equivalent of a dual vector. When n is set to 3, for instance, then we get

$$(000), (001), (010), (011), (100), (101), (110), (111)$$

in this order, and this corresponds to the dual representation of the numbers $0, 1, 2, 3, 4, 5, 6, 7$. It is often named *lexicographic order*. Very often such vectors will be used as the argument vectors of Boolean functions, and it is possible to assign the variables x_1, x_2, x_3 to such vectors.

Now everything is prepared to define *Boolean functions.*

Definition 1.2 Boolean Function. Each unique mapping from $\mathbb{B}^n$ into $\mathbb{B}$ is a *Boolean function* of n variables.

Since $\mathbb{B}^n$ has 2^n elements, there are 2^{2^n} different functions of n variables. Important are the elementary functions for $n = 1$ and $n = 2$. We start with $n = 1$. Then $2^n = 2$ and $2^{2^n} = 4$. Therefore we have two values for the single variable and four different functions. Table 1.3 shows the four different functions of one variable x, $f_i(x)$ will be used for the respective function values:

$$f_0(x) = 0(x)\ , \quad f_1(x) = x\ , \quad f_2(x) = \overline{x}\ , \quad \text{and} \quad f_3(x) = 1(x)\ .$$

Table 1.3: Boolean functions of one variable

x	f_0	f_1	f_2	f_3
0	0	0	1	1
1	0	1	0	1

When we read the function values vertically, then we see the binary vectors of $\mathbb{B}^2$:

$$(f_i(0), f_i(1)) : \quad (00) \quad (01) \quad (10) \quad (11)\ .$$

The functions $0(x)$ and $1(x)$ are constant $= 0$ and $= 1$, resp., the variable x indicates that these functions depend on one variable.

The function $f_2(x)$ converts the two values into each other; it is called **negation** and indicated by a line above the variable, i.e., by $\overline{x}$:

$$\overline{0} = 1\ , \qquad \overline{1} = 0\ . \tag{1.10}$$

Literal can be used as a *common name* for variables and negated variables. The twofold use of the negation reproduces the original value:

$$\overline{\overline{x}} = x\ .$$

The functions of two variables can be seen in Table 1.4.

The set of all possible Boolean functions of two variables is equal to $\mathbb{B}^4$ and satisfies the axioms of a lattice.

There are two *constant* functions: $f_0(x) = 0(x_1, x_2)$ and $f_{15}(x_1, x_2) = 1(x_1, x_2)$. The function $f_1(x_1, x_2)$ is already known as **conjunction**:

$$f_1(x_1, x_2) = x_1 \wedge x_2\ . \tag{1.11}$$

Table 1.4: Boolean functions of two variables

x_1	x_2	f_0	f_1	f_2	f_3	f_4	f_5	f_6	f_7	f_8	f_9	f_{10}	f_{11}	f_{12}	f_{13}	f_{14}	f_{15}
0	0	0	0	0	0	0	0	0	0	1	1	1	1	1	1	1	1
0	1	0	0	0	0	1	1	1	1	0	0	0	0	1	1	1	1
1	0	0	0	1	1	0	0	1	1	0	0	1	1	0	0	1	1
1	1	0	1	0	1	0	1	0	1	0	1	0	1	0	1	0	1

Very often $\wedge$ is omitted, and we simply write $f_1(x_1, x_2) = x_1 x_2$. The main property of this function: it is only equal to 1 if *both* arguments are equal to 1 ($x_1 = x_2 = 1$). This has been seen already when the minimum for $\mathbb{B}$ has been defined.

We are also familiar with $f_7(x_1, x_2)$, the **disjunction**:

$$f_7(x_1, x_2) = x_1 \vee x_2 \; . \tag{1.12}$$

The important property of this function: $f_7(x_1, x_2) = x_1 \vee x_2 = 0$ only holds for $x_1 = 0$ *and* $x_2 = 0$ (remember the maximum for $\mathbb{B}$).

Two functions are very appropriate to indicate the *inequality* and the *equality* of the arguments (see Table 1.4) by their function values:

$$f_6(x_1, x_2) = x_1 \oplus x_2 \quad \textbf{(antivalence)} \tag{1.13}$$

and

$$f_9(x_1, x_2) = x_1 \odot x_2 \quad \textbf{(equivalence)} \; . \tag{1.14}$$

It can be seen that the vector of the antivalence (0110) can be transformed into the vector of the equivalence (1001) and vice versa by means of the negation:

$$\overline{x_1 \odot x_2} = x_1 \oplus x_2 \; , \qquad \overline{x_1 \oplus x_2} = x_1 \odot x_2 \; . \tag{1.15}$$

Two more functions are very often used in circuit design:

$$f_8(x_1, x_2) = \text{NOR}(x_1, x_2) = \overline{x_1 \vee x_2} \tag{1.16}$$

and

$$f_{14}(x_1, x_2) = \text{NAND}(x_1, x_2) = \overline{x_1 \wedge x_2} \; . \tag{1.17}$$

Each reader must ensure that he knows all these functions and can use them properly. It can be taken into consideration that the length of binary vectors representing a Boolean function always is a power of 2: $2^1 = 2,\ 2^2 = 4,\ 2^3 = 8,\ \ldots$

The main problem that must be addressed for many applications is the *exponential complexity*: for $n = 3$ there are already $2^{2^3} = 2^8 = 256$ functions of 3 variables, etc.

In order to cope with these problems formulas will be used. Each correct formula must be built by means of the following steps.

Definition 1.3 Construction of Formulas.

1. The constants 0 and 1 and the single variables $x_1, \ldots, x_n$ are formulas.
2. If F is a formula then $\overline{F}$ is also a formula.
3. If F_1 and F_2 are formulas then

$$(F_1 \wedge F_2)\,, \quad (F_1 \vee F_2)\,, \quad (F_1 \oplus F_2)\,, \quad \text{and} \quad (F_1 \odot F_2)$$

   are also formulas.
4. Each formula can be built when the second and the third rule are applied finitely many times.

Example 1.4 Construction of a Formula of Four Variables.

- We start with x_1, x_2, x_3 and x_4 .
- Thereafter, we build $F_1 = (x_1 \vee x_2)$ and $F_2 = (x_3 \oplus x_4)$.
- Next, we set $F_3 = \overline{F}_2$.
- Finally:

$$f(x_1, x_2, x_3, x_4) = (F_1 \wedge F_3) = (x_1 \vee x_2) \wedge \overline{(x_3 \oplus x_4)}\ .$$

The formula is a nice abbreviation, however, it does not really show the function values. In order to find these values we must go back to the 16 possible combinations of values and calculate each value of $F(x_1, x_2, x_3, x_4)$ by means of the respective parts of the formula. And this brings back the problem of very large tables; see Table 1.5 for the example.

Example 1.5 Lattice of Boolean Functions. The functions $f_1(x_1, x_2, x_3)$, $f_2(x_1, x_2, x_3)$, and $f_3(x_1, x_2, x_3)$ will be given as follows:

$$\begin{aligned}
f_1(x_1, x_2, x_3): &\quad (1\ \ 0\ \ 1\ \ 0\ \ 0\ \ 1\ \ 0\ \ 0)\,,\\
f_2(x_1, x_2, x_3): &\quad (1\ \ 0\ \ 1\ \ 0\ \ 1\ \ 1\ \ 0\ \ 0)\,,\\
f_3(x_1, x_2, x_3): &\quad (1\ \ 0\ \ 1\ \ 1\ \ 1\ \ 1\ \ 0\ \ 0)\,.
\end{aligned}$$

Table 1.5: The function $f = (x_1 \vee x_2) \wedge \overline{(x_3 \oplus x_4)}$

x_1	x_2	x_3	x_4	$x_1 \vee x_2$	$\overline{x_3 \oplus x_4}$	$f(x_1, x_2 x_3, x_4)$
0	0	0	0	0	1	0
0	0	0	1	0	0	0
0	0	1	0	0	0	0
0	0	1	1	0	1	0
0	1	0	0	1	1	1
0	1	0	1	1	0	0
0	1	1	0	1	0	0
0	1	1	1	1	1	1
1	0	0	0	1	1	1
1	0	0	1	1	0	0
1	0	1	0	1	0	0
1	0	1	1	1	1	1
1	1	0	0	1	1	1
1	1	0	1	1	0	0
1	1	1	0	1	0	0
1	1	1	1	1	1	1

It can be seen that $f_1 \leq f_2$, $f_1 \leq f_3$ and $f_2 \leq f_3$. This means that

$$\min(f_1, f_2, f_3) = f_1 , \quad \max(f_1, f_2, f_3) = f_3 .$$

The $\wedge$ and the $\vee$ of any of these three functions is again one of these three functions. The set $\{f_1, f_2, f_3\}$ is *closed* with regard to these two operations and therefore a lattice.

Another possibility is the following *ternary vector*:

$$(011 - 0 - 10) .$$

Sometimes it is also written as

$$(011\ \Phi\ 0\ \Phi\ 10) .$$

The dash (–) or the Φ can be arbitrarily replaced by 0 or 1, in this way we get four functions that satisfy the axioms of a lattice. In circuit design such a lattice is generated by *don't-care conditions*.

Now we split the algebraic background into two directions: *Boolean Algebras* and *Boolean Rings*.

1.2 BOOLEAN ALGEBRAS

We have already considered the four properties of a lattice, now we can add some more laws. The verification of these laws follows the same procedure as before, we enumerate all possibilities and check the left and the right side of the equations. The variables of the next formulas always take their values from $\mathbb{B}$, but all of them can be replaced by vectors when the operations are performed component by component. The 0 must be replaced by the vector $\mathbf{0} = (00\ldots0)$ and the 1 by $\mathbf{1} = (11\ldots1)$ when vector operations are computed:

1. **Distributivity:**
$$x_1 \vee (x_2 \wedge x_3) = (x_1 \vee x_2) \wedge (x_1 \vee x_3)\ , \tag{1.18}$$
$$x_1 \wedge (x_2 \vee x_3) = (x_1 \wedge x_2) \vee (x_1 \wedge x_3)\ , \tag{1.19}$$

2. **Neutral elements:**
$$0 \vee x = x\ , \tag{1.20}$$
$$1 \wedge x = x\ , \tag{1.21}$$

3. **Complement:**
$$x \vee \overline{x} = 1\ , \tag{1.22}$$
$$x \wedge \overline{x} = 0\ , \tag{1.23}$$

4. **De Morgan's Laws:**
$$\overline{x_1 \wedge x_2} = \overline{x}_1 \vee \overline{x}_2\ , \tag{1.24}$$
$$\overline{x_1 \vee x_2} = \overline{x}_1 \wedge \overline{x}_2\ . \tag{1.25}$$

Both these four laws and the four laws of a lattice (see page 4) remain true when the operations $\vee$ and $\wedge$ are exchanged. When these four laws are added to the laws of a lattice then all the axioms of a **Boolean Algebra** are satisfied. As a summary we get the following.

Theorem 1.6 Boolean Algebras. *The structures*

$$(\mathbb{B}, \wedge, \vee, ^{-}, 0, 1),\ (\mathbb{B}, \vee, \wedge, ^{-}, 1, 0),\ (\mathbb{B}^n, \wedge, \vee, ^{-}, \mathbf{0}, \mathbf{1}),\ \textit{and}\ (\mathbb{B}^n, \vee, \wedge, ^{-}, \mathbf{1}, \mathbf{0})$$

are Boolean Algebras.

It is possible that different formulas describe the same function. This is already indicated by the transformation rules from above. It will be later on a problem to find formulas that satisfy a given criterion.

Up until now we reached the following point:

- a Boolean function can be defined by a table; and
- we can find the function table when a formula has been given.

We still need a possibility to find a formula for a function which is given by a table. We go back to Table 1.5 and consider the lines with the function value 1 for

$$f = (x_1 \vee x_2) \wedge \overline{(x_3 \oplus x_4)}\ .$$

Table 1.6: The disjunctive normal form of f

x_1	x_2	x_3	x_4	$f(x_1, x_2, x_3, x_4)$	Conjunction
0	1	0	0	1	$\overline{x}_1\, x_2\, \overline{x}_3\, \overline{x}_4$
0	1	1	1	1	$\overline{x}_1\, x_2\, x_3\, x_4$
1	0	0	0	1	$x_1\, \overline{x}_2\, \overline{x}_3\, \overline{x}_4$
1	0	1	1	1	$x_1\, \overline{x}_2\, x_3\, x_4$
1	1	0	0	1	$x_1\, x_2\, \overline{x}_3\, \overline{x}_4$
1	1	1	1	1	$x_1\, x_2\, x_3\, x_4$

We use, for instance, the first row $(x_1x_2x_3x_4) = (0100)$ and assign a negated variable to the value 0 and a non-negated variable to the value 1. These (negated or non-negated) variables are combined by $\wedge$. For the given vector we get $\overline{x}_1x_2\overline{x}_3\overline{x}_4$. In this way the conjunction will result in the value 1 exactly for the given vector; for the other vectors of $\mathbb{B}^4$ the conjunction results in 0. This will be done for all the given lines of the table, the resulting six conjunctions are combined by $\vee$, the resulting disjunction of conjunctions is called the **disjunctive normal form**. It is *uniquely defined* except the order of the conjunctions. This, however, does not matter because of the commutativity of $\vee$. The other vectors which are not in the table result in $f = 0$. In this way we get a formula for the representation of a Boolean function which is given by Table 1.6:

$$\begin{aligned} f(x_1, x_2, x_3, x_4) = {} & \overline{x}_1x_2\overline{x}_3\overline{x}_4 \vee \overline{x}_1x_2x_3x_4 \vee x_1\overline{x}_2\overline{x}_3\overline{x}_4 \vee \\ & x_1\overline{x}_2x_3x_4 \vee x_1x_2\overline{x}_3\overline{x}_4 \vee x_1x_2x_3x_4 \, . \end{aligned} \tag{1.26}$$

Special denominations are sometimes necessary for the constant functions which are equal to 0 or 1 for all possible vectors $\mathbf{x}$. If it is necessary to show the variables that are important in a given context then we write $0(x_1, x_2, x_3, x_4)$ or $1(x_1, x_2, x_3, x_4)$; if the context is clear we might simply use **0** or **1** for these functions.

A second approach results in the **conjunctive normal form**. We select the binary vectors which are assigned to the function value 0 (see Tables 1.5 and 1.7). Here the non-negated variables are used for the value 0 in the vector, the negated variables are used when the component of the vector is equal to 1. The resulting literals are combined by $\vee$. Each disjunction is equal to 0 for the respective binary vector. The resulting disjunctions are combined by $\wedge$. Since the conjunction with 0 always results in 0, finally the conjunction of all these disjunctions produces all the values 0 of f. In this way we have built the **conjunctive normal form**. It is also *uniquely defined* except the order of the disjunctions:

$$\begin{aligned} f(x_1, x_2, x_3, x_4) = {} & (x_1 \vee x_2 \vee x_3 \vee x_4)(x_1 \vee x_2 \vee x_3 \vee \overline{x}_4)(x_1 \vee x_2 \vee \overline{x}_3 \vee x_4) \\ & (x_1 \vee x_2 \vee \overline{x}_3 \vee \overline{x}_4)(x_1 \vee \overline{x}_2 \vee x_3 \vee \overline{x}_4)(x_1 \vee \overline{x}_2 \vee \overline{x}_3 \vee x_4) \\ & (\overline{x}_1 \vee x_2 \vee x_3 \vee \overline{x}_4)(\overline{x}_1 \vee x_2 \vee \overline{x}_3 \vee x_4)(\overline{x}_1 \vee \overline{x}_2 \vee x_3 \vee \overline{x}_4) \\ & (\overline{x}_1 \vee \overline{x}_2 \vee \overline{x}_3 \vee x_4) \, . \end{aligned} \tag{1.27}$$

If there is a given Boolean function and we specify the value of *one* variable the result is a **subfunction**.

Table 1.7: The conjunctive normal form of f

x_1	x_2	x_3	x_4	$f(x_1, x_2, x_3, x_4)$	Disjunction
0	0	0	0	0	$(x_1 \vee x_2 \vee x_3 \vee x_4)$
0	0	0	1	0	$(x_1 \vee x_2 \vee x_3 \vee \overline{x}_4)$
0	0	1	0	0	$(x_1 \vee x_2 \vee \overline{x}_3 \vee x_4)$
0	0	1	1	0	$(x_1 \vee x_2 \vee \overline{x}_3 \vee \overline{x}_4)$
0	1	0	1	0	$(x_1 \vee \overline{x}_2 \vee x_3 \vee \overline{x}_4)$
0	1	1	0	0	$(x_1 \vee \overline{x}_2 \vee \overline{x}_3 \vee x_4)$
1	0	0	1	0	$(\overline{x}_1 \vee x_2 \vee x_3 \vee \overline{x}_4)$
1	0	1	0	0	$(\overline{x}_1 \vee x_2 \vee \overline{x}_3 \vee x_4)$
1	1	0	1	0	$(\overline{x}_1 \vee \overline{x}_2 \vee x_3 \vee \overline{x}_4)$
1	1	1	0	0	$(\overline{x}_1 \vee \overline{x}_2 \vee \overline{x}_3 \vee x_4)$

Definition 1.7 Cofactors. Let $f(x_1, x_2, \ldots, x_n) = f(x_i, \mathbf{x}_1)$ be a function of n variables. Then the *negative cofactor*

$$f^0_{x_i}(\mathbf{x}_1) = f(x_i = 0, \mathbf{x}_1) \tag{1.28}$$

and the *positive cofactor*

$$f^1_{x_i}(\mathbf{x}_1) = f(x_i = 1, \mathbf{x}_1) \tag{1.29}$$

are the two subfunctions with regard to x_i.

The resulting subfunctions do not depend on the respective variable anymore.

Example 1.8 Cofactors With Regard to x_1. We use the function (1.27) as an example and set $x_1 = 0$:

$$f_{x_1}^0(\mathbf{x}_1) = f(x_1 = 0, x_2, x_3, x_4) = (x_2 \vee x_3 \vee x_4)(x_2 \vee x_3 \vee \overline{x}_4)(x_2 \vee \overline{x}_3 \vee x_4)$$
$$(x_2 \vee \overline{x}_3 \vee \overline{x}_4)(\overline{x}_2 \vee x_3 \vee \overline{x}_4)(\overline{x}_2 \vee \overline{x}_3 \vee x_4) .$$

For $x_1 = 1$ the following function comes into existence:

$$f_{x_1}^1(\mathbf{x}_1) = f(x_1 = 1, x_2, x_3, x_4) = (x_2 \vee x_3 \vee \overline{x}_4)(x_2 \vee \overline{x}_3 \vee x_4)(\overline{x}_2 \vee x_3 \vee \overline{x}_4)$$
$$(\overline{x}_2 \vee \overline{x}_3 \vee x_4) .$$

It can be seen that some vectors of the function table are selected, all the vectors with a special value in the first component.

We go back to the disjunctive normal form (see (1.26)) and use the distributive law after the collection of the disjunctions of the normal form according to x_1:

$$f(x_1, x_2, x_3, x_4) = \overline{x}_1(x_2\overline{x}_3\overline{x}_4 \vee x_2x_3x_4) \vee$$
$$x_1(\overline{x}_2\overline{x}_3\overline{x}_4 \vee \overline{x}_2x_3x_4 \vee x_2\overline{x}_3\overline{x}_4 \vee x_2x_3x_4) . \tag{1.30}$$

This can be abbreviated by

$$f(x_1, x_2, x_3, x_4) = \overline{x}_1 f_{x_1}^0 \vee x_1 f_{x_1}^1 = \overline{x}_1 f_{x_1}^0 \oplus x_1 f_{x_1}^1 \tag{1.31}$$

and has been introduced by *C. Shannon*. $f_{x_1}^0$ and $f_{x_1}^1$ are the *two* cofactors from above. Each Boolean function can be decomposed using the following.

Theorem 1.9 Shannon Decomposition. *Each Boolean function $f(\mathbf{x}) = f(x_i, \mathbf{x}_1)$ can be represented by*

$$f(x_i, \mathbf{x}_1) = \overline{x}_i f(x_i = 0, \mathbf{x}_1) \vee x_i f(x_i = 1, \mathbf{x}_1) \tag{1.32}$$
$$= \overline{x}_i f(x_i = 0, \mathbf{x}_1) \oplus x_i f(x_i = 1, \mathbf{x}_1) \tag{1.33}$$

for any $x_i \in \mathbf{x}$, where $\mathbf{x}_1 = \mathbf{x} \setminus x_i$.

The cofactors $f_{x_1}^0(\mathbf{x}_1) = f_{x_1}^0(x_2, x_3, x_4)$ and $f_{x_1}^1(\mathbf{x}_1) = f_{x_1}^1(x_2, x_3, x_4)$ of (1.31) depend on x_2, x_3, and x_4. As a border case it can happen that $f_{x_1}^0(\mathbf{x}_1) = f_{x_1}^1(\mathbf{x}_1)$. This is an example for the *independence of a function of a variable* because we get step-by-step

$$\begin{aligned} f(x_1, x_2, x_3, x_4) &= \overline{x}_1 f_{x_1}^0(\mathbf{x}_1) \vee x_1 f_{x_1}^1(\mathbf{x}_1) \\ &= \overline{x}_1 f_{x_1}^0(\mathbf{x}_1) \vee x_1 f_{x_1}^0(\mathbf{x}_1) \\ &= (\overline{x}_1 \vee x_1) f_{x_1}^0(\mathbf{x}_1) \\ &= f_{x_1}^0(x_2, x_3, x_4) . \end{aligned} \tag{1.34}$$

This independence can also be seen when we go back to Table 1.4. Here we get for $f_3(x_1, x_2)$:

$$f_3(x_1, x_2) = x_1\overline{x}_2 \vee x_1x_2 = x_1(\overline{x}_2 \vee x_2) = x_1 \; .$$

We get a formula without x_2, nevertheless the function f_3 is a function of x_1 and x_2.

Functions also can be embedded into larger spaces using one additional variable. The reverse application of the Shannon decomposition realizes this expansion. We can assign the function $f(x_1, \ldots, x_n)$ of n variables to both the negative and the positive cofactors of a Shannon decomposition with regard to x_{n+1}

$$f(x_1, \ldots, x_n, x_{n+1} = 0) = f(x_1, \ldots, x_n) \; ,$$
$$f(x_1, \ldots, x_n, x_{n+1} = 1) = f(x_1, \ldots, x_n) \; ,$$

and get as expansion

$$f(x_1, \ldots, x_n, x_{n+1}) = \overline{x}_{n+1} f(x_1, \ldots, x_n) \vee x_{n+1} f(x_1, \ldots, x_n) \tag{1.35}$$

the function $f(x_1, \ldots, x_n, x_{n+1})$ of $n + 1$ variables.

1.3 BOOLEAN RINGS

The algebraic structure of a ring requires an *addition* and a *multiplication*. For each Boolean space $\mathbb{B}^n$ there are two Boolean Rings. For the first Boolean Ring, as the addition the antivalence $\oplus$ will be taken, the conjunction $\wedge$ is used as multiplication. The second (dual) Boolean Ring uses as addition the equivalence $\odot$ and the disjunctions $\vee$ as multiplication. The following axioms must be satisfied.

1. **Commutativity:**
$$x_1 \oplus x_2 = x_2 \oplus x_1 \; , \tag{1.36}$$
$$x_1 \odot x_2 = x_2 \odot x_1 \; , \tag{1.37}$$

2. **Associativity:**
$$x_1 \oplus (x_2 \oplus x_3) = (x_1 \oplus x_2) \oplus x_3 = x_1 \oplus x_2 \oplus x_3 \; , \tag{1.38}$$
$$x_1 \odot (x_2 \odot x_3) = (x_1 \odot x_2) \odot x_3 = x_1 \odot x_2 \odot x_3 \; , \tag{1.39}$$

3. **Zero element:**
$$x \oplus 0 = x \; , \tag{1.40}$$
$$x \odot 1 = x \; , \tag{1.41}$$

4. **Inverse element:**
$$x \oplus x = 0 \; , \tag{1.42}$$
$$x \odot x = 1 \; , \tag{1.43}$$

5. **Unit element:**
$$x \wedge 1 = x \; , \tag{1.44}$$
$$x \vee 0 = x \; , \tag{1.45}$$

6. **Distributivity:**
$$x_1 \wedge (x_2 \oplus x_3) = (x_1 \wedge x_2) \oplus (x_1 \wedge x_3) \; , \tag{1.46}$$
$$x_1 \vee (x_2 \odot x_3) = (x_1 \vee x_2) \odot (x_1 \vee x_3) \; , \tag{1.47}$$

4. **Idempotence:**
$$x \wedge x = x \; , \tag{1.48}$$
$$x \vee x = x \; . \tag{1.49}$$

The idempotence characterizes *Boolean Rings*. The comparison of the upper line in above axioms (valid for the first Boolean Ring) and the associated line below (valid for the second Boolean Ring) show both the duality of the operations ($\wedge \Leftrightarrow \vee$) as well as ($\oplus \Leftrightarrow \odot$) and the opposite roles of the values 0 and 1.

Theorem 1.10 Boolean Rings. *The structures*

$$(\mathbb{B}, \oplus, \wedge, 0, 1),\ (\mathbb{B}, \odot, \vee, 1, 0),\ (\mathbb{B}^n, \oplus, \wedge, \mathbf{0}, \mathbf{1}),\ \textit{and}\ (\mathbb{B}^n, \odot, \vee, \mathbf{1}, \mathbf{0})$$

are Boolean Rings.

The rule (to be checked as before)

$$x_1 \vee x_2 = x_1 \oplus x_2 \oplus x_1\, x_2 \tag{1.50}$$

connects antivalence and disjunction. The complement can be eliminated by

$$\overline{x} = 1 \oplus x\,. \tag{1.51}$$

A next normal form, the **antivalence normal form**, can be constructed by means of the disjunctive normal form. Since two conjunctions are always different, it is possible to find at least one variable x_i which appears non-negated in one conjunction C_1 and negated in C_2:

$$C_1 = x_i C' \text{ and } C_2 = \overline{x}_i C''\,.$$

$C_1 = 1$ and $C_2 = 1$ hold for different vectors, therefore $C_1 \wedge C_2 = 0$; such conjunctions are **orthogonal** to each other. Subsequently this concept of **orthogonality** will also be used for sets of vectors.

The property $C_1\, C_2 = 0$ simplifies the rule (1.50)

$$C_1 \vee C_2 = C_1 \oplus C_2 \oplus C_1\, C_2$$

to

$$C_1 \vee C_2 = C_1 \oplus C_2$$

when the variables are replaced by conjunctions. This relationship is generalized by the following.

Theorem 1.11 Orthogonality of Conjunctions. *Let C_i, $i = 1, \ldots, k$, be k conjunctions. If*

$$C_i \wedge C_j = 0 \qquad \forall\, i \neq j,\ 1 \leq i, j \leq k \tag{1.52}$$

then

$$\bigvee_{i=1}^{k} C_i = \bigoplus_{i=1}^{k} C_i\,. \tag{1.53}$$

All conjunctions (minterms) of a disjunctive normal form are orthogonal to each other. Hence, due to Theorem 1.11 the disjunction sign of the disjunctive normal form (1.26) simply can be replaced by the antivalence without any change of the specified function:

$$\begin{aligned} f(x_1, x_2, x_3, x_4) = {} & \overline{x}_1 x_2 \overline{x}_3 \overline{x}_4 \oplus \overline{x}_1 x_2 x_3 x_4 \oplus x_1 \overline{x}_2 \overline{x}_3 \overline{x}_4 \oplus \\ & x_1 \overline{x}_2 x_3 x_4 \oplus x_1 x_2 \overline{x}_3 \overline{x}_4 \oplus x_1 x_2 x_3 x_4 \ . \end{aligned} \tag{1.54}$$

This antivalence normal form is also *uniquely defined* except the order of the conjunctions.

Finally, in this normal form the negation can be eliminated by using the rule (1.51), and the distributivity law (1.46) will be applied thereafter. For instance:

$$\overline{x}_1 x_2 x_3 x_4 = (1 \oplus x_1) x_2 x_3 x_4 = x_2 x_3 x_4 \oplus x_1 x_2 x_3 x_4 \ .$$

Further simplifications might be possible with other conjunctions. If there are, for instance, two equal conjunctions then we get

$$C_1 \oplus C_1 = 0 \quad \text{and} \quad F \oplus 0 = F \ .$$

Hence, two equal conjunctions can be deleted.

In this way, the antivalence normal form (1.54) can be expressed by the positive polarity Reed-Muller polynomial:

$$\begin{aligned} f(x_1, x_2, x_3, x_4) = {} & x_1 \oplus x_2 \oplus \\ & x_1 x_2 \oplus x_1 x_3 \oplus x_1 x_4 \oplus x_2 x_3 \oplus x_2 x_4 \oplus \\ & x_1 x_2 x_3 \oplus x_1 x_2 x_4 \ . \end{aligned} \tag{1.55}$$

It can be seen that the conjunctions consist of one variable, two variables, and three variables. It is a property of the transformed function $f(x_1, x_2, x_3, x_4)$ that the conjunction of four variables does not occur in the expression (1.55).

There is also the possibility to switch between Boolean Rings and Boolean Algebras by eliminating $\oplus$:

$$x_1 \oplus x_2 = \overline{x}_1 x_2 \vee x_1 \overline{x}_2 \ . \tag{1.56}$$

In order to transfer the conjunctive normal form into an **equivalence normal form**, we need the following two rules:

$$D_1 \wedge D_2 = D_1 \odot D_2 \odot (D_1 \vee D_2) \ , \tag{1.57}$$

$$\overline{x} = 0 \odot x \ . \tag{1.58}$$

In the same way as has been done for the antivalence normal form it can be concluded that for two different disjunctions D_i and D_j always $D_i \vee D_j = 1$ holds, and we get

$$D_1 \wedge D_2 = D_1 \odot D_2 \ ,$$

and generalized the

Theorem 1.12 Orthogonality of Disjunctions. *Let D_i, $i = 1, \dots, k$, be k disjunctions. If*

$$D_i \vee D_j = 0 \qquad \forall\, i \neq j,\ 1 \leq i, j \leq k \tag{1.59}$$

then

$$\bigwedge_{i=1}^{k} D_i = \bigodot_{i=1}^{k} D_i \,. \tag{1.60}$$

Remaining negated variables can be eliminated using (1.58) and the distributive law (1.47).

This second Boolean Ring $(\mathbb{B}, \odot, \vee, 1, 0)$ has the same properties as the ring with the antivalence and the conjunction $(\mathbb{B}, \oplus, \wedge, 0, 1)$. Therefore only the first ring will be used.

1.4 BOOLEAN EQUATIONS AND INEQUALITIES

We start this complex by an example which is easy to understand and shows the problem.

Example 1.13 Solving a Boolean Equation. Let the equation

$$x_1 \vee x_2 = x_3 \wedge x_4 \tag{1.61}$$

be given. Since we are dealing with Boolean functions, the equality can only have the format

$$0 = 0 \quad \text{or} \quad 1 = 1 \,.$$

$x_1 \vee x_2 = 0$ holds for $x_1 = 0, x_2 = 0$. $x_3 \wedge x_4 = 0$ holds for $x_3 = 0,\ x_4 = 0$; $x_3 = 0,\ x_4 = 1$; $x_3 = 1,\ x_4 = 0$; hence, we have the following set of solution vectors with the components (x_1, x_2, x_3, x_4):

$$\{(0000), (0001), (0010)\} \,.$$

Now the identity $1 = 1$ has to be explored, and, according to the definition of $\vee$ and $\wedge$ we get the following solutions:

$$\{(0111), (1011), (1111)\} \,.$$

Altogether, this equation has six solution vectors.

Definition 1.14 Boolean Equation. Let $\mathbf{x} = (x_1, \dots, x_n)$, $f(\mathbf{x})$ and $g(\mathbf{x})$ be two Boolean functions, then

$$f(\mathbf{x}) = g(\mathbf{x}) \tag{1.62}$$

is a *Boolean equation* of n variables. The vector $\mathbf{b} = (b_1, \ldots, b_n)$ is a *solution* of this equation if $f(\mathbf{b}) = g(\mathbf{b})$ (i.e., $f(\mathbf{b}) = g(\mathbf{b}) = 0$ or $f(\mathbf{b}) = g(\mathbf{b}) = 1$).

It is easily possible to reduce the considerations to homogeneous equations.

Theorem 1.15 Homogeneous Equations. *The equation $f(\mathbf{x}) = g(\mathbf{x})$ is equivalent to the following two equations:*

$$\text{homogeneous restrictive equation}: \quad f(\mathbf{x}) \oplus g(\mathbf{x}) = 0\,, \tag{1.63}$$
$$\text{homogeneous characteristic equation}: \quad f(\mathbf{x}) \odot g(\mathbf{x}) = 1\,. \tag{1.64}$$

The original equation and the two homogeneous equations have the same solution set.

It is very easy to eliminate *inequalities*. The inequality

$$f(\mathbf{x}) \leq g(\mathbf{x}) \tag{1.65}$$

is equivalent to the two equations

$$f(\mathbf{x}) \wedge \overline{g(\mathbf{x})} = 0\,, \tag{1.66}$$
$$\overline{f(\mathbf{x})} \vee g(\mathbf{x}) = 1\,. \tag{1.67}$$

As done before, these equivalent transformations can be applied to binary vectors and particularly to Boolean functions.

1.5 LISTS OF TERNARY VECTORS (TVL)

The use of ternary vectors started in the 1970s by Zakrevskij [1975] and was further developed in order to create numerical methods for Boolean problems. They will be defined as a pooling of Boolean vectors.

Let two conjunctions $x_1\, x_2\, x_3\, x_4$ and $\overline{x}_1\, x_2\, x_3\, x_4$ be given; they are equal to 1 for the two binary vectors (1111) and (0111). This can be abbreviated by (−111). The symbol − can be replaced by 0 and by 1, the ternary vector represents two binary vectors.

In the background the laws of a Boolean Algebra are used:

$$x_1\, x_2\, x_3\, x_4 \vee \overline{x}_1\, x_2\, x_3\, x_4 = (x_1 \vee \overline{x}_1)\, x_2\, x_3\, x_4 = x_2\, x_3\, x_4\,. \tag{1.68}$$

The resulting conjunction is equal to 1 for $x_2 = x_3 = x_4 = 1$ and any value of x_1.

When a function is given by a disjunctive form (i.e., several conjunctions are combined by $\vee$), then each conjunction will be translated into a ternary vector, and the different ternary vectors are collected in a list or matrix (TVL).

Example 1.16 TVL of a Boolean Function in Disjunctive Form. The TVL

$$\mathrm{D}(f) = \begin{array}{ccc} x_1 & x_2 & x_3 \\ \hline 1 & - & 0 \\ - & 1 & 1 \\ 1 & 0 & 1 \\ \hline \end{array}$$

represents the disjunctive form of the function

$$f(x_1, x_2, x_3) = x_1\,\overline{x}_3 \vee x_2\,x_3 \vee x_1\,\overline{x}_2\,x_3 \; .$$

The set of all Boolean vectors $\mathbf{x}$ with $f(\mathbf{x}) = 1$ is collected in this list.

When we look at the function $f(x_1, x_2, x_3, x_4) = x_1 \vee x_2 \vee x_3$ in $\mathbb{B}^4$ then we get in a first step three ternary vectors, the respective sets of Boolean vectors, however, overlap each other.

$$\mathrm{D}(f) = \begin{array}{cccc} x_1 & x_2 & x_3 & x_4 \\ \hline 1 & - & - & - \\ - & 1 & - & - \\ - & - & 1 & - \\ \hline \end{array} \tag{1.69}$$

This is an uncomfortable situation and can be avoided when we introduce the **orthogonality** of ternary vectors.

Definition 1.17 Orthogonality. Let $S(\mathbf{t})$ be the set of all $\mathbf{x} \in \mathbb{B}^n$ which can be generated by means of the ternary vector $\mathbf{t}$. Two vectors $\mathbf{t}_1$ and $\mathbf{t}_2$ are *orthogonal* to each other ($\mathbf{t}_1 \perp \mathbf{t}_2$) if $S(\mathbf{t}_1) \cap S(\mathbf{t}_2) = \emptyset$.

Theorem 1.18 Orthogonality. *The property* $\mathbf{t}_1 \perp \mathbf{t}_2$ *holds if and only if for at least one component* i *the combination*

$$t_{1,i} = 0, t_{2,i} = 1 \quad \text{or} \quad t_{1,i} = 1, t_{2,i} = 0 \tag{1.70}$$

exists.

This very useful property can easily be established and tested. The orthogonal TVL

$$\mathrm{ODA}(f) = \begin{array}{cccc} x_1 & x_2 & x_3 & x_4 \\ \hline 1 & - & - & - \\ 0 & 1 & - & - \\ 0 & 0 & 1 & - \\ \hline \end{array} \tag{1.71}$$

generates the same set of Boolean vectors as (1.69). The form predicate ODA indicates that this TVL can, due to Theorem 1.11, alternatively be used to describe a Boolean function in disjunctive form $\mathrm{D}(f)$ or in antivalence form $\mathrm{A}(f)$.

SUMMARY

In this chapter we discussed very briefly some important terms of Boolean Algebras. The Boolean spaces $\mathbb{B}^n$ are built from the elements 0 and 1 or vectors of these elements. Boolean functions are unique mappings from $\mathbb{B}^n$ into $\mathbb{B}$. Boolean expressions of variables and operations describe in a compact manner Boolean functions. A Boolean equation will be built from two Boolean functions and has a set of Boolean vectors as solution. The Boolean Algebra is widely used to specify, design, analyze, and test digital circuits and systems.

EXERCISES

1.1 Use the table method to prove the associative law (1.4).

1.2 Determine the function table of the function:

$$f(x_1, x_2, x_3, x_4) = (x_1 \vee \overline{x}_2) \wedge ((x_1 \oplus x_3) \vee \overline{(x_2 \oplus x_4)}) . \tag{1.72}$$

1.3 A lattice of Boolean functions is specified by Table 1.8 which contains three don't-care values Φ.

(a) How many functions belong to this lattice?

(b) Describe each function of this lattice by a formula in disjunctive normal form.

(c) Apply the idempotence law (1.5) and the absorption law (1.8) to find simplified disjunctive forms of these functions.

(d) Find all pairs of functions of this lattice for which the conjunction results in the minimum and the disjunction in the maximum of all these functions.

(e) Which function of this lattice has the simplest formula in disjunctive form?

Table 1.8: Incompletely specified function that generates a lattice of Boolean functions

x_1	x_2	x_3	$f(x_1, x_2, x_3)$
0	0	0	1
0	0	1	1
0	1	0	0
0	1	1	Φ
1	0	0	0
1	0	1	Φ
1	1	0	Φ
1	1	1	1

1.4 Which binary vectors belong to the solution of the Boolean equation:

$$(x_1 \vee \overline{x}_3) \wedge x_2 = (x_3 \oplus x_4) \vee x_1 . \tag{1.73}$$

Specify the solution set S:

(a) by a list of binary vectors (BVL), and

(b) by an orthogonal list of ternary vectors (TVL).

1.5 The following TVL is not orthogonal.

x_1	x_2	x_3	x_4
1	–	0	0
–	1	1	–
0	–	0	–
–	–	1	1

(1.74)

(a) Which ternary vectors of the TVL (1.74) are not orthogonal to each other?

(b) Assume that the TVL (1.74) represents a function in disjunctive form. Construct the associated ODA-form.

(c) Assume that the TVL (1.74) represents a function in antivalence form. Construct the associated ODA-form.

CHAPTER 2

Derivative Operations of Boolean Functions

2.1 VECTORIAL DERIVATIVE OPERATIONS

Vectorial derivative operations explore the change of function values for pairs of points in the Boolean space. This requires to distinguish between

- the *change of the value of selected variables* to describe the direction of change in the Boolean space and
- the *change of the function values* between the chosen pairs.

Example 2.1 demonstrates all possible directions of change of $\mathbb{B}^3$.

Example 2.1 All Directions of Change in the Boolean Space $\mathbb{B}^3$. Taking one point of $\mathbb{B}^3$, then $2^3 - 1 = 7$ other points remain to specify a direction of change in this Boolean space. Figure 2.1 shows these seven possible directions for the selected point $(x_1, x_2, x_3) = (011)$ of $\mathbb{B}^3$.

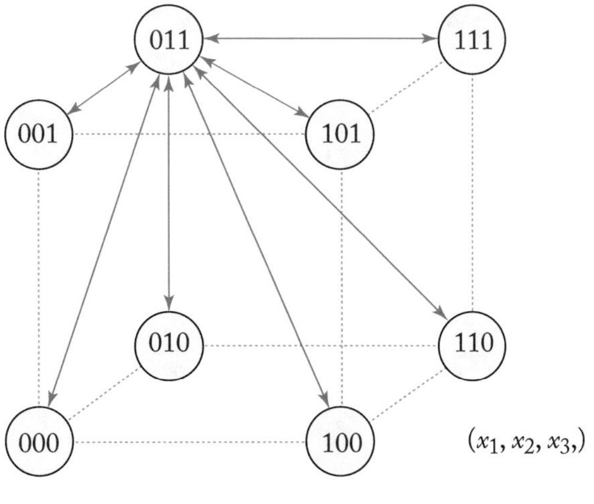

Figure 2.1: Directions of change in B^3 for the selected point $(x_1, x_2, x_3) = (011)$.

There are:

- $\binom{3}{1} = 3$ possibilities to change the value of one of the three variables; the value of x_2, for instance, changes between the points $(x_1, x_2, x_3) = (011)$ and $(x_1, x_2, x_3) = (001)$;
- $\binom{3}{2} = 3$ possibilities to change the value of two of the three variables; e.g., the values of x_2 and x_3 change between the pair of points $(x_1, x_2, x_3) = (011)$ and $(x_1, x_2, x_3) = (000)$; and
- $\binom{3}{3} = 1$ possibility to change the values of all three variables x_1, x_2 and x_3 between the pair of points $(x_1, x_2, x_3) = (011)$ and $(x_1, x_2, x_3) = (100)$.

We extend this idea to $\mathbb{B}^n$ and split the set of n variables $\mathbf{x} = (x_1, x_2, \ldots, x_n)$ into two subsets $\mathbf{x}_0$ and $\mathbf{x}_1 = \mathbf{x} \setminus \mathbf{x}_0$, where $\mathbf{x}_0$ is a nonempty subset of variables that describes the direction of change between the pair of points $(\mathbf{x}_0, \mathbf{x}_1)$ and $(\overline{\mathbf{x}}_0, \mathbf{x}_1)$. All the values of the variables of $\mathbf{x}_0$ must *simultaneously* change to switch between the pair of points in the Boolean space.

As shown in Figure 2.1, the subset of variables $\mathbf{x}_0$ specifies a direction of change that can be depicted as a *vector* in the Boolean space. This motivates the name *vectorial* for this kind of derivative operations.

Vectorial derivative operations explore the changes for all pairs of function values with the same direction of change in the Boolean space. These pairs of function values are characterized by $f(\mathbf{x}_0, \mathbf{x}_1)$ and $f(\overline{\mathbf{x}}_0, \mathbf{x}_1)$. Different change properties are evaluated using the operations $\oplus$ or $\odot$ of Boolean Rings or the operations $\wedge$ or $\vee$ of Boolean Algebras.

Definition 2.2 Vectorial Derivative Operations.
Let $\mathbf{x}_0 = (x_1, \ldots, x_k)$, $\mathbf{x}_1 = \mathbf{x} \setminus \mathbf{x}_0$ be two disjoint sets of Boolean variables, $1 \leq k \leq n$, and $f(\mathbf{x}_0, \mathbf{x}_1) = f(x_1, \ldots, x_n) = f(\mathbf{x})$ be a Boolean function of n variables, then

$$\frac{\partial f(\mathbf{x})}{\partial \mathbf{x}_0} = f(\mathbf{x}_0, \mathbf{x}_1) \oplus f(\overline{\mathbf{x}}_0, \mathbf{x}_1) \tag{2.1}$$

is the **vectorial derivative**,

$$\overline{\frac{\partial f(\mathbf{x})}{\partial \mathbf{x}_0}} = f(\mathbf{x}_0, \mathbf{x}_1) \odot f(\overline{\mathbf{x}}_0, \mathbf{x}_1) \tag{2.2}$$

is the **negation of the vectorial derivative**,

$$\min_{\mathbf{x}_0} f(\mathbf{x}) = f(\mathbf{x}_0, \mathbf{x}_1) \wedge f(\overline{\mathbf{x}}_0, \mathbf{x}_1) \tag{2.3}$$

is the **vectorial minimum**, and

$$\max_{\mathbf{x}_0} f(\mathbf{x}) = f(\mathbf{x}_0, \mathbf{x}_1) \vee f(\overline{\mathbf{x}}_0, \mathbf{x}_1) \tag{2.4}$$

is the **vectorial maximum**

of the Boolean function $f(\mathbf{x})$ with regard to the variables $\mathbf{x}_0$.

The results of these four vectorial derivative operations are again Boolean functions. They generally depend on the same variables $\mathbf{x} = (\mathbf{x}_0, \mathbf{x}_1)$ as the given function $f(\mathbf{x})$. Commonly, all vectorial derivative operations compare the values of a function f in the two points $(\mathbf{x}_0, \mathbf{x}_1)$ and $(\overline{\mathbf{x}}_0, \mathbf{x}_1)$. The difference in their meaning originates from the used Boolean operations in Definition 2.2.

Table 2.1: Comparison of function values in pairs of points

$f(\mathbf{x}_0, \mathbf{x}_1)$	$f(\overline{\mathbf{x}}_0, \mathbf{x}_1)$	$\frac{\partial f(\mathbf{x})}{\partial \mathbf{x}_0}$	$\overline{\frac{\partial f(\mathbf{x})}{\partial \mathbf{x}_0}}$	$\min_{\mathbf{x}_0} f(\mathbf{x})$	$\max_{\mathbf{x}_0} f(\mathbf{x})$
0	0	0	1	0	0
0	1	1	0	0	1
1	0	1	0	0	1
1	1	0	1	1	1

Table 2.1 helps to remember the interpretation of the vectorial derivative operations.

- The *vectorial derivative* of $f(\mathbf{x})$ with regard to $\mathbf{x}_0$ is equal to 1 if the simultaneous change of all variables of the subset $\mathbf{x}_0 \subseteq \mathbf{x}$ changes the value of the given function.
- The *negation of the vectorial derivative* of $f(\mathbf{x})$ with regard to $\mathbf{x}_0$ is equal to 1 if the simultaneous change of all variables of the subset $\mathbf{x}_0 \subseteq \mathbf{x}$ does not change the value of the given function.
- The *vectorial minimum* of $f(\mathbf{x})$ with regard to $\mathbf{x}_0$ is equal to 1 if the simultaneous change of all variables of the subset $\mathbf{x}_0 \subseteq \mathbf{x}$ lets the value of the given function unchanged equal to 1.
- The *vectorial maximum* of $f(\mathbf{x})$ with regard to $\mathbf{x}_0$ is equal to 0 if the simultaneous change of all variables of the subset $\mathbf{x}_0 \subseteq \mathbf{x}$ lets the value of the given function unchanged equal to 0.

The negation of the vectorial derivative with regard to $\mathbf{x}_0$ (2.2) can be calculated using the negation operation ($\overline{\ }$) applied to the vectorial derivative with regard to the same variables (2.1). Therefore, we omit the negation of the vectorial derivative (2.2) in the following discussions of the details of the vectorial derivative operations.

We mentioned already that the vectorial derivative of $f(\mathbf{x}_0, \mathbf{x}_1)$ with regard to $\mathbf{x}_0$ is a function that depends on the same variables $(\mathbf{x}_0, \mathbf{x}_1)$. Such a vectorial derivative can be calculated for each function $f(\mathbf{x}_0, \mathbf{x}_1)$. However, only a subset of functions $f(\mathbf{x}_0, \mathbf{x}_1)$ can be a vectorial derivative.

This property can be verified as follows. Assume that $\frac{\partial f(\mathbf{x}_0,\mathbf{x}_1)}{\partial \mathbf{x}_0}$ is equal to 1 for $(\mathbf{c}_0,\mathbf{c}_1)$. That means: the simultaneous change of values of $\mathbf{c}_0$ causes the value change of the given function $f(\mathbf{x}_0,\mathbf{x}_1)$. The change of all values of $\mathbf{c}_0$ changes the evaluated vector from $(\mathbf{c}_0,\mathbf{c}_1)$ to $(\overline{\mathbf{c}}_0,\mathbf{c}_1)$. The reverse change from $(\overline{\mathbf{c}}_0,\mathbf{c}_1)$ to $(\mathbf{c}_0,\mathbf{c}_1)$ changes the value of the given function $f(\mathbf{x}_0,\mathbf{x}_1)$ as well. Hence, the vectorial derivative of $f(\mathbf{x}_0,\mathbf{x}_1)$ with regard to $\mathbf{x}_0$ is equal to 1 also for $(\overline{\mathbf{c}}_0,\mathbf{c}_1)$. An analog property is satisfied in the case that $\frac{\partial f(\mathbf{x}_0,\mathbf{x}_1)}{\partial \mathbf{x}_0}$ is equal to 0 for $(\mathbf{c}_0,\mathbf{c}_1)$; this vectorial derivative is also equal to 0 for $(\overline{\mathbf{c}}_0,\mathbf{c}_1)$. In summary we get the property:

each vectorial derivative has an even number of function values 1.

This is, however, only a necessary condition for a vectorial derivative.

A sufficient condition that a function

$$g(\mathbf{x}_0,\mathbf{x}_1) = \frac{\partial f(\mathbf{x}_0,\mathbf{x}_1)}{\partial \mathbf{x}_0}$$

is a vectorial derivative looks as follows:

$$\frac{\partial g(\mathbf{x}_0,\mathbf{x}_1)}{\partial \mathbf{x}_0} = 0 \,.$$

Hence, we have the rule:

$$\frac{\partial}{\partial \mathbf{x}_0}\left(\frac{\partial f(\mathbf{x}_0,\mathbf{x}_1)}{\partial \mathbf{x}_0}\right) = 0 \,. \tag{2.5}$$

Using Defintion (2.1) this can be seen by simple transformations:

$$\begin{aligned}
\frac{\partial}{\partial \mathbf{x}_0}\left(\frac{\partial f(\mathbf{x}_0,\mathbf{x}_1)}{\partial \mathbf{x}_0}\right) &= \frac{\partial}{\partial \mathbf{x}_0}\left(f(\mathbf{x}_0,\mathbf{x}_1) \oplus f(\overline{\mathbf{x}}_0,\mathbf{x}_1)\right)\\
&= \left(f(\mathbf{x}_0,\mathbf{x}_1) \oplus f(\overline{\mathbf{x}}_0,\mathbf{x}_1)\right) \oplus \left(f(\overline{\mathbf{x}}_0,\mathbf{x}_1) \oplus f(\overline{\overline{\mathbf{x}}}_0,\mathbf{x}_1)\right)\\
&= f(\mathbf{x}_0,\mathbf{x}_1) \oplus f(\overline{\mathbf{x}}_0,\mathbf{x}_1) \oplus f(\overline{\mathbf{x}}_0,\mathbf{x}_1) \oplus f(\mathbf{x}_0,\mathbf{x}_1)\\
&= f(\mathbf{x}_0,\mathbf{x}_1) \oplus f(\mathbf{x}_0,\mathbf{x}_1) \oplus f(\overline{\mathbf{x}}_0,\mathbf{x}_1) \oplus f(\overline{\mathbf{x}}_0,\mathbf{x}_1)\\
&= 0 \oplus 0\\
&= 0 \,.
\end{aligned}$$

Besides the analysis of simultaneous changes of values in a combinational circuit, the vectorial derivative helps to analyze properties of Boolean functions. A function is *self-dual* if

$$f(\mathbf{x}) = \overline{f(\overline{\mathbf{x}})} \,. \tag{2.6}$$

Equivalent transformations of this equation lead to:

$$\begin{aligned}
f(\mathbf{x}) &= f(\overline{\mathbf{x}}) \oplus 1\\
f(\mathbf{x}) \oplus f(\overline{\mathbf{x}}) &= f(\overline{\mathbf{x}}) \oplus f(\overline{\mathbf{x}}) \oplus 1\\
f(\mathbf{x}) \oplus f(\overline{\mathbf{x}}) &= 0 \oplus 1\\
\frac{\partial f(\mathbf{x})}{\partial \mathbf{x}} &= 1 \,.
\end{aligned} \tag{2.7}$$

Hence, a Boolean function $f(\mathbf{x})$ is self-dual if its vectorial derivative with regard to all variables $\mathbf{x}$ is equal to 1.

Another property of a Boolean function $f(x_i, x_j, \mathbf{x}_1)$ is the symmetry between x_i and x_j. The exchange of the input values for x_i and x_j does not change the function value of such a function. A Boolean function that satisfies

$$(x_i \oplus x_j) \wedge \frac{\partial f(x_i, x_j, \mathbf{x}_1)}{\partial(x_i, x_j)} = 0 \tag{2.8}$$

is symmetric with regard to x_i and x_j. This property can be satisfied for several pairs of variables. A Boolean function $f(x_i, \mathbf{x}_1)$, $x_j \in \mathbf{x}_1$ is symmetric with regard to all pairs of variables if

$$\bigvee_{x_j \in \mathbf{x}_1} \left((x_i \oplus x_j) \wedge \frac{\partial f(x_i, \mathbf{x}_1)}{\partial(x_i, x_j)} \right) = 0\,. \tag{2.9}$$

Due to the 2^n points of $\mathbb{B}^n$ there is a total of $2^n - 1$ different vectorial derivatives of a given function of n variables. However, maximally n of them are linearly independent. The following example demonstrates the linear dependence of three vectorial derivatives.

Example 2.3 Three Linearly Dependent Vectorial Derivatives.
Each function $f(x_i, x_j, x_k, \mathbf{x}_1)$ satisfies:

$$\frac{\partial}{\partial(x_j, x_k)} \left(\frac{\partial}{\partial(x_i, x_k)} \left(\frac{\partial f(x_i, x_j, x_k, \mathbf{x}_1)}{\partial(x_i, x_j)} \right) \right) = 0\,. \tag{2.10}$$

Using Definition (2.1) we get:

$$\begin{aligned}
&\frac{\partial}{\partial(x_j, x_k)} \left(\frac{\partial}{\partial(x_i, x_k)} \left(\frac{\partial f(x_i, x_j, x_k, \mathbf{x}_1)}{\partial(x_i, x_j)} \right) \right) = \\
&\quad = \frac{\partial}{\partial(x_j, x_k)} \left(\frac{\partial}{\partial(x_i, x_k)} \left(f(x_i, x_j, x_k, \mathbf{x}_1) \oplus f(\overline{x}_i, \overline{x}_j, x_k, \mathbf{x}_1) \right) \right) \\
&\quad = \frac{\partial}{\partial(x_j, x_k)} \left(f(x_i, x_j, x_k, \mathbf{x}_1) \oplus f(\overline{x}_i, \overline{x}_j, x_k, \mathbf{x}_1) \oplus \right. \\
&\qquad\qquad \left. f(\overline{x}_i, x_j, \overline{x}_k, \mathbf{x}_1) \oplus f(x_i, \overline{x}_j, \overline{x}_k, \mathbf{x}_1) \right) \\
&\quad = f(x_i, x_j, x_k, \mathbf{x}_1) \oplus f(\overline{x}_i, \overline{x}_j, x_k, \mathbf{x}_1) \oplus f(\overline{x}_i, x_j, \overline{x}_k, \mathbf{x}_1) \oplus f(x_i, \overline{x}_j, \overline{x}_k, \mathbf{x}_1) \oplus \\
&\qquad f(\overline{x}_i, x_j, \overline{x}_k, \mathbf{x}_1) \oplus f(x_i, \overline{x}_j, \overline{x}_k, \mathbf{x}_1) \oplus f(x_i, x_j, x_k, \mathbf{x}_1) \oplus f(\overline{x}_i, \overline{x}_j, x_k, \mathbf{x}_1) \\
&\quad = f(x_i, x_j, x_k, \mathbf{x}_1) \oplus f(x_i, x_j, x_k, \mathbf{x}_1) \oplus f(\overline{x}_i, \overline{x}_j, x_k, \mathbf{x}_1) \oplus f(\overline{x}_i, \overline{x}_j, x_k, \mathbf{x}_1) \oplus \\
&\qquad f(\overline{x}_i, x_j, \overline{x}_k, \mathbf{x}_1) \oplus f(\overline{x}_i, x_j, \overline{x}_k, \mathbf{x}_1) \oplus f(x_i, \overline{x}_j, \overline{x}_k, \mathbf{x}_1) \oplus f(x_i, \overline{x}_j, \overline{x}_k, \mathbf{x}_1) \\
&\quad = 0\,.
\end{aligned}$$

The sequence of three vectorial derivatives (2.10) results in the EXOR of four pairs of identical functions and describes therefore the function $0(\mathbf{x}_1)$ for each given function $f(x_i, x_j, x_k, \mathbf{x}_1)$.

Just like the vectorial derivative, both the vectorial minimum and maximum of $f(\mathbf{x})$ with regard to $\mathbf{x}_0$ depend on all variables $\mathbf{x} = (\mathbf{x}_0, \mathbf{x}_1)$ of the given function $f(\mathbf{x})$. However, function values of $(\mathbf{c}_0, \mathbf{c}_1)$ and $(\overline{\mathbf{c}}_0, \mathbf{c}_1)$ are the same for each vectorial derivative operation. Hence, both the vectorial minimum and maximum do not depend on the simultaneous change of $\mathbf{x}_0$ anymore:

$$\frac{\partial}{\partial \mathbf{x}_0}\left(\min_{\mathbf{x}_0} f(\mathbf{x})\right) = 0 \,, \tag{2.11}$$

$$\frac{\partial}{\partial \mathbf{x}_0}\left(\max_{\mathbf{x}_0} f(\mathbf{x})\right) = 0 \,, \tag{2.12}$$

so that only $2^{2^{n-1}}$ of all 2^{2^n} functions of n variables can be a vectorial minimum or maximum. The proof of the properties (2.11) and (2.12) directly follows from Definition 2.2 using the commutativity (1.2) and (1.1) and the inverse element (1.42):

$$\begin{aligned}
\frac{\partial}{\partial \mathbf{x}_0}\left(\min_{\mathbf{x}_0} f(\mathbf{x})\right) &= \frac{\partial}{\partial \mathbf{x}_0}\left(f(\mathbf{x}_0, \mathbf{x}_1) \wedge f(\overline{\mathbf{x}}_0, \mathbf{x}_1)\right) \\
&= \left(f(\mathbf{x}_0, \mathbf{x}_1) \wedge f(\overline{\mathbf{x}}_0, \mathbf{x}_1)\right) \oplus \left(f(\overline{\mathbf{x}}_0, \mathbf{x}_1) \wedge f(\mathbf{x}_0, \mathbf{x}_1)\right) \\
&= 0 \,, \\
\frac{\partial}{\partial \mathbf{x}_0}\left(\max_{\mathbf{x}_0} f(\mathbf{x})\right) &= \frac{\partial}{\partial \mathbf{x}_0}\left(f(\mathbf{x}_0, \mathbf{x}_1) \vee f(\overline{\mathbf{x}}_0, \mathbf{x}_1)\right) \\
&= \left(f(\mathbf{x}_0, \mathbf{x}_1) \vee f(\overline{\mathbf{x}}_0, \mathbf{x}_1)\right) \oplus \left(f(\overline{\mathbf{x}}_0, \mathbf{x}_1) \vee f(\mathbf{x}_0, \mathbf{x}_1)\right) \\
&= 0 \,.
\end{aligned}$$

The names of these vectorial derivative operations were chosen due to the inequality:

$$\min_{\mathbf{x}_0} f(\mathbf{x}_0, \mathbf{x}_1) \leq f(\mathbf{x}_0, \mathbf{x}_1) \leq \max_{\mathbf{x}_0} f(\mathbf{x}_0, \mathbf{x}_1) \,. \tag{2.13}$$

Again, Definition 2.2, the equivalence between an inequality and Equation (1.66), and the rules of Boolean Algebra confirm the inequality of the vectorial minimum:

$$\begin{aligned}
\min_{\mathbf{x}_0} f(\mathbf{x}_0, \mathbf{x}_1) &\leq f(\mathbf{x}_0, \mathbf{x}_1) \\
\min_{\mathbf{x}_0} f(\mathbf{x}_0, \mathbf{x}_1) \wedge \overline{f(\mathbf{x}_0, \mathbf{x}_1)} &= 0 \\
f(\mathbf{x}_0, \mathbf{x}_1) \wedge f(\overline{\mathbf{x}}_0, \mathbf{x}_1) \wedge \overline{f(\mathbf{x}_0, \mathbf{x}_1)} &= 0 \\
0 &= 0 \,,
\end{aligned}$$

as well as the inequality of the vectorial maximum:

$$\begin{aligned}
f(\mathbf{x}_0, \mathbf{x}_1) &\leq \max_{\mathbf{x}_0} f(\mathbf{x}_0, \mathbf{x}_1) \\
f(\mathbf{x}_0, \mathbf{x}_1) \wedge \overline{\max_{\mathbf{x}_0} f(\mathbf{x}_0, \mathbf{x}_1)} &= 0 \\
f(\mathbf{x}_0, \mathbf{x}_1) \wedge \overline{\left(f(\mathbf{x}_0, \mathbf{x}_1) \vee f(\overline{\mathbf{x}}_0, \mathbf{x}_1)\right)} &= 0 \\
f(\mathbf{x}_0, \mathbf{x}_1) \wedge \overline{f(\mathbf{x}_0, \mathbf{x}_1)} \wedge \overline{f(\overline{\mathbf{x}}_0, \mathbf{x}_1)} &= 0 \\
0 &= 0 \,.
\end{aligned}$$

The three restrictions

$$\min_{\mathbf{x}_0} f(\mathbf{x}) \wedge \overline{f(\mathbf{x})} = 0 \,, \tag{2.14}$$

$$\min_{\mathbf{x}_0} f(\mathbf{x}) \wedge \overline{\max_{\mathbf{x}_0} f(\mathbf{x})} = 0 \,, \tag{2.15}$$

$$f(\mathbf{x}) \wedge \overline{\max_{\mathbf{x}_0} f(\mathbf{x})} = 0 \tag{2.16}$$

follow from (2.13) and can be used to simplify expressions.

Figure 2.2 shows how the three vectorial derivative operations are calculated for a small function using Karnaugh-maps. The arrows in the Karnaugh-map of the given function $f(x_1, x_2, x_3)$ emphasize the pairs of function values which determine the vectorial derivative operations with regard to (x_1, x_3).

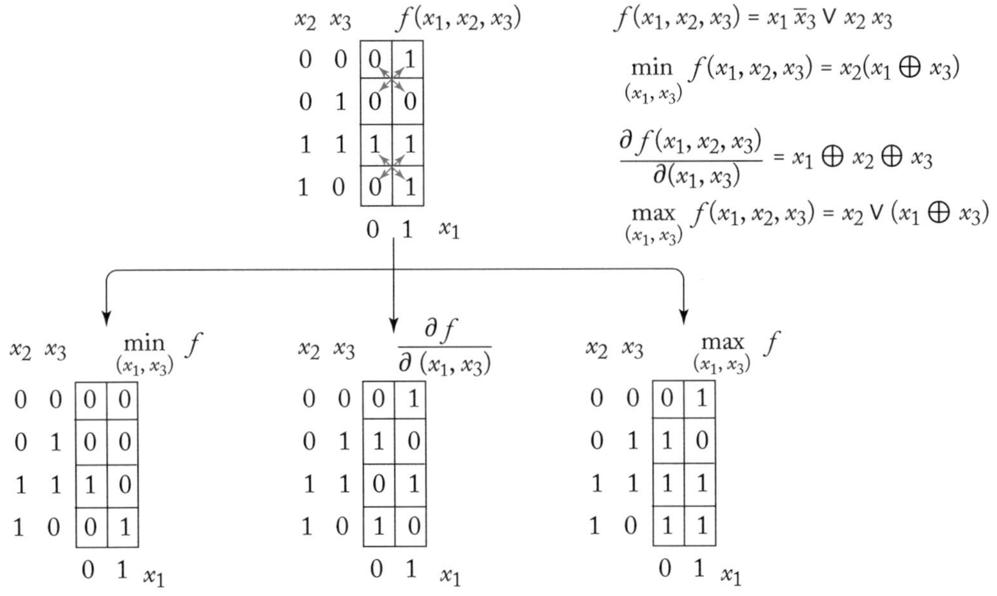

Figure 2.2: Karnaugh-maps of a function and all three associated vectorial derivative operations with regard to (x_1, x_3).

Definition 2.2 can directly be used to confirm the following relations:

$$\min_{\mathbf{x}_0} f(\mathbf{x}) \wedge \frac{\partial f(\mathbf{x})}{\partial \mathbf{x}_0} = 0 \,, \tag{2.17}$$

$$\frac{\partial f(\mathbf{x})}{\partial \mathbf{x}_0} \wedge \overline{\max_{\mathbf{x}_0} f(\mathbf{x})} = 0 \,. \tag{2.18}$$

For the same reason, three vectorial derivative operations satisfy the equation:

$$\min_{\mathbf{x}_0} f(\mathbf{x}) \oplus \max_{\mathbf{x}_0} f(\mathbf{x}) \oplus \frac{\partial f(\mathbf{x})}{\partial \mathbf{x}_0} = 0 \,, \tag{2.19}$$

which can be verified using Definition 2.2:

$$\begin{aligned}
(f(\mathbf{x}_0,\mathbf{x}_1) \wedge f(\overline{\mathbf{x}}_0,\mathbf{x}_1)) \oplus (f(\mathbf{x}_0,\mathbf{x}_1) \vee f(\overline{\mathbf{x}}_0,\mathbf{x}_1)) \oplus (f(\mathbf{x}_0,\mathbf{x}_1) \oplus f(\overline{\mathbf{x}}_0,\mathbf{x}_1)) &= 0 \\
(f(\mathbf{x}_0,\mathbf{x}_1) \wedge f(\overline{\mathbf{x}}_0,\mathbf{x}_1)) \oplus (f(\mathbf{x}_0,\mathbf{x}_1) \oplus f(\overline{\mathbf{x}}_0,\mathbf{x}_1) \oplus (f(\mathbf{x}_0,\mathbf{x}_1) \wedge f(\overline{\mathbf{x}}_0,\mathbf{x}_1)))\oplus & \\
(f(\mathbf{x}_0,\mathbf{x}_1) \oplus f(\overline{\mathbf{x}}_0,\mathbf{x}_1)) &= 0 \\
0 &= 0 \,.
\end{aligned}$$

Equation (2.19) can be transformed such that each of the three vectorial derivative operations is expressed by the two other operations:

$$\min_{\mathbf{x}_0} f(\mathbf{x}) = \max_{\mathbf{x}_0} f(\mathbf{x}) \oplus \frac{\partial f(\mathbf{x})}{\partial \mathbf{x}_0} \,, \tag{2.20}$$

$$\max_{\mathbf{x}_0} f(\mathbf{x}) = \min_{\mathbf{x}_0} f(\mathbf{x}) \oplus \frac{\partial f(\mathbf{x})}{\partial \mathbf{x}_0} \,, \tag{2.21}$$

$$\frac{\partial f(\mathbf{x})}{\partial \mathbf{x}_0} = \min_{\mathbf{x}_0} f(\mathbf{x}) \oplus \max_{\mathbf{x}_0} f(\mathbf{x}) \,. \tag{2.22}$$

These rules can be transformed into expressions that do not contain the $\oplus$-operation explicitly:

$$\min_{\mathbf{x}_0} f(\mathbf{x}) = \max_{\mathbf{x}_0} f(\mathbf{x}) \wedge \overline{\frac{\partial f(\mathbf{x})}{\partial \mathbf{x}_0}} \,, \tag{2.23}$$

$$\max_{\mathbf{x}_0} f(\mathbf{x}) = \min_{\mathbf{x}_0} f(\mathbf{x}) \vee \frac{\partial f(\mathbf{x})}{\partial \mathbf{x}_0} \,, \tag{2.24}$$

$$\frac{\partial f(\mathbf{x})}{\partial \mathbf{x}_0} = \overline{\min_{\mathbf{x}_0} f(\mathbf{x})} \wedge \max_{\mathbf{x}_0} f(\mathbf{x}) \,. \tag{2.25}$$

The easiest way to prove that these transformations are correct: use again Definition 2.2 and calculate the left and the right side of the equations. This demonstrates by their equality that the equations are valid.

Vectorial derivative operations of negated functions can be replaced by vectorial derivative operations of the non-negated function as follows:

$$\frac{\partial \overline{f(\mathbf{x})}}{\partial \mathbf{x}_0} = \frac{\partial f(\mathbf{x})}{\partial \mathbf{x}_0} \,, \tag{2.26}$$

$$\min_{\mathbf{x}_0} \overline{f(\mathbf{x})} = \overline{\max_{\mathbf{x}_0} f(\mathbf{x})} \,, \tag{2.27}$$

$$\max_{\mathbf{x}_0} \overline{f(\mathbf{x})} = \overline{\min_{\mathbf{x}_0} f(\mathbf{x})} \,. \tag{2.28}$$

The validity of (2.26) can be seen, e.g., by

$$
\begin{aligned}
\frac{\partial f(\mathbf{x})}{\partial \mathbf{x}_0} &= f(\mathbf{x}_0, \mathbf{x}_1) \oplus f(\overline{\mathbf{x}}_0, \mathbf{x}_1) \\
&= (f(\mathbf{x}_0, \mathbf{x}_1) \oplus 1) \oplus (f(\overline{\mathbf{x}}_0, \mathbf{x}_1) \oplus 1) \\
&= \overline{f(\mathbf{x}_0, \mathbf{x}_1)} \oplus \overline{f(\overline{\mathbf{x}}_0, \mathbf{x}_1)} \\
&= \frac{\partial \overline{f(\mathbf{x})}}{\partial \mathbf{x}_0} \; .
\end{aligned}
$$

Equations (2.27) and (2.28) can be confirmed using Definition 2.2 and the laws of De Morgan.

If the argument of a vectorial derivative operation consists of two Boolean functions, which are connected by the same operation as used for the definition of the vectorial derivative operation, a separation is possible due to the commutative laws:

$$
\frac{\partial (f(\mathbf{x}) \oplus g(\mathbf{x}))}{\partial \mathbf{x}_0} = \frac{\partial f(\mathbf{x})}{\partial \mathbf{x}_0} \oplus \frac{\partial g(\mathbf{x})}{\partial \mathbf{x}_0} \; , \tag{2.29}
$$

$$
\min_{\mathbf{x}_0}(f(\mathbf{x}) \wedge g(\mathbf{x})) = \min_{\mathbf{x}_0} f(\mathbf{x}) \wedge \min_{\mathbf{x}_0} g(\mathbf{x}) \; , \tag{2.30}
$$

$$
\max_{\mathbf{x}_0}(f(\mathbf{x}) \vee g(\mathbf{x})) = \max_{\mathbf{x}_0} f(\mathbf{x}) \vee \max_{\mathbf{x}_0} g(\mathbf{x}) \; . \tag{2.31}
$$

Next, we assume that the function $g(\mathbf{x}_0, \mathbf{x}_1)$ does not depend on the simultaneous change of all variables of $\mathbf{x}_0$:

$$
\frac{\partial g(\mathbf{x}_0, \mathbf{x}_1)}{\partial \mathbf{x}_0} = 0 \; . \tag{2.32}
$$

All functions $g(\mathbf{x}_1)$ belong to a subset of functions that satisfy (2.32). Vectorial derivative operations of a conjunction between a function $g(\mathbf{x}_0, \mathbf{x}_1)$ that satisfies (2.32) and an arbitrary function $f(\mathbf{x}_0, \mathbf{x}_1)$ can be simplified as follows:

$$
\frac{\partial (g(\mathbf{x}_0, \mathbf{x}_1) \wedge f(\mathbf{x}_0, \mathbf{x}_1))}{\partial \mathbf{x}_0} = g(\mathbf{x}_0, \mathbf{x}_1) \wedge \frac{\partial f(\mathbf{x}_0, \mathbf{x}_1)}{\partial \mathbf{x}_0} \; , \tag{2.33}
$$

$$
\min_{\mathbf{x}_0}(g(\mathbf{x}_0, \mathbf{x}_1) \wedge f(\mathbf{x}_0, \mathbf{x}_1)) = g(\mathbf{x}_0, \mathbf{x}_1) \wedge \min_{\mathbf{x}_0} f(\mathbf{x}_0, \mathbf{x}_1) \; , \tag{2.34}
$$

$$
\max_{\mathbf{x}_0}(g(\mathbf{x}_0, \mathbf{x}_1) \wedge f(\mathbf{x}_0, \mathbf{x}_1)) = g(\mathbf{x}_0, \mathbf{x}_1) \wedge \max_{\mathbf{x}_0} f(\mathbf{x}_0, \mathbf{x}_1) \; . \tag{2.35}
$$

When $g(\mathbf{x}_0, \mathbf{x}_1)$ satisfies (2.32) and the $\wedge$-operations on the left-hand side of (2.33), ..., (2.35) are replaced by $\vee$-operations we get:

$$
\frac{\partial (g(\mathbf{x}_0, \mathbf{x}_1) \vee f(\mathbf{x}_0, \mathbf{x}_1))}{\partial \mathbf{x}_0} = \overline{g(\mathbf{x}_0, \mathbf{x}_1)} \wedge \frac{\partial f(\mathbf{x}_0, \mathbf{x}_1)}{\partial \mathbf{x}_0} \; , \tag{2.36}
$$

$$
\min_{\mathbf{x}_0}(g(\mathbf{x}_0, \mathbf{x}_1) \vee f(\mathbf{x}_0, \mathbf{x}_1)) = g(\mathbf{x}_0, \mathbf{x}_1) \vee \min_{\mathbf{x}_0} f(\mathbf{x}_0, \mathbf{x}_1) \; , \tag{2.37}
$$

$$
\max_{\mathbf{x}_0}(g(\mathbf{x}_0, \mathbf{x}_1) \vee f(\mathbf{x}_0, \mathbf{x}_1)) = g(\mathbf{x}_0, \mathbf{x}_1) \vee \max_{\mathbf{x}_0} f(\mathbf{x}_0, \mathbf{x}_1) \; , \tag{2.38}
$$

and the replacement by $\oplus$-operations leads to:

$$\frac{\partial(g(\mathbf{x}_0,\mathbf{x}_1)\oplus f(\mathbf{x}_0,\mathbf{x}_1))}{\partial\mathbf{x}_0}=\frac{\partial f(\mathbf{x}_0,\mathbf{x}_1)}{\partial\mathbf{x}_0}\,, \tag{2.39}$$

$$\min_{\mathbf{x}_0}(g(\mathbf{x}_0,\mathbf{x}_1)\oplus f(\mathbf{x}_0,\mathbf{x}_1))=\overline{g(\mathbf{x}_0,\mathbf{x}_1)}\wedge\min_{\mathbf{x}_0} f(\mathbf{x}_0,\mathbf{x}_1)\vee g(\mathbf{x}_0,\mathbf{x}_1)\wedge\min_{\mathbf{x}_0}\overline{f(\mathbf{x}_0,\mathbf{x}_1)}\,, \tag{2.40}$$

$$\max_{\mathbf{x}_0}(g(\mathbf{x}_0,\mathbf{x}_1)\oplus f(\mathbf{x}_0,\mathbf{x}_1))=\overline{g(\mathbf{x}_0,\mathbf{x}_1)}\wedge\max_{\mathbf{x}_0} f(\mathbf{x}_0,\mathbf{x}_1)\vee g(\mathbf{x}_0,\mathbf{x}_1)\wedge\max_{\mathbf{x}_0}\overline{f(\mathbf{x}_0,\mathbf{x}_1)}\,. \tag{2.41}$$

If it is known that a Boolean function $f(\mathbf{x}_0,\mathbf{x}_1)$ is independent of simultaneously changing the values of all variables $x_i\in\mathbf{x}_0$, the calculation of the vectorial derivative operations is strongly simplified by the following formulas:

$$\frac{\partial f(\mathbf{x}_0,\mathbf{x}_1)}{\partial\mathbf{x}_0}=0\,, \tag{2.42}$$

$$\min_{\mathbf{x}_0} f(\mathbf{x}_0,\mathbf{x}_1)=f(\mathbf{x}_0,\mathbf{x}_1)\,, \tag{2.43}$$

$$\max_{\mathbf{x}_0} f(\mathbf{x}_0,\mathbf{x}_1)=f(\mathbf{x}_0,\mathbf{x}_1)\,. \tag{2.44}$$

2.2 SINGLE DERIVATIVE OPERATIONS

In this section we want to explore the effect of changing the value of one variable. This is a special case of the vectorial derivative operations where the set of variables $\mathbf{x}_0$ contains only the single variable x_i.

Figure 2.3 reuses Example 2.1 and emphasizes the directions of change with regard to single variables in $\mathbb{B}^3$ by solid vectors. The dashed vectors describe directions of change where the values of more than one variable have been changed simultaneously.

Very welcome features justify the autonomous definition of derivative operations with regard to a single variable (as shown in Definition 2.4) and the discussion of special properties. Due to the restriction of changing a single variable the cofactors $f_{x_i}^0=f(x_i=0,\mathbf{x}_1)$ and $f_{x_i}^1=f(x_i=1,\mathbf{x}_1)$ of the Shannon decomposition (1.33) specify the pairs of function values that are evaluated.

Definition 2.4 Single Derivative Operations.
Let $f(\mathbf{x})=f(x_i,\mathbf{x}_1)$ be a Boolean function of n variables, then

$$\frac{\partial f(\mathbf{x})}{\partial x_i}=f(x_i=0,\mathbf{x}_1)\oplus f(x_i=1,\mathbf{x}_1) \tag{2.45}$$

is the (single) **derivative**,

$$\overline{\frac{\partial f(\mathbf{x})}{\partial x_i}}=f(x_i=0,\mathbf{x}_1)\odot f(x_i=1,\mathbf{x}_1) \tag{2.46}$$

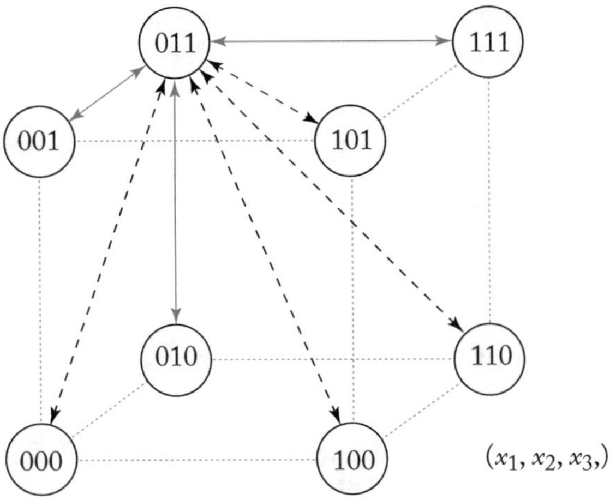

Figure 2.3: Directions of change with regard to single variables in B^3 for the selected point $(x_1, x_2, x_3) = (011)$.

is the **negation of the** (single) **derivative**,

$$\min_{x_i} f(\mathbf{x}) = f(x_i = 0, \mathbf{x}_1) \wedge f(x_i = 1, \mathbf{x}_1) \tag{2.47}$$

is the (single) **minimum**, and

$$\max_{x_i} f(\mathbf{x}) = f(x_i = 0, \mathbf{x}_1) \vee f(x_i = 1, \mathbf{x}_1) \tag{2.48}$$

is the (single) **maximum**

of the Boolean function $f(\mathbf{x})$ with regard to the variable x_i.

For a clear separation of this subset of vectorial derivative operations we omit the term *vectorial* when the set of variables $\mathbf{x}_0$ only contains one variable x_i. The term *single* emphasizes that the value of a single variable will change. This term is used to distinguish the single derivative operations from vectorial derivative operations. In general, the term *single* is omitted. The results of these four single derivative operations are again Boolean functions. Due to the constant values of x_i in the used cofactors, they only depend, however, on the variables of $\mathbf{x}_1$. The interpretation of the single derivative operations can be restricted to the evaluation of the change behavior caused by the change of the single variable x_i.

- The derivative of $f(\mathbf{x})$ with regard to x_i is equal to 1 for such assignments $\mathbf{x}_1 = \mathbf{c}$ where the change of the variable $x_i \in \mathbf{x}$ changes the value of the given function.

- The negation of the derivative of $f(\mathbf{x})$ with regard to x_i is equal to 1 for such assignments $\mathbf{x}_1 = \mathbf{c}$ where the change of the variable $x_i \in \mathbf{x}$ does not change the value of the given function.
- The minimum of $f(\mathbf{x})$ with regard to x_i is equal to 1 for such assignments $\mathbf{x}_1 = \mathbf{c}$ where the change of the variable $x_i \in \mathbf{x}$ lets the value of the given function unchanged equal to 1. Hence, the minimum expresses the *for all* relation ($\forall$) of $f(\mathbf{x})$ with regard to the variable x_i.
- The maximum of $f(\mathbf{x})$ with regard to x_i is equal to 1 if at least once the value 1 of the given function appears for the assignment $\mathbf{x}_1 = \mathbf{c}$. Hence, the maximum expresses the *exists* relation ($\exists$) of $f(\mathbf{x})$ with regard to the variable x_i.

Theorem 2.5 states that the single derivative operations of Definition 2.4 are equivalent to the vectorial derivative operations of Definition 2.2 when the set of variables $\mathbf{x}_0$ contains only the change variable x_i.

Theorem 2.5 Single Derivative Operations—Subset of Vectorial Derivatives. *If the set of variables $\mathbf{x}_0$ that specifies the direction of change of the vectorial derivative operations of the function $f(\mathbf{x}) = f(\mathbf{x}_0, \mathbf{x}_1)$ only contains a single variable x_i, then each vectorial derivative operation of Definition 2.2 is equivalent to the correspondent single derivative operation of Definition 2.4:*

$$\frac{\partial f(\mathbf{x})}{\partial x_i} = f(x_i, \mathbf{x}_1) \oplus f(\overline{x}_i, \mathbf{x}_1) = f(x_i = 0, \mathbf{x}_1) \oplus f(x_i = 1, \mathbf{x}_1) , \tag{2.49}$$

$$\overline{\frac{\partial f(\mathbf{x})}{\partial x_i}} = f(x_i, \mathbf{x}_1) \odot f(\overline{x}_i, \mathbf{x}_1) = f(x_i = 0, \mathbf{x}_1) \odot f(x_i = 1, \mathbf{x}_1) , \tag{2.50}$$

$$\min_{x_i} f(\mathbf{x}) = f(x_i, \mathbf{x}_1) \wedge f(\overline{x}_i, \mathbf{x}_1) = f(x_i = 0, \mathbf{x}_1) \wedge f(x_i = 1, \mathbf{x}_1) , \tag{2.51}$$

$$\max_{x_i} f(\mathbf{x}) = f(x_i, \mathbf{x}_1) \vee f(\overline{x}_i, \mathbf{x}_1) = f(x_i = 0, \mathbf{x}_1) \vee f(x_i = 1, \mathbf{x}_1) . \tag{2.52}$$

Theorem 2.5 can be proven by applying the Shannon decomposition (1.33) to the left-hand side of the equations in Theorem 2.5. Based on Definition 2.2 and transformations using the rules to simplify Boolean expressions lead to the right-hand side taken from Definition 2.4, e.g., for the derivative with regard to x_i we get:

$$\begin{aligned}
\frac{\partial f(x_i, \mathbf{x}_1)}{\partial x_i} &= f(x_i, \mathbf{x}_1) \oplus f(\overline{x}_i, \mathbf{x}_1) \\
&= (\overline{x}_i f(x_i = 0, \mathbf{x}_1) \oplus x_i f(x_i = 1, \mathbf{x}_1)) \oplus (\overline{x}_i f(x_i = 1, \mathbf{x}_1) \oplus x_i f(x_i = 0, \mathbf{x}_1)) \\
&= \overline{x}_i (f(x_i = 0, \mathbf{x}_1) \oplus f(x_i = 1, \mathbf{x}_1)) \oplus x_i (f(x_i = 1, \mathbf{x}_1) \oplus f(x_i = 0, \mathbf{x}_1)) \\
&= (\overline{x}_i \oplus x_i)(f(x_i = 0, \mathbf{x}_1) \oplus f(x_i = 1, \mathbf{x}_1)) \\
&= f(x_i = 0, \mathbf{x}_1) \oplus f(x_i = 1, \mathbf{x}_1) .
\end{aligned}$$

Similar transformations can be used for the proof of the other single derivative operations.

Due to the same arguments as in the case of the more general negated vectorial derivative, we omit the negation of the derivative (2.46) in the following discussions of the details of the single derivative operations.

For the derivative of $f(\mathbf{x})$ with regard to x_i two border cases are possible. The first one is:

$$\frac{\partial f(\mathbf{x})}{\partial x_i} = f(x_i = 0, \mathbf{x}_1) \oplus f(x_i = 1, \mathbf{x}_1) = 0 \, . \tag{2.53}$$

In this case the negative cofactor $f(x_i = 0, \mathbf{x}_1)$ and the positive cofactor $f(x_i = 1, \mathbf{x}_1)$ are equal to each other and the change of the value of x_i cannot cause a value change of $f(\mathbf{x})$. Hence, the function $f(\mathbf{x})$ of the Boolean space $\mathbb{B}^n$ is independent of the variable x_i and can be expressed by a Boolean function $f(\mathbf{x}_1)$ over $\mathbb{B}^{n-1}$. Such a simplification is welcome in circuit design because it allows to reduce the number of gates and consequently the power consumption.

The other extreme case is:

$$\frac{\partial f(\mathbf{x})}{\partial x_i} = f(x_i = 0, \mathbf{x}_1) \oplus f(x_i = 1, \mathbf{x}_1) = 1 \, . \tag{2.54}$$

In this case the negative cofactor $f(x_i = 0, \mathbf{x}_1)$ is the negation of the positive cofactor $f(x_i = 1, \mathbf{x}_1)$. Hence, the change of x_i causes the change of the function value of $f(\mathbf{x})$ for all assignments to the variables $\mathbf{x}_1$. This property can also be utilized in circuit design.

Example 2.6 Linear Separation of x_i.
If (2.54) is satisfied, the function $f(\mathbf{x})$ is *linear* in x_i and can be expressed by:

$$f(\mathbf{x}) = x_i \oplus g(\mathbf{x}_1) \, . \tag{2.55}$$

The substitution of (2.55) into (2.45) confirms the correctness of the linearity condition (2.54):

$$\begin{aligned}
\frac{\partial f(\mathbf{x})}{\partial x_i} &= f(x_i = 0, \mathbf{x}_1) \oplus f(x_i = 1, \mathbf{x}_1) \\
\frac{\partial (x_i \oplus g(\mathbf{x}_1))}{\partial x_i} &= 0 \oplus g(\mathbf{x}_1) \oplus 1 \oplus g(\mathbf{x}_1) \\
\frac{\partial (x_i \oplus g(\mathbf{x}_1))}{\partial x_i} &= 1 \, .
\end{aligned}$$

Definition (2.45) directly reveals another important property of all single derivatives. The derivative of $f(\mathbf{x})$ with regard to the variable x_i is independent of x_i:

$$\frac{\partial}{\partial x_i}\left(\frac{\partial f(\mathbf{x})}{\partial x_i}\right) = 0 \, . \tag{2.56}$$

Using Definition (2.45), the abbreviations (1.28) and (1.29) of the cofactors, and basic rules for simplifications, the law (2.56) can be confirmed by:

$$\begin{aligned}\frac{\partial}{\partial x_i}\left(\frac{\partial f(\mathbf{x})}{\partial x_i}\right) &= \frac{\partial}{\partial x_i}\left(f^0_{x_i}(\mathbf{x}_1) \oplus f^1_{x_i}(\mathbf{x}_1)\right) \\ &= \left(f^0_{x_i}(\mathbf{x}_1) \oplus f^1_{x_i}(\mathbf{x}_1)\right) \oplus \left(f^0_{x_i}(\mathbf{x}_1) \oplus f^1_{x_i}(\mathbf{x}_1)\right) \\ &= 0\,.\end{aligned}$$

The number of values 1 in a single derivative can be used as a measure of linearity of $f(\mathbf{x})$ with regard to the selected variable x_i.

Definition 2.7 Degree of Linearity. A Boolean function $f(x_i, \mathbf{x}_1)$ of n variables has a *degree of linearity* with regard to the variable x_i within the interval $[0, \ldots, 1]$ defined by

$$\mathbf{degree}^{lin}_{x_i} f(x_i, \mathbf{x}_1) = \frac{1}{2^{n-1}} * \left|\frac{\partial f(x_i, \mathbf{x}_1)}{\partial x_i}\right| , \tag{2.57}$$

where $|g(\mathbf{x})|$ is the number of values 1 of the evaluated function $g(\mathbf{x})$.

In the extreme case of Example 2.6 we have:

$$\mathbf{degree}^{lin}_{x_i} f(\mathbf{x}) = 1 \quad \text{and} \quad \mathbf{degree}^{lin}_{x_i} g(\mathbf{x}_1) = 0\,.$$

Generally, the variable x_i can be separated from each function $f(x_i, \mathbf{x}_1)$ using an EXOR-gate:

$$f(x_i, \mathbf{x}_1) = x_i \oplus g(x_i, \mathbf{x}_1)\,, \tag{2.58}$$

where

$$\mathbf{degree}^{lin}_{x_i} f(x_i, \mathbf{x}_1) = 1 - \mathbf{degree}^{lin}_{x_i} g(x_i, \mathbf{x}_1)\,. \tag{2.59}$$

Similar to the derivative with regard to x_i, both the minimum and the maximum of $f(\mathbf{x})$ with regard to x_i are defined by means of the positive and negative cofactors of this function. Hence, these two single derivative operations do also not depend on the variable x_i anymore:

$$\frac{\partial}{\partial x_i}\left(\min_{x_i} f(\mathbf{x})\right) = 0\,, \tag{2.60}$$

$$\frac{\partial}{\partial x_i}\left(\max_{x_i} f(\mathbf{x})\right) = 0\,, \tag{2.61}$$

so that subsequent calculations become easier.

Figure 2.4 shows how the three single derivative operations are calculated for a small function with regard to x_2 using Karnaugh-maps. The arrows in the Karnaugh-map of the given function $f(x_1, x_2, x_3)$ emphasize the pairs of function values which determine the single derivative operations with regard to x_2.

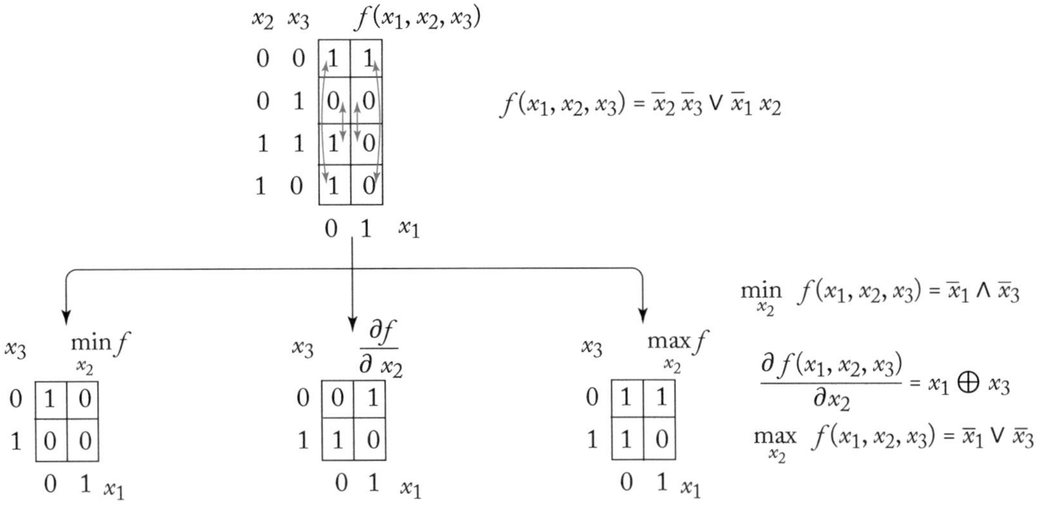

Figure 2.4: Karnaugh-maps of a function and all three associated single derivative operations with regard to x_2.

As special case of (2.13) we get for the single minimum and maximum the inequality:

$$\min_{x_i} f(\mathbf{x}) \leq f(\mathbf{x}) \leq \max_{x_i} f(\mathbf{x}) \,, \tag{2.62}$$

which motivates the chosen names of these derivative operations. The truth of this inequality can be shown in the same way as demonstrated for (2.13). From (2.62) follow three restrictions which can be used to simplify expressions:

$$\min_{x_i} f(\mathbf{x}) \wedge \overline{f(\mathbf{x})} = 0 \,, \tag{2.63}$$

$$\min_{x_i} f(\mathbf{x}) \wedge \overline{\max_{x_i} f(\mathbf{x})} = 0 \,, \tag{2.64}$$

$$f(\mathbf{x}) \wedge \overline{\max_{x_i} f(\mathbf{x})} = 0 \,. \tag{2.65}$$

Definition 2.4 of the single derivative operations can directly be used to confirm the following relations:

$$\min_{x_i} f(\mathbf{x}) \wedge \overline{\frac{\partial f(\mathbf{x})}{\partial x_i}} = 0 \,, \tag{2.66}$$

$$\frac{\partial f(\mathbf{x})}{\partial x_i} \wedge \overline{\max_{x_i} f(\mathbf{x})} = 0 \,. \tag{2.67}$$

We noticed already that a Boolean function $f(x_i, \mathbf{x}_1)$ of $\mathbb{B}^n$ can be independent of the variable x_i and provided the condition (2.53) to check whether the function satisfies this property. The maximum with regard to x_i can be used to map such a function from $\mathbb{B}^n$ to $\mathbb{B}^{n-1}$:

$$f(\mathbf{x}_1) = \max_{x_i} f(x_i, \mathbf{x}_1) \; . \tag{2.68}$$

The three single derivative operations satisfy the equation:

$$\min_{x_i} f(\mathbf{x}) \oplus \max_{x_i} f(\mathbf{x}) \oplus \frac{\partial f(\mathbf{x})}{\partial x_i} = 0 \; , \tag{2.69}$$

because the associated definitions only differ in the operation between the two cofactors. Using Definition 2.4 and the abbreviations of the cofactors we get for (2.69):

$$\begin{aligned}
(f_{x_i}^0(\mathbf{x}_1) \wedge f_{x_i}^1(\mathbf{x}_1)) \oplus (f_{x_i}^0(\mathbf{x}_1) \vee f_{x_i}^1(\mathbf{x}_1)) \oplus (f_{x_i}^0(\mathbf{x}_1) \oplus f_{x_i}^1(\mathbf{x}_1)) &= 0 \\
(f_{x_i}^0(\mathbf{x}_1) \wedge f_{x_i}^1(\mathbf{x}_1)) \oplus (f_{x_i}^0(\mathbf{x}_1) \oplus f_{x_i}^1(\mathbf{x}_1) \oplus f_{x_i}^0(\mathbf{x}_1) \wedge f_{x_i}^1(\mathbf{x}_1)) \oplus (f_{x_i}^0(\mathbf{x}_1) \oplus f_{x_i}^1(\mathbf{x}_1)) &= 0 \\
0 &= 0 \; .
\end{aligned}$$

Equation (2.69) can be transformed such that each single derivative operation is expressed by the other two operations:

$$\min_{x_i} f(\mathbf{x}) = \max_{x_i} f(\mathbf{x}) \oplus \frac{\partial f(\mathbf{x})}{\partial x_i} \; , \tag{2.70}$$

$$\max_{x_i} f(\mathbf{x}) = \min_{x_i} f(\mathbf{x}) \oplus \frac{\partial f(\mathbf{x})}{\partial x_i} \; , \tag{2.71}$$

$$\frac{\partial f(\mathbf{x})}{\partial x_i} = \min_{x_i} f(\mathbf{x}) \oplus \max_{x_i} f(\mathbf{x}) \; . \tag{2.72}$$

These rules can be transformed into expressions that do not contain the $\oplus$-operation:

$$\min_{x_i} f(\mathbf{x}) = \max_{x_i} f(\mathbf{x}) \wedge \overline{\frac{\partial f(\mathbf{x})}{\partial x_i}} \; , \tag{2.73}$$

$$\max_{x_i} f(\mathbf{x}) = \min_{x_i} f(\mathbf{x}) \vee \frac{\partial f(\mathbf{x})}{\partial x_i} \; , \tag{2.74}$$

$$\frac{\partial f(\mathbf{x})}{\partial x_i} = \overline{\min_{x_i} f(\mathbf{x})} \wedge \max_{x_i} f(\mathbf{x}) \; . \tag{2.75}$$

The transformation from (2.70) to (2.73) uses the rule (1.56) and the restriction (2.67):

$$\begin{aligned}
\min_{x_i} f(\mathbf{x}) &= \max_{x_i} f(\mathbf{x}) \oplus \frac{\partial f(\mathbf{x})}{\partial x_i} \; , \\
&= \overline{\max_{x_i} f(\mathbf{x})} \wedge \frac{\partial f(\mathbf{x})}{\partial x_i} \vee \max_{x_i} f(\mathbf{x}) \wedge \overline{\frac{\partial f(\mathbf{x})}{\partial x_i}} \; , \\
&= \max_{x_i} f(\mathbf{x}) \wedge \overline{\frac{\partial f(\mathbf{x})}{\partial x_i}} \; .
\end{aligned}$$

The transformation from (2.71) to (2.74) can be executed in a similar manner:

$$
\begin{aligned}
\max_{x_i} f(\mathbf{x}) &= \min_{x_i} f(\mathbf{x}) \oplus \frac{\partial f(\mathbf{x})}{\partial x_i} \\
&= \left(\min_{x_i} f(\mathbf{x}) \vee \frac{\partial f(\mathbf{x})}{\partial x_i}\right) \oplus \left(\min_{x_i} f(\mathbf{x}) \wedge \frac{\partial f(\mathbf{x})}{\partial x_i}\right) \\
&= \left(\min_{x_i} f(\mathbf{x}) \vee \frac{\partial f(\mathbf{x})}{\partial x_i}\right) \oplus 0 \\
&= \min_{x_i} f(\mathbf{x}) \vee \frac{\partial f(\mathbf{x})}{\partial x_i} \; .
\end{aligned}
$$

The last transformation from (2.72) to (2.75) uses again the rule (1.56), but now the restriction (2.64):

$$
\begin{aligned}
\frac{\partial f(\mathbf{x})}{\partial x_i} &= \min_{x_i} f(\mathbf{x}) \oplus \max_{x_i} f(\mathbf{x}) \\
&= \overline{\min_{x_i} f(\mathbf{x})} \wedge \max_{x_i} f(\mathbf{x}) \vee \min_{x_i} f(\mathbf{x}) \wedge \overline{\max_{x_i} f(\mathbf{x})} \\
&= \overline{\min_{x_i} f(\mathbf{x})} \wedge \max_{x_i} f(\mathbf{x}) \; .
\end{aligned}
$$

Next, we explore several cases where Boolean operations occur within the argument of a single derivative operation. By using the rules of the Boolean Algebra the complement operation can be eliminated from the argument of all single derivative operations:

$$
\frac{\partial \overline{f(\mathbf{x})}}{\partial x_i} = \frac{\partial f(\mathbf{x})}{\partial x_i} \; , \tag{2.76}
$$

$$
\min_{x_i} \overline{f(\mathbf{x})} = \overline{\max_{x_i} f(\mathbf{x})} \; , \tag{2.77}
$$

$$
\max_{x_i} \overline{f(\mathbf{x})} = \overline{\min_{x_i} f(\mathbf{x})} \; . \tag{2.78}
$$

These and the following formulas are special cases of the vectorial derivative operation, where the set $\mathbf{x}_0$ only contains the variable x_i.

The argument of a derivative operation can consist of more than one Boolean function. In the case of two functions which are connected by the same operation as used for the definition of the single derivative operation, a separation is easy:

$$
\frac{\partial (f(\mathbf{x}) \oplus g(\mathbf{x}))}{\partial x_i} = \frac{\partial f(\mathbf{x})}{\partial x_i} \oplus \frac{\partial g(\mathbf{x})}{\partial x_i} \; , \tag{2.79}
$$

$$
\min_{x_i}(f(\mathbf{x}) \wedge g(\mathbf{x})) = \min_{x_i} f(\mathbf{x}) \wedge \min_{x_i} g(\mathbf{x}) \; , \tag{2.80}
$$

$$
\max_{x_i}(f(\mathbf{x}) \vee g(\mathbf{x})) = \max_{x_i} f(\mathbf{x}) \vee \max_{x_i} g(\mathbf{x}) \; . \tag{2.81}
$$

If the function $f(\mathbf{x}) = f(x_i, \mathbf{x}_1)$ depends on x_i, but the function $g(\mathbf{x}_1)$ is independent of x_i, we get for the connection by a $\wedge$-operation:

$$\frac{\partial(g(\mathbf{x}_1) \wedge f(x_i, \mathbf{x}_1))}{\partial x_i} = g(\mathbf{x}_1) \wedge \frac{\partial f(x_i, \mathbf{x}_1)}{\partial x_i} , \tag{2.82}$$

$$\min_{x_i}(g(\mathbf{x}_1) \wedge f(x_i, \mathbf{x}_1)) = g(\mathbf{x}_1) \wedge \min_{x_i} f(x_i, \mathbf{x}_1) , \tag{2.83}$$

$$\max_{x_i}(g(\mathbf{x}_1) \wedge f(x_i, \mathbf{x}_1)) = g(\mathbf{x}_1) \wedge \max_{x_i} f(x_i, \mathbf{x}_1) . \tag{2.84}$$

When the $\wedge$-operation on the left-hand side of (2.82), ..., (2.84) will be replaced by a $\vee$-operation, we get:

$$\frac{\partial(g(\mathbf{x}_1) \vee f(x_i, \mathbf{x}_1))}{\partial x_i} = \overline{g(\mathbf{x}_1)} \wedge \frac{\partial f(x_i, \mathbf{x}_1)}{\partial x_i} , \tag{2.85}$$

$$\min_{x_i}(g(\mathbf{x}_1) \vee f(x_i, \mathbf{x}_1)) = g(\mathbf{x}_1) \vee \min_{x_i} f(x_i, \mathbf{x}_1) , \tag{2.86}$$

$$\max_{x_i}(g(\mathbf{x}_1) \vee f(x_i, \mathbf{x}_1)) = g(\mathbf{x}_1) \vee \max_{x_i} f(x_i, \mathbf{x}_1) , \tag{2.87}$$

and the replacement by $\oplus$-operations leads to:

$$\frac{\partial(g(\mathbf{x}_1) \oplus f(x_i, \mathbf{x}_1))}{\partial x_i} = \frac{\partial f(x_i, \mathbf{x}_1)}{\partial x_i} , \tag{2.88}$$

$$\min_{x_i}(g(\mathbf{x}_1) \oplus f(x_i, \mathbf{x}_1)) = \overline{g(\mathbf{x}_1)} \wedge \min_{x_i} f(x_i, \mathbf{x}_1) \vee g(\mathbf{x}_1) \wedge \min_{x_i} \overline{f(x_i, \mathbf{x}_1)} , \tag{2.89}$$

$$\max_{x_i}(g(\mathbf{x}_1) \oplus f(x_i, \mathbf{x}_1)) = \overline{g(\mathbf{x}_1)} \wedge \max_{x_i} f(x_i, \mathbf{x}_1) \vee g(\mathbf{x}_1) \wedge \max_{x_i} \overline{f(x_i, \mathbf{x}_1)} . \tag{2.90}$$

It can occur that a derivative operation with regard to x_i has to be calculated for a function $f(\mathbf{x}_1)$ that does not depend on x_i, i.e., $x_i \notin \mathbf{x}_1$. In such a case we get the following results:

$$\frac{\partial f(\mathbf{x}_1)}{\partial x_i} = 0 , \tag{2.91}$$

$$\min_{x_i} f(\mathbf{x}_1) = f(\mathbf{x}_1) , \tag{2.92}$$

$$\max_{x_i} f(\mathbf{x}_1) = f(\mathbf{x}_1) . \tag{2.93}$$

A positive polarity Reed-Muller polynomial is an antivalence form of the function f in which only non-negated variables occur. Using the associativity (1.38), the commutativity (1.36), and the distributivity (1.46), such a polynomial can be transformed into:

$$f(x_i, \mathbf{x}_1) = x_i \wedge g(\mathbf{x}_1) \oplus h(\mathbf{x}_1) . \tag{2.94}$$

The rules (2.79), (2.82), and (2.91) simplify the calculation of the derivative of (2.94) with regard to x_i. Using (2.79) and (2.91) we get:

$$\frac{\partial(x_i \wedge g(\mathbf{x}_1) \oplus h(\mathbf{x}_1))}{\partial x_i} = \frac{\partial(x_i \wedge g(\mathbf{x}_1))}{\partial x_i} \oplus \frac{\partial h(\mathbf{x}_1)}{\partial x_i}$$
$$\frac{\partial(x_i \wedge g(\mathbf{x}_1) \oplus h(\mathbf{x}_1))}{\partial x_i} = \frac{\partial(x_i \wedge g(\mathbf{x}_1))}{\partial x_i} \oplus 0$$
$$\frac{\partial(x_i \wedge g(\mathbf{x}_1) \oplus h(\mathbf{x}_1))}{\partial x_i} = \frac{\partial(x_i \wedge g(\mathbf{x}_1))}{\partial x_i},$$

and the consecutive application of (2.82) leads finally to:

$$\frac{\partial(x_i \wedge g(\mathbf{x}_1) \oplus h(\mathbf{x}_1))}{\partial x_i} = g(\mathbf{x}_1) \wedge \frac{\partial x_i}{\partial x_i}$$
$$\frac{\partial(x_i \wedge g(\mathbf{x}_1) \oplus h(\mathbf{x}_1))}{\partial x_i} = g(\mathbf{x}_1) \wedge 1$$
$$\frac{\partial(x_i \wedge g(\mathbf{x}_1) \oplus h(\mathbf{x}_1))}{\partial x_i} = g(\mathbf{x}_1) .$$

Similar simplifications exist for the single minimum and maximum.

2.3 m-FOLD DERIVATIVE OPERATIONS

Commonly, each vectorial and as subset each single derivative operation evaluates the change behavior of a given function of n variables for certain pairs of points in the Boolean space $\mathbb{B}^n$. These pairs of points specify Boolean subspaces $\mathbb{B}^1$ in the case of a single derivative operation. Hence, the single derivative operations of the Boolean function $f(\mathbf{x}) = f(x_i, \mathbf{x}_1)$ with regard to x_i describe the change behavior of $f(x_i, \mathbf{x}_1)$ in 2^{n-1} subspaces $\mathbb{B}^1 = \{x_i | x_i \in \{0, 1\}\}$.

Knowing that the result of a single derivative operation is again a Boolean function, a consecutive calculation of several single derivative operations of the same type is possible. Taking, for example, the variables x_i and x_j of the function $f(\mathbf{x}) = f(x_i, x_j, \mathbf{x}_1)$ as variables of the consecutively calculated single derivative operations of the same type, the solution describes the change behavior of $f(x_i, x_j, \mathbf{x}_1)$ in 2^{n-2} subspaces $\mathbb{B}^2 = \{(x_i, x_j) | x_i, x_j \in \{0, 1\}\}$. Hence, the consecutive calculation of several single derivative operations of the same type provide an insight into the change behavior of certain subspaces. m-fold derivative operations describe this more global quality of the change behavior.

Definition 2.8 ***m*-fold Derivative.** Let $\mathbf{x}_0 = (x_1, x_2, \ldots, x_m)$, $\mathbf{x}_1 = (x_{m+1}, x_{m+2}, \ldots, x_n)$ be two disjoint sets of Boolean variables, and $f(\mathbf{x}_0, \mathbf{x}_1) = f(x_1, x_2, \ldots, x_n) = f(\mathbf{x})$ be a Boolean function of n variables, then

$$\frac{\partial^m f(\mathbf{x}_0, \mathbf{x}_1)}{\partial x_1 \partial x_2 \ldots \partial x_m} = \frac{\partial}{\partial x_m}\left(\cdots\left(\frac{\partial}{\partial x_2}\left(\frac{\partial f(\mathbf{x}_0, \mathbf{x}_1)}{\partial x_1}\right)\right)\cdots\right) \tag{2.95}$$

is the m-**fold derivative** of the Boolean function $f(\mathbf{x}_0, \mathbf{x}_1)$ with regard to the subset of variables $\mathbf{x}_0$.

The consecutive calculation using Definition (2.95) does not explicitly reveal the relation between the function values in the 2^m points of a subspace

$$\mathbb{B}^m = \{(x_1, \ldots, x_m) | x_i, \in \{0, 1\}\}, \forall i = 1, \ldots, m$$

and the associated Boolean value of the m-fold derivative. The detailed calculation of the m-fold derivative using m-times Definition (2.45) of the derivative leads to both an alternative formula to calculate the m-fold derivative and the understanding of their meaning.

Example 2.9 2-fold Derivative: Alternative Rule.
The 2-fold derivative of $f(x_1, x_2, \mathbf{x}_1)$ with regard to x_1 and x_2 can be calculated as follows:

$$\begin{aligned}
\frac{\partial^2 f(x_1, x_2, \mathbf{x}_1)}{\partial x_1 \partial x_2} &= \frac{\partial}{\partial x_2}\left(\frac{\partial f(x_1, x_2, \mathbf{x}_1)}{\partial x_1}\right) \\
&= \frac{\partial}{\partial x_2}\left(f(x_1 = 0, x_2, \mathbf{x}_1) \oplus f(x_1 = 1, x_2, \mathbf{x}_1)\right) \\
&= f(x_1 = 0, x_2 = 0, \mathbf{x}_1) \oplus f(x_1 = 1, x_2 = 0, \mathbf{x}_1) \oplus \\
&\quad\; f(x_1 = 0, x_2 = 1, \mathbf{x}_1) \oplus f(x_1 = 1, x_2 = 1, \mathbf{x}_1) \\
&= \bigoplus_{(c_1, c_2) \in \mathbb{B}^2} f(x_1 = c_1, x_2 = c_2, \mathbf{x}_1) \,.
\end{aligned} \tag{2.96}$$

Obviously, the value of the 2-fold derivative for the subspace $\mathbb{B}^2$ with $\mathbf{x}_1 = \mathbf{c}_1$ results from the EXOR-operation of the four function values of $f(x_1, x_2, \mathbf{x}_1 = \mathbf{c}_1)$. Hence, the 2-fold derivative is equal to 1 if the associated subspace $\mathbb{B}^2$ contains either one or three function values 1.

Formula (2.96) can be generalized for an arbitrary number m of variables $\mathbf{x}_0 = (x_1, x_2, \ldots, x_m)$:

$$\frac{\partial^m f(\mathbf{x}_0, \mathbf{x}_1)}{\partial x_1 \partial x_2 \ldots \partial x_m} = \bigoplus_{\mathbf{c}_0 \in \mathbb{B}^m} f(\mathbf{x}_0 = \mathbf{c}_0, \mathbf{x}_1) \,. \tag{2.97}$$

This formula reveals the meaning of the m-fold derivative of $f(\mathbf{x}_0, \mathbf{x}_1)$ with regard to $\mathbf{x}_0$:

- the m-fold derivative is equal to 1 for for subspaces $\mathbb{B}^m$ with $\mathbf{x}_1 = \mathbf{c}_1$ that contain an *odd* number of function values 1.

The possibility to distinguish between subspaces of odd or even numbers of function values 1 of a given function opens a wide field of applications.

Example 2.10 Evaluation of Subspaces Calculating a 2-fold Derivative in $\mathbb{B}^3$.
The 2-fold derivative of

$$f(x_1, x_2, x_3) = x_1 x_2 \vee x_2 \overline{x}_3 \vee \overline{x}_2 x_3$$

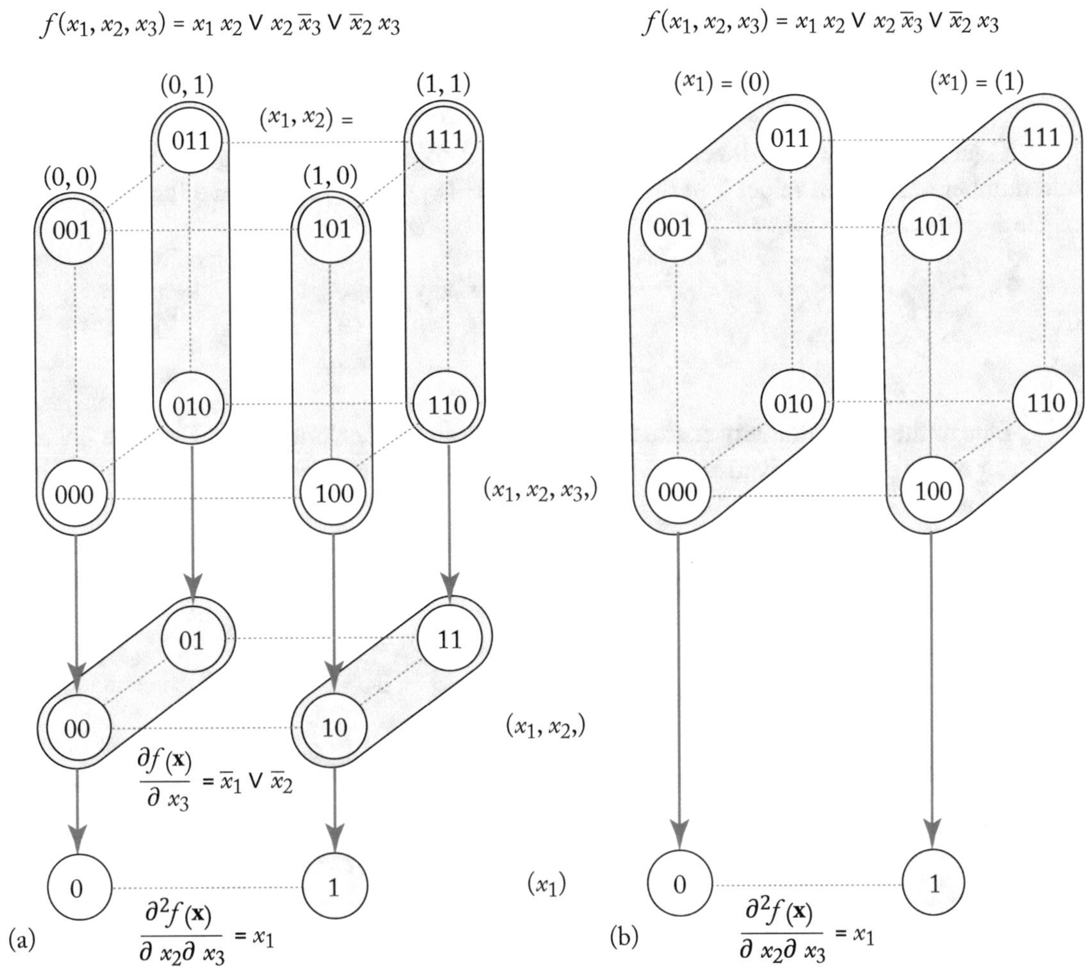

Figure 2.5: Calculation of a 2-fold derivative: (a) consecutively, at first with regard to x_3 and thereafter with regard to x_2, and (b) directly by evaluating the function values 1 in subspaces $\mathbb{B}^2$.

with regard to x_2 and x_3 can be calculated in two different ways.

Figure 2.5a shows the consecutive calculation using Definition (2.95). In the first step the four subspaces $\mathbb{B}^1\ (x_1, x_2) = (c_1, c_2)$ with gray background in the top left part of Figure 2.5 are evaluated: different function values in these subspaces result in function values 1 of the derivative

$$\frac{\partial f(x_1, x_2, x_3)}{\partial x_3} = \overline{x}_1 \vee \overline{x}_2 \ .$$

In the subsequent step the two subspaces $\mathbb{B}^1\ (x_1) = (c_1)$ with gray background in the left middle of Figure 2.5 are evaluated: different function values in these subspaces result in function values

1 of the 2-fold derivative

$$\frac{\partial^2 f(x_1, x_2, x_3)}{\partial x_2 \partial x_3} = x_1 \ .$$

Figure 2.5b shows the direct calculation of this 2-fold derivative using the rule (2.97): an odd number of function values 1 in these subspaces $\mathbb{B}^2\ (x_1) = (c_1)$ results in a function value 1 of the 2-fold derivative

$$\frac{\partial^2 f(x_1, x_2, x_3)}{\partial x_2 \partial x_3} = x_1 \ .$$

Due to the commutativity of the $\oplus$-operation it can be concluded from Example 2.9 that the result of the m-fold derivative does not depend on the order in which the single derivatives are calculated.

Formula (2.97) also shows that the m-fold derivative of $f(\mathbf{x}_0, \mathbf{x}_1)$ of n variables with regard to $\mathbf{x}_0$ is a Boolean function that does not depend on the m variables $x_i \in \mathbf{x}_0$, but only on the $n - m$ variables of $\mathbf{x}_1$:

$$\bigvee_{i=1}^{m} \frac{\partial}{\partial x_i} \left(\frac{\partial^m f(\mathbf{x}_0, \mathbf{x}_1)}{\partial x_1 \partial x_2 \dots \partial x_m} \right) = 0 \ . \tag{2.98}$$

This strong simplification is a very welcome property of the m-fold derivative and suggests to generalize the single minimum and maximum to m-fold derivative operations as well.

Definition 2.11 ***m*-fold Minimum and Maximum.**
Let $\mathbf{x}_0 = (x_1, x_2, \dots, x_m)$, $\mathbf{x}_1 = (x_{m+1}, \dots, x_n)$ be two disjoint sets of Boolean variables, and $f(\mathbf{x}_0, \mathbf{x}_1) = f(x_1, x_2 \dots, x_n) = f(\mathbf{x})$ be a Boolean function of n variables, then

$$\min_{\mathbf{x}_0}{}^m f(\mathbf{x}_0, \mathbf{x}_1) = \min_{x_m} \left(\dots \left(\min_{x_2} \left(\min_{x_1} f(\mathbf{x}_0, \mathbf{x}_1) \right) \right) \dots \right) \tag{2.99}$$

is the m-**fold minimum** and

$$\max_{\mathbf{x}_0}{}^m f(\mathbf{x}_0, \mathbf{x}_1) = \max_{x_m} \left(\dots \left(\max_{x_2} \left(\max_{x_1} f(\mathbf{x}_0, \mathbf{x}_1) \right) \right) \dots \right) \tag{2.100}$$

is the m-**fold maximum**
of the Boolean function $f(\mathbf{x}_0, \mathbf{x}_1)$ with regard to the subset of variables $\mathbf{x}_0$.

We again explore the meaning of these two m-fold derivative operations by means of simple examples.

Example 2.12 2-fold Minimum. The 2-fold minimum of $f(x_1, x_2, \mathbf{x}_1)$ with regard to x_1 and x_2 can be calculated as follows:

$$\begin{aligned}
\min_{\mathbf{x}_0}{}^2 f(x_1, x_2, \mathbf{x}_1) &= \min_{x_2} \left(\min_{x_1} f(x_1, x_2, \mathbf{x}_1) \right) \\
&= \min_{x_2} \left(f(x_1 = 0, x_2, \mathbf{x}_1) \wedge f(x_1 = 1, x_2, \mathbf{x}_1) \right) \\
&= f(x_1 = 0, x_2 = 0, \mathbf{x}_1) \wedge f(x_1 = 1, x_2 = 0, \mathbf{x}_1) \wedge \\
&\quad\; f(x_1 = 0, x_2 = 1, \mathbf{x}_1) \wedge f(x_1 = 1, x_2 = 1, \mathbf{x}_1) \\
&= \bigwedge_{(c_1, c_2) \in \mathbb{B}^2} f(x_1 = c_1, x_2 = c_2, \mathbf{x}_1) \,. \qquad (2.101)
\end{aligned}$$

Formula (2.101) shows that the value of the 2-fold minimum for the subspace $\mathbb{B}^2$ with $\mathbf{x}_1 = \mathbf{c}_1$ results from the $\wedge$-operation of the four function values of $f(x_1, x_2, \mathbf{x}_1 = \mathbf{c}_1)$. Hence, the 2-fold minimum is equal to 1 if the associated subspace $\mathbb{B}^2$ contains for all four points function values 1.

Formula (2.101) can be generalized to an arbitrary number m of variables $\mathbf{x}_0 = (x_1, x_2, \ldots, x_m)$:

$$\min_{\mathbf{x}_0}{}^m f(\mathbf{x}_0, \mathbf{x}_1) = \bigwedge_{\mathbf{c}_0 \in \mathbb{B}^m} f(\mathbf{x}_0 = \mathbf{c}_0, \mathbf{x}_1) \,. \qquad (2.102)$$

This formula reveals the meaning of the m-fold minimum of $f(\mathbf{x}_0, \mathbf{x}_1)$ with regard to $\mathbf{x}_0$:

- the m-fold minimum is equal to 1 for subspaces $\mathbb{B}^m$ with $\mathbf{x}_1 = \mathbf{c}_1$ that contain *for all* points function values 1.

The possibility to detect subspaces containing for all ($\forall$) points function values 1 is a very useful property of the m-fold minimum.

Due to the commutativity of the $\wedge$-operation it can be concluded from Example 2.12 that the result of the m-fold minimum does not depend on the order in which the single minimum operations are used.

Formula (2.102) also shows that the m-fold minimum of $f(\mathbf{x}_0, \mathbf{x}_1)$ of n variables with regard to $\mathbf{x}_0$ is a Boolean function that does not depend on all m variables $x_i \in \mathbf{x}_0$, but only on $n - m$ variables of $\mathbf{x}_1$:

$$\bigvee_{i=1}^{m} \frac{\partial}{\partial x_i} \left(\min_{\mathbf{x}_0}{}^m f(\mathbf{x}_0, \mathbf{x}_1) \right) = 0 \,. \qquad (2.103)$$

Hence, the m-fold minimum simplifies a given function in the same strong manner as the m-fold derivative.

The meaning of the m-fold maximum can also be discovered using an example where $m = 2$.

Example 2.13 2-fold Maximum. The 2-fold maximum of $f(x_1, x_2, \mathbf{x}_1)$ with regard to x_1 and x_2 can be calculated as follows:

$$\begin{aligned}
\max_{\mathbf{x}_0}{}^2 f(x_1, x_2, \mathbf{x}_1) &= \max_{x_2}\left(\max_{x_1} f(x_1, x_2, \mathbf{x}_1)\right) \\
&= \max_{x_2}\left(f(x_1 = 0, x_2, \mathbf{x}_1) \vee f(x_1 = 1, x_2, \mathbf{x}_1)\right) \\
&= f(x_1 = 0, x_2 = 0, \mathbf{x}_1) \vee f(x_1 = 1, x_2 = 0, \mathbf{x}_1) \vee \\
&\quad\; f(x_1 = 0, x_2 = 1, \mathbf{x}_1) \vee f(x_1 = 1, x_2 = 1, \mathbf{x}_1) \\
&= \bigvee_{(c_1,c_2)\in\mathbb{B}^2} f(x_1 = c_1, x_2 = c_2, \mathbf{x}_1)\,.
\end{aligned} \tag{2.104}$$

Hence, the value of the 2-fold maximum for the subspace $\mathbb{B}^2$ with $\mathbf{x}_1 = \mathbf{c}_1$ results from the $\vee$-operation of the four function values of $f(x_1, x_2, \mathbf{x}_1 = \mathbf{c}_1)$. Therefore, the 2-fold maximum is equal to 1 if the function value 1 exists at least once in the associated subspace $\mathbb{B}^2$.

Formula (2.104) can be generalized for an arbitrary number m of variables $\mathbf{x}_0 = (x_1, x_2, \ldots, x_m)$:

$$\max_{\mathbf{x}_0}{}^m f(\mathbf{x}_0, \mathbf{x}_1) = \bigvee_{\mathbf{c}_0\in\mathbb{B}^m} f(\mathbf{x}_0 = \mathbf{c}_0, \mathbf{x}_1)\,. \tag{2.105}$$

This formula reveals the meaning of the m-fold maximum of $f(\mathbf{x}_0, \mathbf{x}_1)$ with regard to $\mathbf{x}_0$:

- the m-fold maximum is equal to 1 for subspaces $\mathbb{B}^m$ with $\mathbf{x}_1 = \mathbf{c}_1$ in which the function value 1 *exists* at least for one point, and

- the m-fold maximum is equal to 0 for subspaces $\mathbb{B}^m$ with $\mathbf{x}_1 = \mathbf{c}_1$ in which the function value 1 *exists* for no point.

The possibility to detect subspaces in which at least once a function value 1 exists ($\exists$) is one of the main applications of the m-fold maximum.

Due to the commutativity of the $\vee$-operation it can be concluded from Example 2.13 that the result of the m-fold maximum does not depend on the order in which the single maximum operations are used.

Formula (2.105) also shows that the m-fold maximum of $f(\mathbf{x}_0, \mathbf{x}_1)$ of n variables with regard to $\mathbf{x}_0$ is a Boolean function that does not depend on the m variables $x_i \in \mathbf{x}_0$, but only on the $n - m$ variables of $\mathbf{x}_1$:

$$\bigvee_{i=1}^{m} \frac{\partial}{\partial x_i}\left(\max_{\mathbf{x}_0}{}^m f(\mathbf{x}_0, \mathbf{x}_1)\right) = 0\,. \tag{2.106}$$

Hence, the m-fold maximum with regard to $\mathbf{x}_0$ also strongly simplifies the given function $f(\mathbf{x}_0, \mathbf{x}_1)$. This property is used in many applications.

A Boolean function $f(\mathbf{x}_0, \mathbf{x}_1)$ of n variables that satisfies

$$\bigvee_{i=1}^{m} \frac{\partial f(\mathbf{x}_0, \mathbf{x}_1)}{\partial x_i} = 0 \tag{2.107}$$

does not depend on all variables $x_i \in \mathbf{x}_0$ and can be mapped from the Boolean space $\mathbb{B}^n$ into the Boolean space $\mathbb{B}^{n-m}$ using the m-fold maximum with regard to $\mathbf{x}_0$:

$$f(\mathbf{x}_1) = \max_{\mathbf{x}_0}{}^m f(\mathbf{x}_0, \mathbf{x}_1) \,. \tag{2.108}$$

An equation similar to (2.19) for the three vectorial derivative operations or (2.69) for the three single derivative operations does not exist for the m-fold derivative, the m-fold minimum, and the m-fold maximum. This statement results from different properties of the Boolean spaces $\mathbb{B}^1$ and $\mathbb{B}^n$ with $n > 1$. A counterexample is an even Boolean function $f(\mathbf{x})$ of n variables, $n > 1$, with two function values 1. Both the m-fold derivative and the m-fold minimum of this function with regard to all variables of $\mathbf{x}$ are equal to 0, but the m-fold maximum with regard to the same variables is equal to 1; hence, an equation similar to (2.69) does not hold for the so far defined three m-fold derivative operations. A fourth m-fold derivative operation closes this gap.

Definition 2.14 **Δ-Operation.** Let $\mathbf{x}_0 = (x_1, x_2, \ldots, x_m)$, $\mathbf{x}_1 = (x_{m+1}, x_{m+2}, \ldots, x_n)$ be two disjoint sets of Boolean variables and $f(\mathbf{x}_0, \mathbf{x}_1) = f(x_1, x_2, \ldots, x_n) = f(\mathbf{x})$ be a Boolean function of n variables, then

$$\Delta_{\mathbf{x}_0} f(\mathbf{x}_0, \mathbf{x}_1) = \min_{\mathbf{x}_0}{}^m f(\mathbf{x}_0, \mathbf{x}_1) \oplus \max_{\mathbf{x}_0}{}^m f(\mathbf{x}_0, \mathbf{x}_1) \tag{2.109}$$

is the Δ-**operation** of the Boolean function $f(\mathbf{x}_0, \mathbf{x}_1)$ with regard to the set of variables $\mathbf{x}_0$.

Due to the properties of the m-fold minimum (2.103) and the m-fold maximum (2.106) it follows from Definition (2.109) that the Δ-operation of $f(\mathbf{x}_0, \mathbf{x}_1)$ of n variables with regard to $\mathbf{x}_0$ is a Boolean function that does not depend on the m variables $x_i \in \mathbf{x}_0$, but only on the $n - m$ variables of $\mathbf{x}_1$:

$$\bigvee_{i=1}^{m} \frac{\partial}{\partial x_i} \left(\Delta_{\mathbf{x}_0} f(\mathbf{x}_0, \mathbf{x}_1)\right) = 0 \,. \tag{2.110}$$

The meaning of the Δ-operation can be derived from Definition (2.109) and the alternative possibilities to calculate the m-fold minimum (2.102) and m-fold maximum (2.105).

- The Δ-operation of $f(\mathbf{x}_0, \mathbf{x}_1)$ with regard to $\mathbf{x}_0$ is equal to 1 for such subspaces $\mathbb{B}^m$ with $\mathbf{x}_1 = \mathbf{c}_1$ that contain both function values 0 and 1 of the given function such that value changes of the given function are possible.

Figure 2.6 shows the calculation of all four 2-fold derivative operations of a Boolean function $f(x_1, x_2, x_3, x_4)$ using Karnaugh-maps. The arrows in the Karnaugh-map of the given function emphasize the pairs of function values which are evaluated in the first step calculating the three derivative operations with regard to x_3. Both the Karnaugh-maps of these single derivative operations and the associated expression do not depend anymore on the variable x_3.

The 2-fold derivative operations with regard to (x_3, x_4) are calculated in the second step by evaluation of the function values in the Boolean spaces $\mathbb{B}^2$ $(x_1, x_2) = (c_1, c_2)$ indicated by arrows in the intermediate Karnaugh-maps. The result functions of the 2-fold derivative operations with regard to (x_3, x_4) show that

- the 2-fold minimum is equal to 1 only in the left subspace where the function $f(\mathbf{x})$ is equal to 1 *for all* patterns of (x_3, x_4);
- the 2-fold derivative is equal to 1 only for the subspace $(x_1, x_2) = (0, 1)$ that contains an odd number of functions values 1;
- the 2-fold maximum is equal to 1 for the three subspaces in which at least one function value 1 *exists*; and
- the Δ-operation is equal to 1 for the two subspaces in which at least one function value 0 and one function value 1 occur.

Figure 2.6 also shows that the $\oplus$-operation of the 2-fold minimum and the 2-fold maximum with regard to the same variables x_3 and x_4 is equal to the Δ-operation but different from the 2-fold derivative with regard to these variables.

Knowing the definitions of all m-fold derivative operations we are exploring their properties. The inequality (2.62) can be generalized for m-fold minima of $f(\mathbf{x}) = f(\mathbf{x}_0, x_k, \mathbf{x}_1)$, $x_i, x_j \in \mathbf{x}_0$ and different values of m using Definition (2.99):

$$\min_{(\mathbf{x}_0, x_k)}{}^{m+1} f(\mathbf{x}) \leq \min_{\mathbf{x}_0}{}^{m} f(\mathbf{x}) \leq \cdots \leq \min_{(x_i, x_j)}{}^{2} f(\mathbf{x}) \leq \min_{x_i} f(\mathbf{x}) \leq f(\mathbf{x}) \,. \tag{2.111}$$

The more variables belong to the subset $\mathbf{x}_0$ of m variables the smaller is the function of the m-fold minimum.

The $\leq$ - relations of (2.111) combine $<$ and $=$. If the condition

$$\frac{\partial \min_{\mathbf{x}_0}{}^{m} f(\mathbf{x}_0, x_k, \mathbf{x}_1)}{\partial x_k} = 0 \tag{2.112}$$

is satisfied, the special case of the equality

$$\min_{(x_k, \mathbf{x}_0)}{}^{m+1} f(\mathbf{x}_0, x_k, \mathbf{x}_1) = \min_{\mathbf{x}_0}{}^{m} f(\mathbf{x}_0, x_k, \mathbf{x}_1) \tag{2.113}$$

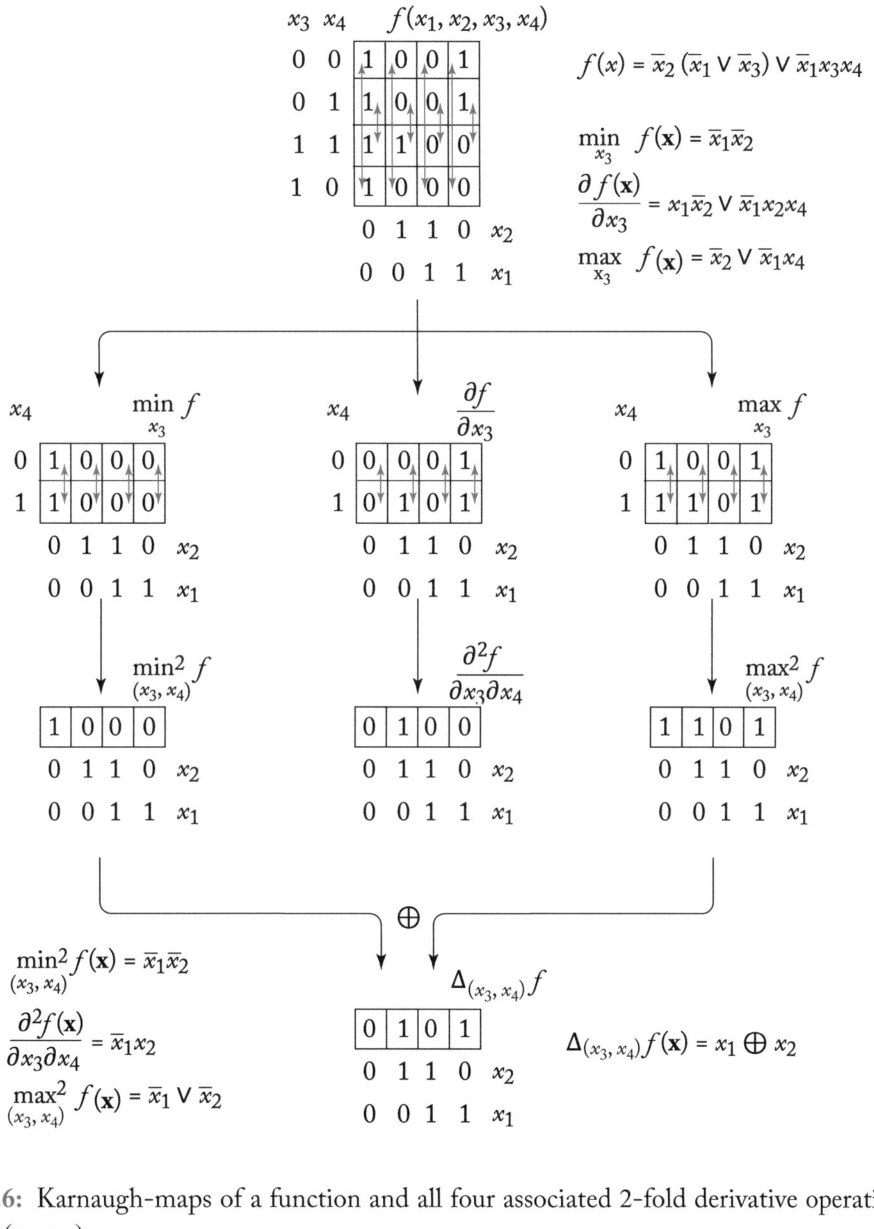

Figure 2.6: Karnaugh-maps of a function and all four associated 2-fold derivative operations with regard to (x_3, x_4).

holds. Using Definition (2.45) we get from (2.112):

$$\begin{aligned}\min_{\mathbf{x}_0}{}^m f(\mathbf{x}_0, x_k = 0, \mathbf{x}_1) \oplus \min_{\mathbf{x}_0}{}^m f(\mathbf{x}_0, x_k = 1, \mathbf{x}_1) = 0 \\ \min_{\mathbf{x}_0}{}^m f(\mathbf{x}_0, x_k = 0, \mathbf{x}_1) = \min_{\mathbf{x}_0}{}^m f(\mathbf{x}_0, x_k = 1, \mathbf{x}_1)\ .\end{aligned} \tag{2.114}$$

Using Definition (2.47) on the left-hand side and the Shannon decomposition (1.32) on the right-hand side of (2.113) we get:

$$\begin{aligned}\min_{\mathbf{x}_0}{}^m f(\mathbf{x}_0, x_k = 0, \mathbf{x}_1) \wedge \min_{\mathbf{x}_0}{}^m f(\mathbf{x}_0, x_k = 1, \mathbf{x}_1) = \\ \overline{x}_k \min_{\mathbf{x}_0}{}^m f(\mathbf{x}_0, x_k = 0, \mathbf{x}_1) \vee x_k \min_{\mathbf{x}_0}{}^m f(\mathbf{x}_0, x_k = 1, \mathbf{x}_1)\end{aligned} \tag{2.115}$$

and the substitution of (2.114) in this formula proves the equality (2.113):

$$\begin{aligned}\min_{\mathbf{x}_0}{}^m f(\mathbf{x}_0, x_k = 1, \mathbf{x}_1) \wedge \min_{\mathbf{x}_0}{}^m f(\mathbf{x}_0, x_k = 1, \mathbf{x}_1) = \\ \overline{x}_k \min_{\mathbf{x}_0}{}^m f(\mathbf{x}_0, x_k = 1, \mathbf{x}_1) \vee x_k \min_{\mathbf{x}_0}{}^m f(\mathbf{x}_0, x_k = 1, \mathbf{x}_1) \\ \min_{\mathbf{x}_0}{}^m f(\mathbf{x}_0, x_k = 1, \mathbf{x}_1) = (\overline{x}_k \vee x_k) \min_{\mathbf{x}_0}{}^m f(\mathbf{x}_0, x_k = 1, \mathbf{x}_1) \\ \min_{\mathbf{x}_0}{}^m f(\mathbf{x}_0, x_k = 1, \mathbf{x}_1) = \min_{\mathbf{x}_0}{}^m f(\mathbf{x}_0, x_k = 1, \mathbf{x}_1)\ .\end{aligned} \tag{2.116}$$

If the condition (2.112) is not satisfied the cofactors of $f(\mathbf{x}_0, x_k, \mathbf{x}_1)$ with regard to x_k are not equal and cause the strong inequality:

$$\min_{(x_k, \mathbf{x}_0)}{}^{m+1} f(\mathbf{x}_0, x_k, \mathbf{x}_1) < \min_{\mathbf{x}_0}{}^m f(\mathbf{x}_0, x_k, \mathbf{x}_1)\ . \tag{2.117}$$

The m-fold minimum with regard to $\mathbf{x}_0$ is a Boolean function which does not depend on the variables $\mathbf{x}_0$. Consequently, the n-fold minimum of a Boolean function $f(\mathbf{x})$ of n variables with regard to all variables $\mathbf{x}$ is either the value 0 or 1. Only the constant function $f(\mathbf{x}) = 1(\mathbf{x})$ has the value 1 as result of the n-fold minimum; the other functions $f(\mathbf{x}) \neq 1(\mathbf{x})$ have the value 0 as result of the n-fold minimum.

Similar properties exist for the m-fold maximum of $f(\mathbf{x}) = f(\mathbf{x}_0, x_k, \mathbf{x}_1)$, $x_i, x_j \in \mathbf{x}_0$:

$$f(\mathbf{x}) \leq \max_{x_i} f(\mathbf{x}) \leq \max_{(x_i, x_j)}{}^2 f(\mathbf{x}) \leq \cdots \leq \max_{\mathbf{x}_0}{}^m f(\mathbf{x}) \leq \max_{(\mathbf{x}_0, x_k)}{}^{m+1} f(\mathbf{x})\ . \tag{2.118}$$

Analog to the m-fold minimum, the $\leq$-relation in (2.118) can be split into an equation and a strong inequality. If the condition

$$\frac{\partial \max_{\mathbf{x}_0}{}^m f(\mathbf{x}_0, x_k, \mathbf{x}_1)}{\partial x_k} = 0 \tag{2.119}$$

is satisfied, the equation

$$\max_{\mathbf{x}_0}{}^m f(\mathbf{x}_0, x_k, \mathbf{x}_1) = \max_{(x_k, \mathbf{x}_0)}{}^{m+1} f(\mathbf{x}_0, x_k, \mathbf{x}_1) \tag{2.120}$$

holds, and otherwise we have the inequality:

$$\max_{\mathbf{x}_0}{}^m f(\mathbf{x}_0, x_k, \mathbf{x}_1) < \max_{(x_k, \mathbf{x}_0)}{}^{m+1} f(\mathbf{x}_0, x_k, \mathbf{x}_1) . \tag{2.121}$$

Reverse properties also exist for the n-fold maximum with regard to all variables of the function $f(\mathbf{x})$ of n variables. Only the constant function $f(\mathbf{x}) = 0(\mathbf{x})$ has the value 0 as result of the n-fold maximum; all other functions $f(\mathbf{x}) \neq 0(\mathbf{x})$ have the value 1 as result of the n-fold maximum.

From the inequalities (2.111), (2.118), and the relation between an inequality and Equation (1.66) follow some helpful equations for simplifications:

$$\min_{\mathbf{x}_0}{}^m f(\mathbf{x}_0, \mathbf{x}_1) \wedge \overline{f(\mathbf{x}_0, \mathbf{x}_1)} = 0 , \tag{2.122}$$

$$\min_{\mathbf{x}_0}{}^m f(\mathbf{x}_0, \mathbf{x}_1) \wedge \overline{\max_{\mathbf{x}_0}{}^m f(\mathbf{x}_0, \mathbf{x}_1)} = 0 , \tag{2.123}$$

$$f(\mathbf{x}_0, \mathbf{x}_1) \wedge \overline{\max_{\mathbf{x}_0}{}^m f(\mathbf{x}_0, \mathbf{x}_1)} = 0 . \tag{2.124}$$

Two more such rules can be derived from the definitions:

$$\min_{\mathbf{x}_0}{}^m f(\mathbf{x}_0, \mathbf{x}_1) \wedge \Delta_{\mathbf{x}_0} f(\mathbf{x}_0, \mathbf{x}_1) = 0 , \tag{2.125}$$

$$\Delta_{\mathbf{x}_0} f(\mathbf{x}_0, \mathbf{x}_1) \wedge \overline{\max_{\mathbf{x}_0}{}^m f(\mathbf{x}_0, \mathbf{x}_1)} = 0 . \tag{2.126}$$

The m-fold minimum is equal to 1 only for subspaces $\mathbf{x}_1 = \mathbf{c}_1$ with $f(\mathbf{x}_0, \mathbf{x}_1 = \mathbf{c}_1) = 1$. For such subspaces the Δ-operation is equal to 0 because no different function values occur within each of these subspaces; hence, we have (2.125). A contradiction to the rule (2.126) requires that the m-fold maximum is equal to 0 for a subspace $\mathbf{x}_1 = \mathbf{c}_1$. This is only possible for a subspace $\mathbf{x}_1 = \mathbf{c}_1$ with constant values $f(\mathbf{x}_0, \mathbf{x}_1 = \mathbf{c}_1) = 0$. However, for such a subspace the Δ-operation is equal to 0 because no different function values occur; hence, we also have (2.126).

Based on Definition (2.109) the m-fold minimum, m-fold maximum, and Δ-operation of a function are related to each other by:

$$\min_{\mathbf{x}_0}{}^m f(\mathbf{x}_0, \mathbf{x}_1) \oplus \max_{\mathbf{x}_0}{}^m f(\mathbf{x}_0, \mathbf{x}_1) \oplus \Delta_{\mathbf{x}_0} f(\mathbf{x}_0, \mathbf{x}_1) = 0 . \tag{2.127}$$

Each of these m-fold derivative operations can be calculated by means of the other two functions. In addition to Definition (2.109) we get:

$$\min_{\mathbf{x}_0}{}^m f(\mathbf{x}_0, \mathbf{x}_1) = \max_{\mathbf{x}_0}{}^m f(\mathbf{x}_0, \mathbf{x}_1) \oplus \Delta_{\mathbf{x}_0} f(\mathbf{x}_0, \mathbf{x}_1) , \tag{2.128}$$

$$\max_{\mathbf{x}_0}{}^m f(\mathbf{x}_0, \mathbf{x}_1) = \min_{\mathbf{x}_0}{}^m f(\mathbf{x}_0, \mathbf{x}_1) \oplus \Delta_{\mathbf{x}_0} f(\mathbf{x}_0, \mathbf{x}_1) . \tag{2.129}$$

Using the relation (1.56) between Boolean Rings and Boolean Algebras these two rules and Definition (2.109) can be transformed into expressions that do not contain the $\oplus$-operation.

The additional utilization of the restrictions (2.123), (2.125), and (2.126) leads to the following simplified expressions:

$$\min_{\mathbf{x}_0}{}^m f(\mathbf{x}_0, \mathbf{x}_1) = \max_{\mathbf{x}_0}{}^m f(\mathbf{x}_0, \mathbf{x}_1) \wedge \overline{\Delta_{\mathbf{x}_0} f(\mathbf{x}_0, \mathbf{x}_1)} \,, \tag{2.130}$$

$$\max_{\mathbf{x}_0}{}^m f(\mathbf{x}_0, \mathbf{x}_1) = \min_{\mathbf{x}_0}{}^m f(\mathbf{x}_0, \mathbf{x}_1) \vee \Delta_{\mathbf{x}_0} f(\mathbf{x}_0, \mathbf{x}_1) \,, \tag{2.131}$$

$$\Delta_{\mathbf{x}_0} f(\mathbf{x}_0, \mathbf{x}_1) = \overline{\min_{\mathbf{x}_0}{}^m f(\mathbf{x}_0, \mathbf{x}_1)} \wedge \max_{\mathbf{x}_0}{}^m f(\mathbf{x}_0, \mathbf{x}_1) \,. \tag{2.132}$$

The m-fold derivative operations of negated functions can by transformed into m-fold derivative operations of the associated non-negated function:

$$\frac{\partial^m \overline{f(\mathbf{x}_0, \mathbf{x}_1)}}{\partial x_1 \partial x_2 \ldots \partial x_m} = \frac{\partial^m f(\mathbf{x}_0, \mathbf{x}_1)}{\partial x_1 \partial x_2 \ldots \partial x_m} \,, \tag{2.133}$$

$$\min_{\mathbf{x}_0}{}^m \overline{f(\mathbf{x}_0, \mathbf{x}_1)} = \overline{\max_{\mathbf{x}_0}{}^m f(\mathbf{x}_0, \mathbf{x}_1)} \,, \tag{2.134}$$

$$\max_{\mathbf{x}_0}{}^m \overline{f(\mathbf{x}_0, \mathbf{x}_1)} = \overline{\min_{\mathbf{x}_0}{}^m f(\mathbf{x}_0, \mathbf{x}_1)} \,, \tag{2.135}$$

$$\Delta_{\mathbf{x}_0} \overline{f(\mathbf{x}_0, \mathbf{x}_1)} = \Delta_{\mathbf{x}_0} f(\mathbf{x}_0, \mathbf{x}_1) \,. \tag{2.136}$$

Due to the even number of function values in each Boolean subspace, the negation of the function does not change the parity of function values in each subspace; hence, the replacement of $\overline{f(\mathbf{x}_0, \mathbf{x}_1)}$ by $f(\mathbf{x}_0, \mathbf{x}_1)$ in (2.133) does not change the result of the m-fold derivative. The rules (2.134) and (2.135) directly follow from the application of the laws of De Morgan to the associated definitions. Finally, the replacement of $\overline{f(\mathbf{x}_0, \mathbf{x}_1)}$ by $f(\mathbf{x}_0, \mathbf{x}_1)$ in (2.136) does not change the property that different function values occur within a subspace; hence, we also have (2.136).

If the argument of an m-fold derivative operation consists of two functions that are connected by the same operation as used for the m-fold derivative operation, a separation into two appropriate derivative operations is possible. The basis operations are $\oplus$ for the m-fold derivative, $\wedge$ for the m-fold minimum, and $\vee$ for the m-fold maximum.

$$\frac{\partial^m (f(\mathbf{x}_0, \mathbf{x}_1) \oplus g(\mathbf{x}_0, \mathbf{x}_1))}{\partial x_1 \partial x_2 \ldots \partial x_m} = \frac{\partial^m f(\mathbf{x}_0, \mathbf{x}_1)}{\partial x_1 \partial x_2 \ldots \partial x_m} \oplus \frac{\partial^m g(\mathbf{x}_0, \mathbf{x}_1)}{\partial x_1 \partial x_2 \ldots \partial x_m} \,, \tag{2.137}$$

$$\min_{\mathbf{x}_0}{}^m (f(\mathbf{x}_0, \mathbf{x}_1) \wedge g(\mathbf{x}_0, \mathbf{x}_1)) = \min_{\mathbf{x}_0}{}^m f(\mathbf{x}_0, \mathbf{x}_1) \wedge \min_{\mathbf{x}_0}{}^m g(\mathbf{x}_0, \mathbf{x}_1) \,, \tag{2.138}$$

$$\max_{\mathbf{x}_0}{}^m (f(\mathbf{x}_0, \mathbf{x}_1) \vee g(\mathbf{x}_0, \mathbf{x}_1)) = \max_{\mathbf{x}_0}{}^m f(\mathbf{x}_0, \mathbf{x}_1) \vee \max_{\mathbf{x}_0}{}^m g(\mathbf{x}_0, \mathbf{x}_1) \,. \tag{2.139}$$

The rule (2.137) holds, because the parity of $f(\mathbf{x}_0, \mathbf{x}_1) \oplus g(\mathbf{x}_0, \mathbf{x}_1)$ is equal to the $\oplus$-operation of the separately calculated parities of $f(\mathbf{x}_0, \mathbf{x}_1)$ and $g(\mathbf{x}_0, \mathbf{x}_1)$ for each subspace $\mathbf{x}_1 = \mathbf{c}_1$. The m-fold minimum of $f(\mathbf{x}_0, \mathbf{x}_1) \wedge g(\mathbf{x}_0, \mathbf{x}_1)$ is equal to 1 only for subspaces $\mathbf{x}_1 = \mathbf{c}_1$ where both $f(\mathbf{x}_0, \mathbf{x}_1 = \mathbf{c}_1) = 1(\mathbf{x}_0)$ and $g(\mathbf{x}_0, \mathbf{x}_1 = \mathbf{c}_1) = 1(\mathbf{x}_0)$. The right-hand side of (2.138) is equal to

1 exactly for these subspaces. Similarly, the m-fold maximum of $f(\mathbf{x}_0, \mathbf{x}_1) \vee g(\mathbf{x}_0, \mathbf{x}_1)$ is equal to 0 only for subspaces $\mathbf{x}_1 = \mathbf{c}_1$ where both $f(\mathbf{x}_0, \mathbf{x}_1 = \mathbf{c}_1) = 0(\mathbf{x}_0)$ and $g(\mathbf{x}_0, \mathbf{x}_1 = \mathbf{c}_1) = 0(\mathbf{x}_0)$. The right-hand side of (2.139) is equal to 0 exactly for these subspaces.

The Δ-operation requires the operation $\wedge$ for the m-fold minimum, $\vee$ for the m-fold maximum, and $\oplus$ for the connection of these intermediate functions. Hence, there is no unique operation in the definition of $\Delta_{\mathbf{x}_0} f(\mathbf{x}_0, \mathbf{x}_1)$ so that a similar separation is not possible.

Next, we assume that the function $g(\mathbf{x}_0, \mathbf{x}_1)$ does not depend on all variables $x_i \in \mathbf{x}_0$:

$$\bigvee_{i=1}^{m} \left(\frac{\partial g(\mathbf{x}_0, \mathbf{x}_1)}{\partial x_i} \right) = 0 \,, \tag{2.140}$$

so that $g(\mathbf{x}_0, \mathbf{x}_1)$ can be replaced by $g(\mathbf{x}_1)$. m-fold derivative operations of a conjunction between a function $g(\mathbf{x}_1)$ and an arbitrary function $f(\mathbf{x}_0, \mathbf{x}_1)$, $\mathbf{x}_0 = (x_1, x_2, \ldots, x_m)$, can be simplified as follows:

$$\frac{\partial^m (g(\mathbf{x}_1) \wedge f(\mathbf{x}_0, \mathbf{x}_1))}{\partial x_1 \partial x_2 \ldots \partial x_m} = g(\mathbf{x}_1) \wedge \frac{\partial^m f(\mathbf{x}_0, \mathbf{x}_1)}{\partial x_1 \partial x_2 \ldots \partial x_m} \,, \tag{2.141}$$

$$\min_{\mathbf{x}_0}{}^m (g(\mathbf{x}_1) \wedge f(\mathbf{x}_0, \mathbf{x}_1)) = g(\mathbf{x}_1) \wedge \min_{\mathbf{x}_0}{}^m f(\mathbf{x}_0, \mathbf{x}_1) \,, \tag{2.142}$$

$$\max_{\mathbf{x}_0}{}^m (g(\mathbf{x}_1) \wedge f(\mathbf{x}_0, \mathbf{x}_1)) = g(\mathbf{x}_1) \wedge \max_{\mathbf{x}_0}{}^m f(\mathbf{x}_0, \mathbf{x}_1) \,, \tag{2.143}$$

$$\Delta_{\mathbf{x}_0} (g(\mathbf{x}_1) \wedge f(\mathbf{x}_0, \mathbf{x}_1)) = g(\mathbf{x}_1) \wedge \Delta_{\mathbf{x}_0} f(\mathbf{x}_0, \mathbf{x}_1) \,. \tag{2.144}$$

The function $g(\mathbf{x}_1)$ operates in the rules (2.141), ..., (2.144) like a filter: only in subspaces with $g(\mathbf{x}_1) = 1$ a value 1 of the used m-fold derivative operation of $f(\mathbf{x}_0, \mathbf{x}_1)$ with regard to $\mathbf{x}_0$ occurs on both sides of these equations; otherwise both sides are equal to 0.

Still assuming that the function $g(\mathbf{x}_1)$ satisfies (2.140), the replacement of the $\wedge$-operations on the left-hand side of (2.141), ..., (2.144) by an $\vee$-operation facilitates the simplifications:

$$\frac{\partial^m (g(\mathbf{x}_1) \vee f(\mathbf{x}_0, \mathbf{x}_1))}{\partial x_1 \partial x_2 \ldots \partial x_m} = \overline{g(\mathbf{x}_1)} \wedge \frac{\partial^m f(\mathbf{x}_0, \mathbf{x}_1)}{\partial x_1 \partial x_2 \ldots \partial x_m} \,, \tag{2.145}$$

$$\min_{\mathbf{x}_0}{}^m (g(\mathbf{x}_1) \vee f(\mathbf{x}_0, \mathbf{x}_1)) = g(\mathbf{x}_1) \vee \min_{\mathbf{x}_0}{}^m f(\mathbf{x}_0, \mathbf{x}_1) \,, \tag{2.146}$$

$$\max_{\mathbf{x}_0}{}^m (g(\mathbf{x}_1) \vee f(\mathbf{x}_0, \mathbf{x}_1)) = g(\mathbf{x}_1) \vee \max_{\mathbf{x}_0}{}^m f(\mathbf{x}_0, \mathbf{x}_1) \,, \tag{2.147}$$

$$\Delta_{\mathbf{x}_0} (g(\mathbf{x}_1) \vee f(\mathbf{x}_0, \mathbf{x}_1)) = \overline{g(\mathbf{x}_1)} \wedge \Delta_{\mathbf{x}_0} f(\mathbf{x}_0, \mathbf{x}_1) \,. \tag{2.148}$$

The disjunction of $g(\mathbf{x}_1)$ and $f(\mathbf{x}_0, \mathbf{x}_1)$ in subspaces with $g(\mathbf{x}_1) = 1$ results in constant functions $f(\mathbf{x}_0, \mathbf{x}_1 = \mathbf{c}_1) = 1(\mathbf{x}_0)$ with an even number of function values 1. Due to this even number (2.145) holds, and this constant function $1(\mathbf{x}_0)$ is the direct cause of (2.146), ..., (2.148).

The replacement of the $\wedge$-operations on the left-hand side of (2.141), ..., (2.144) by an $\oplus$-operation leads under the same condition (2.140) for $g(\mathbf{x}_1)$ to the rules:

$$\frac{\partial^m (g(\mathbf{x}_1) \oplus f(\mathbf{x}_0, \mathbf{x}_1))}{\partial x_1 \partial x_2 \ldots \partial x_m} = \frac{\partial^m f(\mathbf{x}_0, \mathbf{x}_1)}{\partial x_1 \partial x_2 \ldots \partial x_m} , \tag{2.149}$$

$$\min_{\mathbf{x}_0}{}^m (g(\mathbf{x}_1) \oplus f(\mathbf{x}_0, \mathbf{x}_1)) = g(\mathbf{x}_1) \min_{\mathbf{x}_0}{}^m \overline{f(\mathbf{x}_0, \mathbf{x}_1)} \vee \overline{g(\mathbf{x}_1)} \min_{\mathbf{x}_0}{}^m f(\mathbf{x}_0, \mathbf{x}_1) , \tag{2.150}$$

$$\max_{\mathbf{x}_0}{}^m (g(\mathbf{x}_1) \oplus f(\mathbf{x}_0, \mathbf{x}_1)) = g(\mathbf{x}_1) \max_{\mathbf{x}_0}{}^m \overline{f(\mathbf{x}_0, \mathbf{x}_1)} \vee \overline{g(\mathbf{x}_1)} \max_{\mathbf{x}_0}{}^m f(\mathbf{x}_0, \mathbf{x}_1) , \tag{2.151}$$

$$\Delta_{\mathbf{x}_0} (g(\mathbf{x}_1) \oplus f(\mathbf{x}_0, \mathbf{x}_1)) = \Delta_{\mathbf{x}_0} f(\mathbf{x}_0, \mathbf{x}_1) . \tag{2.152}$$

The antivalence of $g(\mathbf{x}_1)$ and $f(\mathbf{x}_0, \mathbf{x}_1)$ in subspaces with $g(\mathbf{x}_1) = 1$ does not change both the parity and the property of unique function values within these subspaces; hence, $g(\mathbf{x}_1)$ does not occur in the result of (2.149) and (2.152). The rules (2.150) and (2.151) are the result of elementary transformations using the analog rules with $\wedge$ and $\vee$.

2.4 DERIVATIVE OPERATIONS OF XBOOLE

Successful applications of the Boolean Differential Calculus require:

1. the knowledge of the theory as presented in this chapter,
2. a model that describes the needed artifacts and steps to find the solution, and
3. efficient tools to calculate the solution.

Chapter 3 demonstrates such models (item 2) for a wide field of applications. Here, we explain efficient algorithms (item 3) that calculate derivative operations. These algorithms are implemented in the XBOOLE library (see Steinbach [1992]) using the programming language C. This library is used in the XBOOLE-Monitor that is available for free on the web page

`http://www.informatik.tu-freiberg.de/xboole`

and facilitates in an easy way the computation of the solutions of problems using both operations of the Boolean Algebra and the Boolean Differential Calculus. Many examples and exercises that use the XBOOLE-Monitor are given in the book Steinbach and Posthoff [2009], and special methods to use the XBOOLE-Monitor to solve Boolean differential equations are explained in the book Steinbach and Posthoff [2013a].

The following algorithms assume that the function is given as a ternary vector list (TVL) in orthogonal disjunctive or antivalence form (ODA), see Section 1.5, and the dedicated variables of the derivative operations are specified as a set of variables (SV). Tables 2.2 and 2.3 explain the used XBOOLE-operations.

XBOOLE provides functions that calculate the vectorial derivative `DERV(f,sv)`, the vectorial minimum `MINV(f,sv)`, and the vectorial maximum `MAXV(f,sv)`. The algorithms for these

Table 2.2: Selected XBOOLE-operations of set operations

XBOOLE	Name	Set Operation	Boolean Operation
CPL (f1)	Complement	$\overline{F_1}$	$\overline{f_1}$
ISC (f1, f2)	Intersection	$F_1 \cap F_2$	$f_1 \wedge f_2$
UNI (f1, f2)	Union	$F_1 \cup F_2$	$f_1 \vee f_2$
SYD (f1, f2)	Symmetric difference	$F_1 \Delta F_2$	$f_1 \oplus f_2$

Table 2.3: Selected further XBOOLE-operations

XBOOLE	Name	Result
ORTH(f)	Orthogonalization	TVL in ODA-form from D-form or A-form
SFORM(f, form)	Set form	Assigns a form (D, A, K, E) to f
CEL(f, sv, cs)	Change elements	Replaces elements in columns sv of f: 0 by cs[1], 1 by cs[2], – by cs[3]
DCO(f, sv)	Delete columns	columns of sv are deleted in f, D-form
OBB(f)	Orthogonal block building	reduces the number of rows in ODA-form

derivative operations implement the associated definitions. The transformation of $f(\mathbf{x}_0, \mathbf{x}_1)$ to $f(\overline{\mathbf{x}}_0, \mathbf{x}_1)$ is realized by the XBOOLE-operation CEL(f, x0, "10-").

Example 2.15 Complement of Selected Columns. A given TVL f in ODA-form represents the Boolean function $f(x_1, x_2, x_3, x_4) = f(\mathbf{x}_0, \mathbf{x}_1)$ where $\mathbf{x}_0 = (x_2, x_4)$. These variables are specified in the set of variables sv=<x2,x4>. Figure 2.7 shows how the wanted TVL of $f(\overline{\mathbf{x}}_0, \mathbf{x}_1)$ is calculated using the XBOOLE-operation CEL(f, sv, cs) with cs="10-".

$$
\begin{array}{cccc} x_1 & x_2 & x_3 & x_4 \\ \hline 0 & 0 & - & 1 \\ 1 & - & 0 & 1 \\ - & 1 & 1 & - \\ \hline \end{array}
$$

$$
\text{ODA}\,(f(\mathbf{x}_0, \mathbf{x}_1)) = \begin{array}{cccc} 0 & 0 & - & 1 \\ 1 & - & 0 & 1 \\ - & 1 & 1 & - \end{array} \longrightarrow \text{ODA}\,(f(\overline{\mathbf{x}}_0, \mathbf{x}_1)) = \begin{array}{cccc} 0 & 1 & - & 0 \\ 1 & - & 0 & 0 \\ - & 0 & 1 & - \end{array}
$$

Figure 2.7: Complement of selected columns: CEL(f, <x2,x4>, "10-").

Algorithms 2.1, 2.2, and 2.3 of the three vectorial derivative operations of $f(\mathbf{x}_0, \mathbf{x}_1)$ with regard to $\mathbf{x}_0$ transform $f(\mathbf{x}_0, \mathbf{x}_1)$ into $f(\overline{\mathbf{x}}_0, \mathbf{x}_1)$ as shown in Example 2.15 and use the XBOOLE-operations SYD, ISC, or UNI according to Definitions (2.1), (2.3), or (2.4).

Algorithm 2.1 Vectorial derivative of $f(\mathbf{x}_0, \mathbf{x}_1)$ with regard to $\mathbf{x}_0 = (x_{01}, x_{02}, \ldots, x_{0k})$

Input : $f = f(\mathbf{x}_0, \mathbf{x}_1)$ in ODA-form and $\mathbf{x}_0 \in \mathbf{x}$,
Output : TVL of $g = g(\mathbf{x}_0, \mathbf{x}_1) = \dfrac{\partial f(\mathbf{x}_0, \mathbf{x}_1)}{\partial \mathbf{x}_0}$ in ODA-form

1: $h \leftarrow \text{CEL}(f, \mathbf{x}_0, "10-")$
2: $g \leftarrow \text{SYD}(f, h)$

Algorithm 2.2 Vectorial minimum of $f(\mathbf{x}_0, \mathbf{x}_1)$ with regard to $\mathbf{x}_0 = (x_{01}, x_{02}, \ldots, x_{0k})$

Input : $f = f(\mathbf{x}_0, \mathbf{x}_1)$ in ODA-form and $\mathbf{x}_0 \in \mathbf{x}$,
Output : TVL of $g = g(\mathbf{x}_0, \mathbf{x}_1) = \min\limits_{\mathbf{x}_0} f(\mathbf{x}_0, \mathbf{x}_1)$ in ODA-form

1: $h \leftarrow \text{CEL}(f, \mathbf{x}_0, "10-")$
2: $g \leftarrow \text{ISC}(f, h)$

Algorithm 2.3 Vectorial maximum of $f(\mathbf{x}_0, \mathbf{x}_1)$ with regard to $\mathbf{x}_0 = (x_{01}, x_{02}, \ldots, x_{0k})$

Input : $f = f(\mathbf{x}_0, \mathbf{x}_1)$ in ODA-form and $\mathbf{x}_0 \in \mathbf{x}$,
Output : TVL of $g = g(\mathbf{x}_0, \mathbf{x}_1) = \max\limits_{\mathbf{x}_0} f(\mathbf{x}_0, \mathbf{x}_1)$ in ODA-form

1: $h \leftarrow \text{CEL}(f, \mathbf{x}_0, "10-")$
2: $g \leftarrow \text{UNI}(f, h)$

Algorithm 2.4 specifies a very efficient procedure to calculate the m-fold derivative. Using (2.97) instead of Definition (2.95), it is not necessary to execute a sequence of single derivatives. Algorithm 2.4 utilizes the property that the m-fold derivative is equal to 1 in subspaces where an odd number of function values 1 exists.

A ternary vector that contains a dash (–) in one column of $\mathbf{x}_0$ describes an even number of function values. Hence, these vectors do not contribute to the solution and are deleted by the function DRCD$(f, \mathbf{x}_0)$ *delete rows containing dashes in columns* $\mathbf{x}_0$ in line 1 of Algorithm 2.4. The solution function $g(\mathbf{x}_1)$ does not depend on $\mathbf{x}_0$. Hence, the function DCO$(h, \mathbf{x}_0)$ deletes the columns $\mathbf{x}_0$. The XBOOLE-operation SFORM assigns the antivalence form to the result TVL

Algorithm 2.4 m-fold derivative of $f(\mathbf{x}_0, \mathbf{x}_1)$ with regard to $\mathbf{x}_0 = (x_{01}, x_{02}, \ldots, x_{0m})$

Input : $f = f(\mathbf{x}_0, \mathbf{x}_1)$ in ODA-form and $\mathbf{x}_0 \in \mathbf{x}$,

Output : TVL of $g = g(\mathbf{x}_1) = \dfrac{\partial^m f(\mathbf{x}_0, \mathbf{x}_1)}{\partial x_{01} \partial x_{02} \ldots \partial x_{0m}}$ in ODA-form

1: $h \leftarrow \text{DRCD}(f, \mathbf{x}_0)$
2: $h \leftarrow \text{SFORM}(\text{DCO}(h, \mathbf{x}_0), \text{A-form})$
3: $g \leftarrow \text{ORTH}(h)$

in line 2 of Algorithm 2.4. The orthogonalization of this antivalence form using the XBOOLE-operation ORTH in line 3 of Algorithm 2.4 results in the wanted m-fold derivative.

Example 2.16 2-fold Derivative. The 2-fold derivative of the function

$$f(\mathbf{x}) = x_1 x_2 x_3 \overline{x}_5 \vee x_3 x_4 x_5 \vee x_1 x_3 \overline{x}_4 x_5 \vee \overline{x}_1 x_2 x_4 \overline{x}_5 \vee \overline{x}_1 \overline{x}_3 \overline{x}_4 x_5 \vee \overline{x}_1 x_2 \overline{x}_3 \overline{x}_4 \overline{x}_5 \tag{2.153}$$

with regard to $\mathbf{x}_0 = (x_1, x_3)$ has to be calculated.

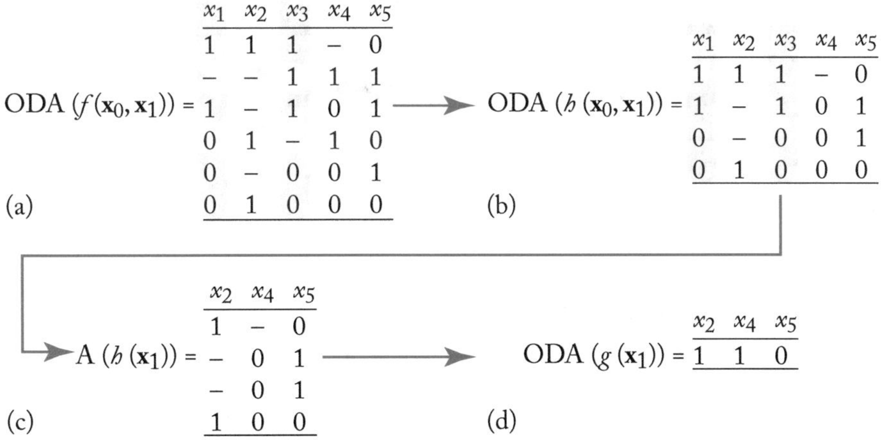

Figure 2.8: Calculation of a 2-fold derivative using Algorithm 2.4: (a) given function f; (b) result h of step 1; (c) result h of step 2; and (d) wanted 2-fold derivative g, result of step 3.

Figure 2.8 summarizes all steps of Algorithm 2.4 to solve this task. Figure 2.8a shows the TVL of the given function f and emphasizes the columns of $\mathbf{x}_0 = (x_1, x_3)$ by light blue backgrounds with regard to which the 2-fold derivative has to be calculated.

There are dashes in the second and the fourth row of these columns. These rows specify even numbers of function values 1. Hence, the function DRCD($f, \mathbf{x}_0$) of line 1 of Algorithm 2.4 removes these rows, and we get the intermediate function h of Figure 2.8b.

The 2-fold derivative of f with regard to $\mathbf{x}_0 = (x_1, x_3)$ does not depend on the variables of $\mathbf{x}_0$. Hence, the columns with light blue background in Figure 2.8b are deleted due to line 2 of Algorithm 2.4, and we get the next intermediate function h of Figure 2.8c.

The XBOOLE-operation `ORTH` of line 3 of Algorithm 2.4 deletes the identical rows two and three and removes the common binary vector (100) from row number one and four. The result of this calculation is the 2-fold derivative of f with regard to $\mathbf{x}_0 = (x_1, x_3)$ shown in Figure 2.8d:

$$g(x_2, x_4, x_5) = \frac{\partial^2 f(x_1, x_2, x_3, x_4, x_5)}{\partial x_1 \partial x_3} = x_2 x_4 \overline{x}_5 \; .$$

Algorithm 2.5 calculates the m-fold maximum of $f(\mathbf{x}_0, \mathbf{x}_1)$ with regard to $\mathbf{x}_0 = (x_{01}, x_{02}, \ldots, x_{0m})$. Algorithm 2.5 is still easier than Algorithm 2.4. It is also not necessary to execute a sequence of maxima as defined in (2.100). Due to (2.105), the m-fold maximum is equal to 1 in subspaces where at least one function value 1 exists. Hence, all columns of $\mathbf{x}_0$ can be removed. The remaining rows indicate subspaces for which the m-fold maximum is equal to 1.

Algorithm 2.5 m-fold maximum of $f(\mathbf{x}_0, \mathbf{x}_1)$ with regard to $\mathbf{x}_0 = (x_1, x_2, \ldots, x_m)$

Input : $f = f(\mathbf{x}_0, \mathbf{x}_1)$ in ODA-form and $\mathbf{x}_0 \in \mathbf{x}$,
Output : TVL of $g = g(\mathbf{x}_1) = \max\limits_{\mathbf{x}_0}{}^m f(\mathbf{x}_0, \mathbf{x}_1)$ in ODA-form

1: $h \leftarrow$ SFORM(DCO($f, \mathbf{x}_0$), D-form)
2: $g \leftarrow$ ORTH(h)

Example 2.17 2-fold Maximum. We take the same function (2.153) of Example 2.16 and calculate the 2-fold maximum with regard to $\mathbf{x}_0 = (x_1, x_3)$.

Figure 2.9 summarizes all steps of Algorithm 2.5 to calculate the 2-fold maximum. The columns with light blue background in Figure 2.9a emphasizes variables $\mathbf{x}_0 = (x_1, x_3)$ with regard to which the 2-fold maximum of this function f has to be calculated.

The 2-fold maximum of f with regard to $\mathbf{x}_0 = (x_1, x_3)$ does not depend on the variables of $\mathbf{x}_0$. Hence, the column with light blue background in Figure 2.9a are deleted due to line 1 of Algorithm 2.5. A TVL losses its orthogonality when some columns are removed. Due to (2.105), the restricted TVL in Figure 2.9b has a D-form.

The orthogonalization in line 2 of Algorithm 2.5 generates the wanted 2-fold maximum in ODA-form as shown in Figure 2.9c. In a post-processing step, the last two rows are merged

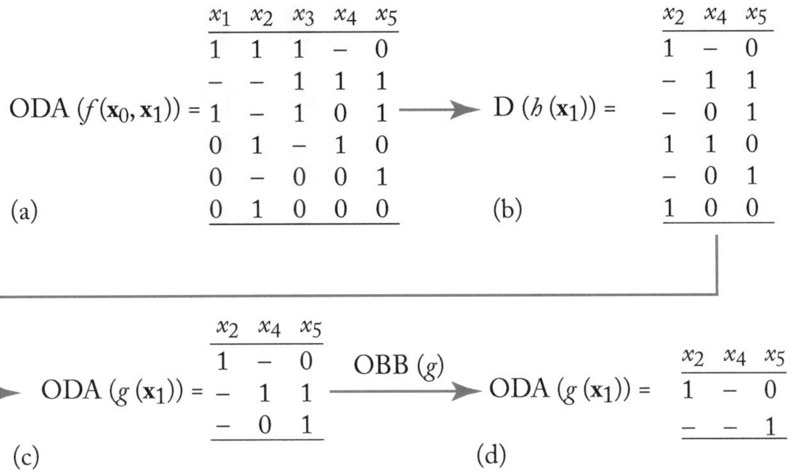

(a) $\mathrm{ODA}\,(f(\mathbf{x}_0, \mathbf{x}_1)) =$

x_1	x_2	x_3	x_4	x_5
1	1	1	–	0
–	–	1	1	1
1	–	1	0	1
0	1	–	1	0
0	–	0	0	1
0	1	0	0	0

→ (b) $\mathrm{D}\,(h\,(\mathbf{x}_1)) =$

x_2	x_4	x_5
1	–	0
–	1	1
–	0	1
1	1	0
–	0	1
1	0	0

→ (c) $\mathrm{ODA}\,(g\,(\mathbf{x}_1)) =$

x_2	x_4	x_5
1	–	0
–	1	1
–	0	1

— OBB (g) → (d) $\mathrm{ODA}\,(g\,(\mathbf{x}_1)) =$

x_2	x_4	x_5
1	–	0
–	–	1

Figure 2.9: Calculation of a 2-fold maximum using Algorithm 2.5: (a) given function f; (b) result h of step 1; (c) wanted 2-fold maximum g, of step 2; and (d) reduced number of rows of the 2-fold maximum g.

into a single ternary vector by means of the XBOOLE-operation `OBB`. Figure 2.9d shows the found TVL of the 2-fold maximum of $f(\mathbf{x})$ (2.153) with regard to $\mathbf{x}_0 = (x_1, x_3)$:

$$g(x_2, x_4, x_5) = \max_{(x_1,x_3)}{}^2 f(x_1, x_2, x_3, x_4, x_5) = x_2 \vee x_5 \,.$$

It is possible to calculate the m-fold minimum by means of the m-fold maximum:

$$\min_{\mathbf{x}_0}{}^m f(\mathbf{x}_0, \mathbf{x}_1) = \overline{\max_{\mathbf{x}_0}{}^m \overline{f(\mathbf{x}_0, \mathbf{x}_1)}} \,. \tag{2.154}$$

However, the inner complement operation is time-consuming because all variables $\mathbf{x} = (\mathbf{x}_0, \mathbf{x}_1)$ must be considered. The calculation of a sequence of single minima as shown in Algorithm 2.6 finds the solution of the m-fold minimum faster, because the number of variables is reduced in each swap of the iteration.

The function $\mathrm{SPLIT}(g, x_{0i}))$ in line 3 of Algorithm 2.6 factorizes the TVL of g in three TVLs h_0, h_1, and h_d controlled by the value 0, 1, or dash (–) in the column x_{0i}. The orthogonality of these three TVLs is neither destroyed by the factorization nor by the implicit removing of the column x_{0i}. A dash in the column x_{0i} indicates that the associated conjunction is equal to 1 for both $x_{0i} = 0$ and $x_{0i} = 1$; hence, these conjunctions can be directly included into the minimum with regard to x_{0i} using the XBOOLE-operation `UNI` in line 4 (here realizing a con-

Algorithm 2.6 m-fold minimum of $f(\mathbf{x}_0, \mathbf{x}_1)$ with regard to $\mathbf{x}_0 = (x_{01}, x_{02}, \ldots, x_{0m})$

Input : $f = f(\mathbf{x}_0, \mathbf{x}_1)$ in ODA-form and $\mathbf{x}_0 \in \mathbf{x}$,
Output : TVL of $g = g(\mathbf{x}_1) = \min\limits_{\mathbf{x}_0}{}^m f(\mathbf{x}_0, \mathbf{x}_1)$ in ODA-form

1: $g \leftarrow f$
2: **for** $i = 1, \ldots, m$ **do**
3: $\quad \langle h_0, h_1, h_d \rangle \leftarrow \mathrm{SPLIT}(g, x_{0i}))$
4: $\quad g \leftarrow \mathrm{UNI}(\mathrm{ISC}(h_0, h_1), h_d)$
5: **end for**

catenation). The XBOOLE-operation `ISC` calculates the minimum for the other conjunctions belonging to both subspaces $x_{0i} = 0$ and $x_{0i} = 1$.

Example 2.18 2-fold Minimum. Again, we take the function (2.153) of Example 2.16 and calculate the 2-fold minimum with regard to $\mathbf{x}_0 = (x_1, x_3)$. Figure 2.10 summarizes all steps of Algorithm 2.6 to calculate the 2-fold minimum.

The columns with light blue background in Figure 2.10a emphasize variables $\mathbf{x}_0 = (x_1, x_3)$ with regard to which the 2-fold minimum of this function f has to be calculated. This TVL is copied to the TVL of g which is transformed in two sweeps into the wanted 2-fold minimum.

In the first sweep of the for-loop of Algorithm 2.6 the TVL of Figure 2.10b is split controlled by x_1 into the TVLs h_0, h_1, and h_d. Figure 2.10c show that h_0 contains the last three rows, h_1 the first and third, and h_d the second row of g. The intersection of h_0 and h_1 results in the single vector of the left TVL of Figure 2.10d which is concatenated with h_d to the intermediate result $\min_{x_1} f$ shown in the right TVL of Figure 2.10d. It can be seen that this first sweep removes one column of the TVL of g and reduces the number of rows.

Analog transformations are executed with regard to x_3 in the second sweep (see Figures 2.10e and f). Due to the empty TVL of h_0, the intersection of h_0 and h_1 results in an empty TVL. The concatenation of this empty TVL and the empty TVL of h_d leads to an empty TVL of the final result:

$$g(x_2, x_4, x_5) = \min_{(x_1, x_3)}{}^2 f(x_1, x_2, x_3, x_4, x_5) = 0 \,.$$

All the other derivative operations can be calculated using the implemented XBOOLE-operations. According to Definition (2.109) of the Δ-operation the XBOOLE-operations `MINK` of Algorithm 2.6, `MAXK` of Algorithm 2.5, and `SYD` for the $\oplus$-operation can be used. All single derivative operations are special cases of the according m-fold derivative operations with $m = 1$. Hence, Algorithms 2.4, 2.5, and 2.6 also efficiently calculate the single derivative operations. A similar statement is true for all derivative operations for lattices of Boolean functions. The

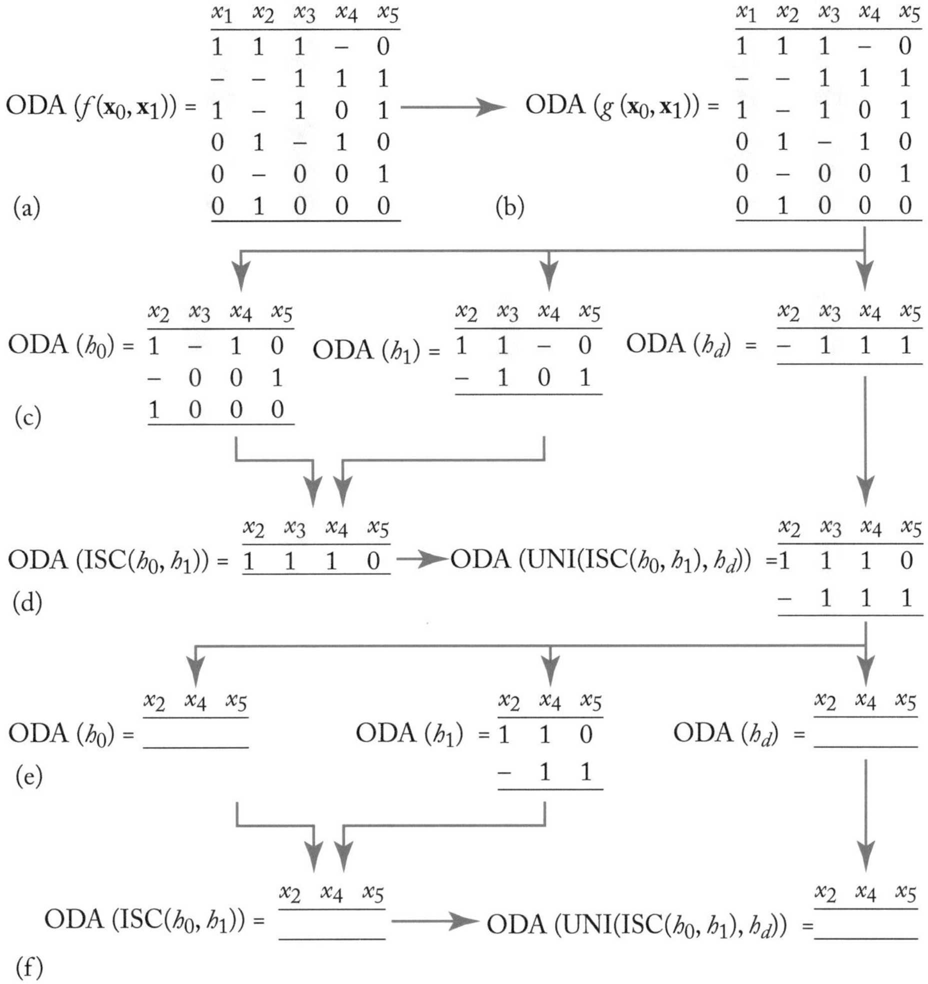

Figure 2.10: Calculation of a 2-fold minimum using Algorithm 2.6: (a) given function f; (b) copied function g (line 1); (c) split functions (line 3, sweep 1); (d) minimum with regard to x_1 (line 4, sweep 1); (e) split functions (line 3, sweep 2); and (f) $g(\mathbf{x}_1) = 0$: 2-fold minimum of f with regard to (x_1, x_3) (line 4, sweep 2).

calculation of all mark functions of such derivative operations only require the implemented Algorithms 2.1–2.6, and the XBOOLE-operations for sets of Table 2.2.

SUMMARY

The Boolean Differential Calculus extends the basics of Boolean calculations such that *changes* of function values are evaluated. These changes can be focused to pairs of function values for vectorial derivative operations. The second function value of each of such a pair is reached from the first one by the simultaneous change of the values for the same selected set of variables. Single derivative operations are a special case of the vectorial derivative operations where the value of only one variable changes. The change behavior of whole subspaces is evaluated by m-fold derivative operations. Boolean derivative operations are calculated for given Boolean functions. The result of each derivative operation is again a Boolean function with special properties. There are very efficient implementations of algorithms for all derivative operations.

EXERCISES

2.1 Calculate the vectorial derivative of

$$f(x_1, x_2, x_3, x_4) = (x_1\overline{x}_2 \oplus \overline{x}_3 x_4) \vee \overline{x}_1 x_4 \tag{2.155}$$

with regard to $\mathbf{x}_0 = (x_1, x_3)$.

2.2 Explore the symmetry properties of the function:

$$f(x_1, x_2, x_3, x_4) = (x_1 x_2 \oplus x_3 x_4) \oplus (x_1 \oplus x_2)(x_3 \oplus x_4)\ . \tag{2.156}$$

(a) How many pairs of variables exist for the function (2.156)?

(b) How many vectorial derivatives are needed in (2.9) to verify that the exchange of the values of each pair of variables does not change the value of the function (2.156)?

(c) Verify by means of (2.9) whether (2.156) is symmetric with regard to all pairs of variables.

(d) An elementary symmetric function $S_n^i(\mathbf{x})$ of n variables is equal to 1 if the value 1 is assigned to exactly i variables. Transform (2.156) into an antivalence normal form and express this form using appropriate elementary symmetric functions. How many variables must be equal to 1 such that the function (2.156) evaluates to 1?

2.3 Verify by means of Definition 2.2 that the rule (2.18) is satisfied for all functions $f(\mathbf{x})$.

2.4 Verify Theorem 2.5 for $\min_{x_i} f(\mathbf{x})$ using Definitions 2.2 and 2.4 as well as the Shannon decomposition (1.32).

2.5 A Boolean function is given by the following expression:

$$f(x_1, x_2, x_3, x_4, x_5) = \overline{x}_1\overline{x}_2 x_3 \vee \overline{x}_1 x_2 x_3 x_4 \vee x_1\overline{x}_4\overline{x}_5 \vee \overline{x}_1 x_2 \overline{x}_4 \vee \overline{x}_4 x_5 \vee \overline{x}_1\overline{x}_2\overline{x}_3\overline{x}_4\ . \tag{2.157}$$

Verify by means of single derivatives with regard to each of the five variables whether the function (2.157) really depends on all variables of the given expression. If possible, simplify this function using the appropriate maximum operation.

2.6 Calculate the degree of linearity with regard to x_1 of the functions:

$$f_1(x_1, x_2, x_3) = x_1 \oplus x_2 \oplus x_3 \ , \tag{2.158}$$
$$f_2(x_1, x_2, x_3) = x_1 \oplus x_2 \oplus x_3 \oplus x_1 x_2 \ , \tag{2.159}$$
$$f_3(x_1, x_2, x_3) = x_1 \oplus x_2 \oplus x_3 \oplus x_1 x_2 x_3 \ . \tag{2.160}$$

2.7 Explore the properties of the function

$$f(x_1, x_2, x_3, x_4, x_5) = x_1 \overline{x}_2 x_5 \vee \overline{x}_1 \overline{x}_3 x_4 x_5 \vee \overline{x}_1 \overline{x}_4 x_5 \vee \overline{x}_2 \overline{x}_3 \overline{x}_5 \vee x_3 \overline{x}_4 \overline{x}_5 \tag{2.161}$$

for all subspaces $(x_1, x_2, x_3) = (c_1, c_2, c_3)$. Which subspace

(a) contains an odd number of function values 1?

(b) contains only function values 1?

(c) contains at least one function value 1?

(d) contains different function values?

2.8 Verify the truth of (2.150). Which rules have to be used in a possible sequence of transformations?

2.9 The 3-fold derivative of the function

$$f(x_1, x_2, x_3, x_4, x_5) = x_1 \overline{x}_2 x_4 \overline{x}_5 \vee \overline{x}_1 \overline{x}_2 x_3 \overline{x}_4 x_5 \vee x_2 \overline{x}_3 x_5 \vee \overline{x}_1 x_2 \overline{x}_3 \overline{x}_5 \vee x_1 \overline{x}_2 x_3 \overline{x}_4 \tag{2.162}$$

with regard to (x_1, x_3, x_5) has to be calculated using Algorithm 2.4.

(a) Check whether all conjunctions of the given expression are orthogonal to each other?

(b) Represent the function (2.162) as TVL in ODA-form and apply Algorithm 2.4 to compute the requested m-fold derivative.

(c) Represent this m-fold derivative as expression in disjunctive form.

CHAPTER 3

Derivative Operations of Lattices of Boolean Functions

3.1 BOOLEAN LATTICES OF BOOLEAN FUNCTIONS

There are applications, e.g., in circuit design, where derivative operations are needed for a set of Boolean functions. Basically, the needed derivative operation can be separately calculated for each function of the set. However, this separate calculation is a time-consuming procedure. Very often these sets of functions satisfy the conditions of a lattice. We will show in this section:

1. each derivative operation for all Boolean functions of a lattice results again in a lattice of Boolean functions;
2. all functions of the lattice can be described using two mark functions and a restriction specified by certain vectorial derivatives; and
3. it is not needed to calculate derivative operations for each function of the lattice separately, but the mark functions as well as the restriction of the new lattice can be directly calculated.

Lattices of Boolean functions were already introduced in Section 1.2. Here, we provide a very general possibility for their description. We also will show that a lattice of Boolean functions can additionally satisfy the laws of a Boolean Algebra; hence, we are talking about Boolean lattices of Boolean functions.

A Boolean function of n variables is a mapping from $\mathbb{B}^n$ into $\mathbb{B}$. Hence, a vector of 2^n elements of $\mathbb{B}$ in a certain order characterizes a Boolean function. All 2^{2^n} such vectors describe all Boolean functions of n variables and are the elements of $\mathbb{B}^{2^n}$. Due to Theorem 1.6 the set of all Boolean functions of n variables also establishes together with the operations $\wedge$, $\vee$, and $^-$ a Boolean Algebra $(\mathbb{B}^{2^n}, \wedge, \vee, ^-, \mathbf{0}, \mathbf{1})$.

If the conjunction and the disjunction of any two elements of $L \subset \mathbb{B}^{2^n}$ are again elements of L then L is a sublattice of $\mathbb{B}^{2^n}$ or, for short, a lattice of Boolean functions. There is a direct relationship between an incompletely specified Boolean function and a lattice of Boolean functions. The set of all 2^n input patterns $\mathbf{x} = (x_1, x_2, \ldots, x_n)$ of an incompletely specified Boolean function can be divided into three disjoint sets:

$$\begin{aligned} \bullet\ \mathbf{x} \in \text{don't-care-set} \quad &\Leftrightarrow f_\varphi(x_1, \ldots, x_n) = 1 \\ &\Leftrightarrow \text{it is allowed to choose the function value of} \\ &\quad\ f(\mathbf{x}) \text{ without any restrictions}\,, \end{aligned} \tag{3.1}$$

- $\mathbf{x} \in$ ON-set $\quad \Leftrightarrow f_q(x_1, \ldots, x_n) = 1$
$$\Leftrightarrow (f_\varphi(x_1, \ldots, x_n) = 0) \wedge (f(x_1, \ldots, x_n) = 1)\,, \tag{3.2}$$

- $\mathbf{x} \in$ OFF-set $\quad \Leftrightarrow f_r(x_1, \ldots, x_n) = 1$
$$\Leftrightarrow (f_\varphi(x_1, \ldots, x_n) = 0) \wedge (f(x_1, \ldots, x_n) = 0)\,. \tag{3.3}$$

These mark functions $f_q(\mathbf{x})$, $f_r(\mathbf{x})$ and $f_\varphi(\mathbf{x})$ cover the whole Boolean space

$$f_q(\mathbf{x}) \vee f_r(\mathbf{x}) \vee f_\varphi(\mathbf{x}) = 1 \tag{3.4}$$

for all vectors $\mathbf{x}$, and they are also mutually disjoint:

$$f_q(\mathbf{x}) \wedge f_r(\mathbf{x}) = 0\,, \tag{3.5}$$
$$f_q(\mathbf{x}) \wedge f_\varphi(\mathbf{x}) = 0\,, \tag{3.6}$$
$$f_r(\mathbf{x}) \wedge f_\varphi(\mathbf{x}) = 0\,. \tag{3.7}$$

A function $f_i(\mathbf{x})$ belongs to the lattice of Boolean functions characterized by the mark functions $f_q(\mathbf{x})$ and $f_r(\mathbf{x})$, in short $f_i(\mathbf{x}) \in L\langle f_q(\mathbf{x}), f_r(\mathbf{x})\rangle$, if

$$f_q(\mathbf{x}) \leq f_i(\mathbf{x}) \leq \overline{f_r(\mathbf{x})}\,. \tag{3.8}$$

The inequality (3.8) can be transformed into the equation

$$f_q(\mathbf{x}) \wedge \overline{f_i(\mathbf{x})} \vee f_i(\mathbf{x}) \wedge f_r(\mathbf{x}) = 0\,. \tag{3.9}$$

An incompletely specified Boolean function with $|f_\varphi(\mathbf{x})|$ don't-cares can be realized by one of $2^{|f_\varphi(\mathbf{x})|}$ functions.

Example 3.1 Relationship: Incompletely Specified Function—Lattice of Functions. The don't-cares of the incompletely specified Boolean function $f_{is}(\mathbf{x}) = f_{is}(x_1, x_2, x_3, x_4)$ are labeled in the Karnaugh-map of Figure 3.1a using the symbol Φ.

Figure 3.1b shows all 16 functions of the associated lattice. The rows and columns of the Karnaugh-maps are arranged in the same way as shown in Figure 3.1a, but the code patterns are omitted. These 16 functions are arranged such that the number of function values 1 decreases by 1 in the next lower level. Fields with light blue background indicate the don't-cares. The 16 functions in Figure 3.1b are different in these positions.

The conjunction of any pair of these 16 functions results in one these functions. The conjunction of two functions of the same row in Figure 3.1b belongs to a row below or is even the smallest function $f_1(\mathbf{x}) = f_q(\mathbf{x})$. Vice versa, the disjunction of any pair of these 16 functions also results in one these functions; and the disjunction of two functions of the same row in Figure 3.1b belongs to a row above or is even the largest function $f_{16}(\mathbf{x}) = \overline{f_r(\mathbf{x})}$.

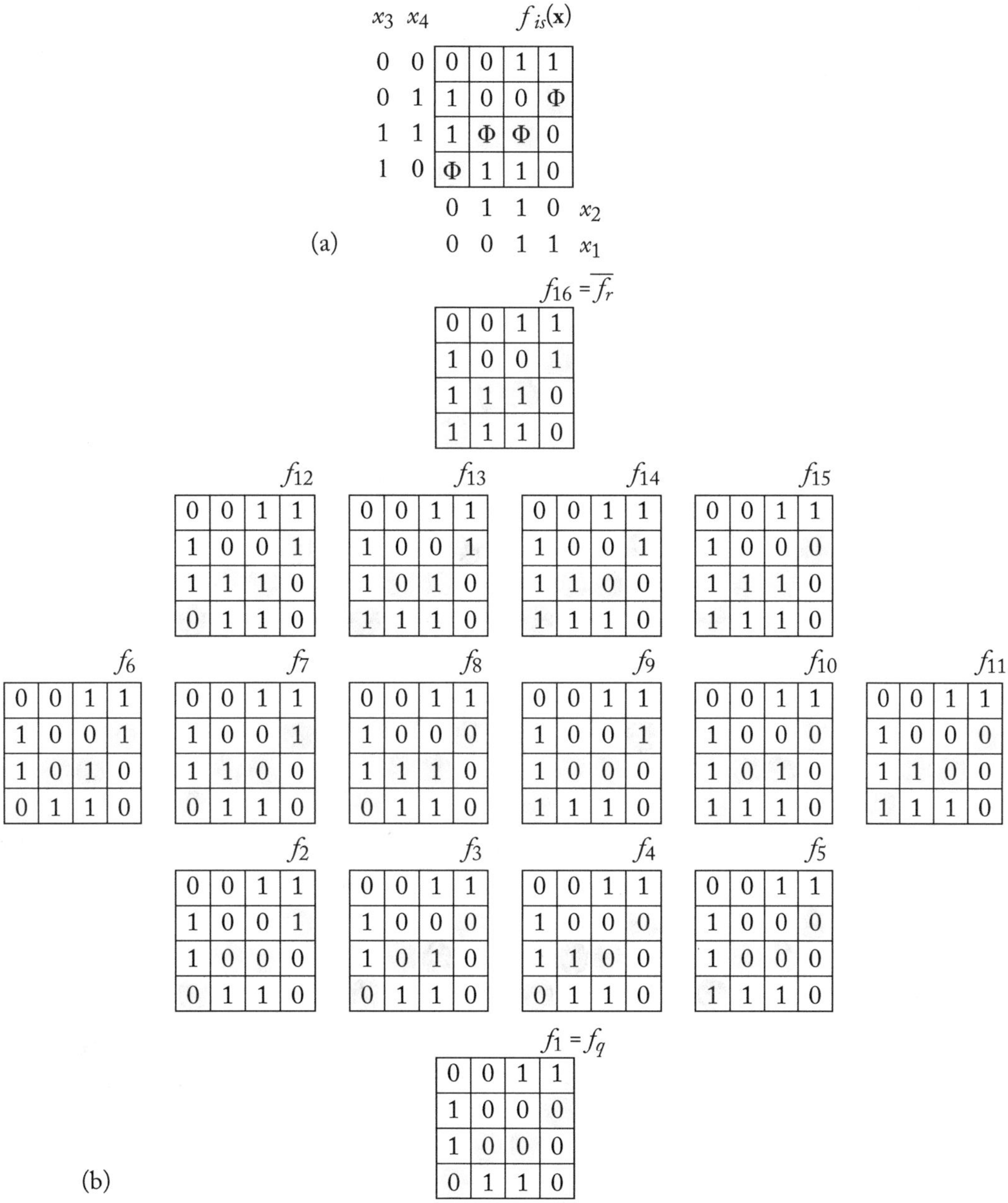

Figure 3.1: Related functions: (a) incompletely specified function and (b) 16 functions of the associated lattice.

The set of functions generated by an incompletely specified function does not only satisfy the axioms of a lattice, but even the axioms of a Boolean Algebra. The distributivity holds for three arbitrarily chosen functions of the lattice. The neutral elements of the lattice L are the smallest function $f_q(\mathbf{x})$ for the disjunction and the largest function $\overline{f_r(\mathbf{x})}$ for the conjunction. Hence, all $f_i(\mathbf{x}) \in L\langle f_q(\mathbf{x}), f_r(\mathbf{x})\rangle$ satisfy:

$$f_q(\mathbf{x}) \vee f_i(\mathbf{x}) = f_i(\mathbf{x}) \, , \tag{3.10}$$
$$\overline{f_r(\mathbf{x})} \wedge f_i(\mathbf{x}) = f_i(\mathbf{x}) \, . \tag{3.11}$$

Using (3.10), (3.11), and the absorption, it can easily be shown that:

$$\begin{aligned} f_i(\mathbf{x}) \wedge f_q(\mathbf{x}) &= f_q(\mathbf{x}) \\ (f_q(\mathbf{x}) \vee f_i(\mathbf{x})) \wedge f_q(\mathbf{x}) &= f_q(\mathbf{x}) \\ f_q(\mathbf{x}) &= f_q(\mathbf{x}) \, , \end{aligned} \tag{3.12}$$

$$\begin{aligned} f_i(\mathbf{x}) \vee \overline{f_r(\mathbf{x})} &= \overline{f_r(\mathbf{x})} \\ (\overline{f_r(\mathbf{x})} \wedge f_i(\mathbf{x})) \vee \overline{f_r(\mathbf{x})} &= \overline{f_r(\mathbf{x})} \\ \overline{f_r(\mathbf{x})} &= \overline{f_r(\mathbf{x})} \, . \end{aligned} \tag{3.13}$$

Using the mark functions $f_q(\mathbf{x})$ and $\overline{f_r(\mathbf{x})}$ of the lattice L a special complement operation for each function $f_i(\mathbf{x}) \in L\langle f_q(\mathbf{x}), f_r(\mathbf{x})\rangle$, indicated by ${}^{L}\overline{f_i(\mathbf{x})}$ can be defined:

$${}^{L}\overline{f_i(\mathbf{x})} = \overline{f_i(\mathbf{x})} \wedge \overline{f_r(\mathbf{x})} \vee f_q(\mathbf{x}) \, . \tag{3.14}$$

Example 3.2 Lattice Complement. Each function $f_i(\mathbf{x})$ of the lattice $L\langle f_q(\mathbf{x}), f_r(\mathbf{x})\rangle$ shown in Figure 3.1b has a complement function ${}^{L}\overline{f_i(\mathbf{x})}$ (3.14) belonging to the same lattice:

$$\begin{aligned}
&{}^{L}\overline{f_1(\mathbf{x})} = f_{16}(\mathbf{x}) \, , && {}^{L}\overline{f_2(\mathbf{x})} = f_{15}(\mathbf{x}) \, , && {}^{L}\overline{f_3(\mathbf{x})} = f_{14}(\mathbf{x}) \, , && {}^{L}\overline{f_4(\mathbf{x})} = f_{13}(\mathbf{x}) \, , \\
&{}^{L}\overline{f_5(\mathbf{x})} = f_{12}(\mathbf{x}) \, , && {}^{L}\overline{f_6(\mathbf{x})} = f_{11}(\mathbf{x}) \, , && {}^{L}\overline{f_7(\mathbf{x})} = f_{10}(\mathbf{x}) \, , && {}^{L}\overline{f_8(\mathbf{x})} = f_{9}(\mathbf{x}) \, , \\
&{}^{L}\overline{f_9(\mathbf{x})} = f_{8}(\mathbf{x}) \, , && {}^{L}\overline{f_{10}(\mathbf{x})} = f_{7}(\mathbf{x}) \, , && {}^{L}\overline{f_{11}(\mathbf{x})} = f_{6}(\mathbf{x}) \, , && {}^{L}\overline{f_{12}(\mathbf{x})} = f_{5}(\mathbf{x}) \, , \\
&{}^{L}\overline{f_{13}(\mathbf{x})} = f_{4}(\mathbf{x}) \, , && {}^{L}\overline{f_{14}(\mathbf{x})} = f_{3}(\mathbf{x}) \, , && {}^{L}\overline{f_{15}(\mathbf{x})} = f_{2}(\mathbf{x}) \, , && {}^{L}\overline{f_{16}(\mathbf{x})} = f_{1}(\mathbf{x}) \, .
\end{aligned}$$

All $f_i(\mathbf{x}) \in L\langle f_q(\mathbf{x}), f_r(\mathbf{x})\rangle$ satisfy:

$$f_i(\mathbf{x}) \wedge {}^{L}\overline{f_i(\mathbf{x})} = f_q(\mathbf{x}) \, , \tag{3.15}$$
$$f_i(\mathbf{x}) \vee {}^{L}\overline{f_i(\mathbf{x})} = \overline{f_r(\mathbf{x})} \, . \tag{3.16}$$

The substitution of (3.14) into (3.15) and the simplification using (3.12) confirm (3.15):

$$
\begin{aligned}
f_i(\mathbf{x}) \wedge {}^L\overline{f_i(\mathbf{x})} &= f_q(\mathbf{x}) \\
f_i(\mathbf{x}) \wedge (\overline{f_i(\mathbf{x})} \wedge \overline{f_r(\mathbf{x})} \vee f_q(\mathbf{x})) &= f_q(\mathbf{x}) \\
f_i(\mathbf{x}) \wedge \overline{f_i(\mathbf{x})} \wedge \overline{f_r(\mathbf{x})} \vee f_i(\mathbf{x}) \wedge f_q(\mathbf{x}) &= f_q(\mathbf{x}) \\
f_i(\mathbf{x}) \wedge f_q(\mathbf{x}) &= f_q(\mathbf{x}) \\
f_q(\mathbf{x}) &= f_q(\mathbf{x}) \, .
\end{aligned}
$$

Similarly, (3.16) can be confirmed using (3.14), (3.10), and (3.13).

The reason that a lattice generated by an incompletely specified function satisfies the axioms of a Boolean Algebra is that such a set of 2^n functions can be mapped to a Boolean space $\mathbb{B}^n$; there is an isomorphism between $(L\langle f_q(\mathbf{x}), f_r(\mathbf{x})\rangle, \wedge, \vee, \overline{}, f_q(\mathbf{x}), \overline{f_r(\mathbf{x})})$ with $|f_\varphi(\mathbf{x})| = k$ and $(\mathbb{B}^k, \wedge, \vee, \overline{}, \mathbf{0}, \mathbf{1})$.

There are sets of Boolean functions that satisfy the axioms of a lattice and even a Boolean Algebra, but cannot be expressed by an incompletely specified function; see Steinbach [2013] and Steinbach and Posthoff [2013b, 2015]. All functions $f_i(\mathbf{x}) \in L$ that satisfy (3.9) are specified by the ON-set $f_q(\mathbf{x})$ and the OFF-set $f_r(\mathbf{x})$; for short we write $L\langle f_q(\mathbf{x}), f_r(\mathbf{x})\rangle$. A subset of functions $g_i(\mathbf{x}) \in L_g \subset L$ can be created such that the ON-set $g_q(\mathbf{x}) = f_q(\mathbf{x})$ and the OFF-set $g_r(\mathbf{x}) = f_r(\mathbf{x})$, i.e., the mark functions of L_g and L are the same, but the functions $g_i(\mathbf{x})$ satisfies the stronger restriction

$$
g_q(\mathbf{x}) \wedge \overline{g_i(\mathbf{x})} \vee g_i(\mathbf{x}) \wedge g_r(\mathbf{x}) \vee \frac{\partial g(\mathbf{x})}{\partial \mathbf{x}_0} = 0 \, . \tag{3.17}
$$

A short characteristic of this lattice is $L_g\left\langle g_q(\mathbf{x}), g_r(\mathbf{x}), \frac{\partial g(\mathbf{x})}{\partial \mathbf{x}_0}\right\rangle$. Due to the vectorial derivative in (3.17) the set of functions $g_i(\mathbf{x}) \in L_g$ is smaller than L, but satisfies the axioms of a lattice and a Boolean Algebra as well.

Example 3.3 General Lattice of Functions. The Karnaugh-map of Figure 3.2a describes an example of a lattice where eight function values, indicated by a light blue background, are not fixed. In case of an incompletely specified function these eight patterns would generate $2^8 = 256$ functions $g_i(\mathbf{x}) \in L_{is}$, but the lattice of Figure 3.2 only consists of 16 functions $g_i(\mathbf{x}) \in L_g$ with $L_g \subset L_{is}$. These 16 functions satisfy the additional condition that each pair of function values, indicated by two diamonds connected by a line in Figure 3.2a, carries either two function values 0 or two function values 1. In this way four independent alternatives and consequently $2^4 = 16$ functions remain. All these functions satisfy the lattice Equation (3.17) for $\mathbf{x} = (x_1, x_2, x_3, x_4)$, $\mathbf{x}_0 = (x_2, x_3, x_4)$ and

$$
g_q(\mathbf{x}) = x_2\overline{x}_3\overline{x}_4 \vee \overline{x}_2 x_3 x_4 \, , \tag{3.18}
$$

$$
\overline{g_r(\mathbf{x})} = \overline{x}_1(x_2 \oplus x_3) \vee x_1(x_2 \oplus x_4) \vee (x_3 \odot x_4) \, . \tag{3.19}
$$

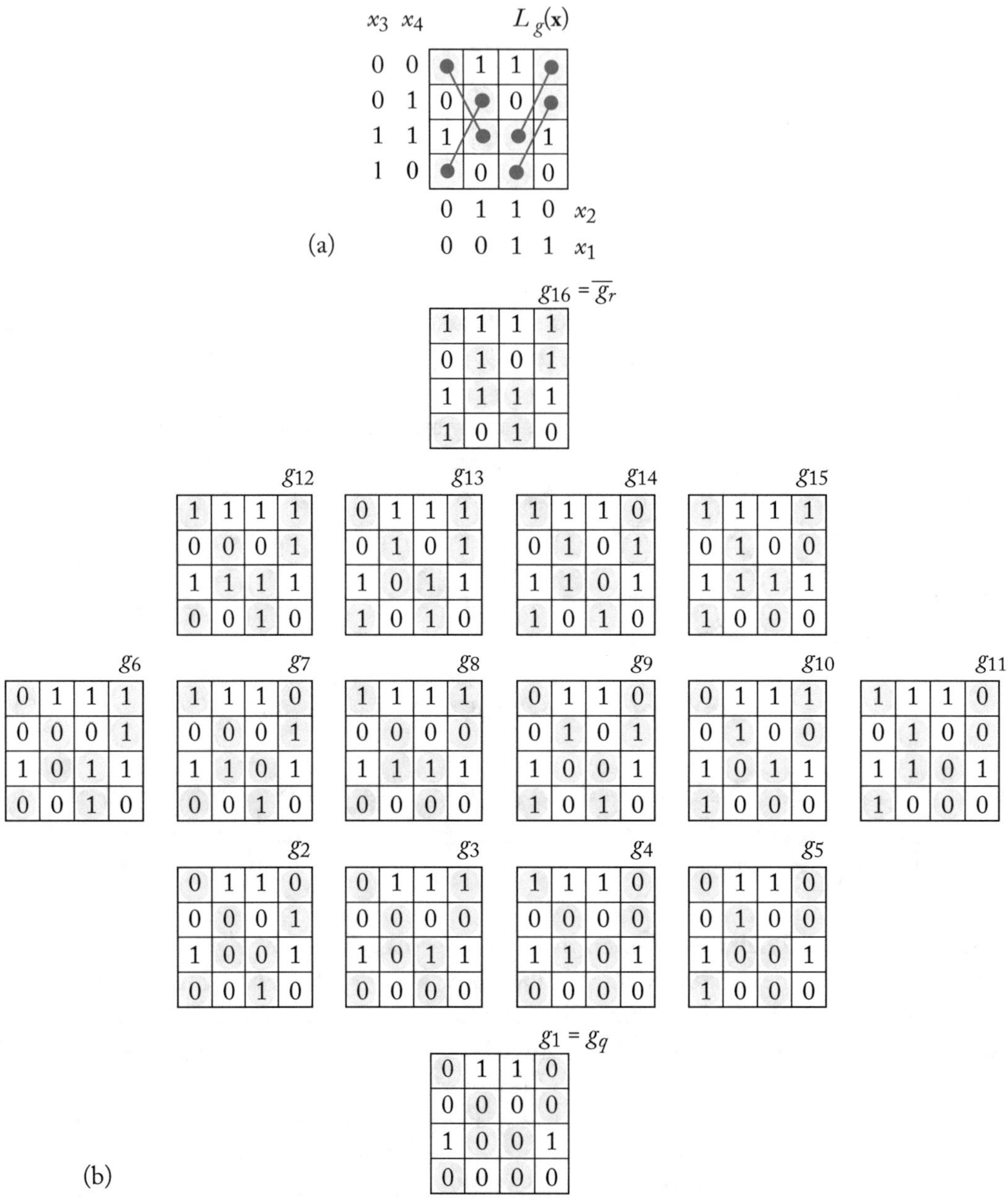

Figure 3.2: General lattice $L_g(\mathbf{x})$: (a) specification of $g_q(\mathbf{x})$, $g_q(\mathbf{x})$, and pairs of equal function values and (b) 16 functions of the lattice.

Figure 3.2b shows all 16 functions of this general lattice. Reusing the same principle of Figure 3.1, the rows and columns of the Karnaugh-maps are arranged in the same way as shown in Figure 3.2a, but the code patterns are omitted. These 16 functions have been arranged such that the number of function values 1 is decreased by 2 in the next lower level. Fields with light blue background indicate patterns not belonging to the ON-set $g_q(\mathbf{x})$ or the OFF-set $g_r(\mathbf{x})$. The 16 functions $g_i(\mathbf{x})$ in Figure 3.2b are different in these positions and satisfy

$$\frac{\partial g(x_1, x_2, x_3, x_4)}{\partial(x_2, x_3, x_4)} = 0 . \qquad (3.20)$$

The conjunction of any pair of these 16 functions results in one of these functions. The conjunction of two functions of the same row in Figure 3.2b belongs to a row below or is even the smallest function $g_1(\mathbf{x}) = g_q(\mathbf{x})$. Vice versa, the disjunction of any pair of these 16 functions also results in one of these functions, and the disjunction of two functions of the same row in Figure 3.2b belongs to a row above or is even the largest function $g_{16}(\mathbf{x}) = \overline{g_r(\mathbf{x})}$.

All functions of the lattice shown in Figure 3.2 satisfy the restriction (3.20). Despite of this restriction the function $g_q(\mathbf{x})$ (3.18) is the neutral element of L_g for the disjunction, and $\overline{g_r(\mathbf{x})}$ (3.19) is the neutral element of this lattice for the conjunctions. The complement function of this more general lattice is also defined by (3.14), where $g(\mathbf{x})$ is renamed back to $f(\mathbf{x})$.

Example 3.4 Lattice Complement $^L\overline{g_i(\mathbf{x})}$ of L_g Shown in Figure 3.2. Each function $g_i(\mathbf{x})$ of the lattice $L_g\left\langle g_q(\mathbf{x}), g_r(\mathbf{x}), \frac{\partial g(\mathbf{x})}{\partial(x_2,x_3,x_4)}\right\rangle$ shown in Figure 3.2b has a complement function $^L\overline{g_i(\mathbf{x})}$ (3.14) belonging to the same lattice:

$$\begin{array}{llll}
^L\overline{g_1(\mathbf{x})} = g_{16}(\mathbf{x}) , & ^L\overline{g_2(\mathbf{x})} = g_{15}(\mathbf{x}) , & ^L\overline{g_3(\mathbf{x})} = g_{14}(\mathbf{x}) , & ^L\overline{g_4(\mathbf{x})} = g_{13}(\mathbf{x}) , \\
^L\overline{g_5(\mathbf{x})} = g_{12}(\mathbf{x}) , & ^L\overline{g_6(\mathbf{x})} = g_{11}(\mathbf{x}) , & ^L\overline{g_7(\mathbf{x})} = g_{10}(\mathbf{x}) , & ^L\overline{g_8(\mathbf{x})} = g_9(\mathbf{x}) , \\
^L\overline{g_9(\mathbf{x})} = g_8(\mathbf{x}) , & ^L\overline{g_{10}(\mathbf{x})} = g_7(\mathbf{x}) , & ^L\overline{g_{11}(\mathbf{x})} = g_6(\mathbf{x}) , & ^L\overline{g_{12}(\mathbf{x})} = g_5(\mathbf{x}) , \\
^L\overline{g_{13}(\mathbf{x})} = g_4(\mathbf{x}) , & ^L\overline{g_{14}(\mathbf{x})} = g_3(\mathbf{x}) , & ^L\overline{g_{15}(\mathbf{x})} = g_2(\mathbf{x}) , & ^L\overline{g_{16}(\mathbf{x})} = g_1(\mathbf{x}) .
\end{array}$$

We learned from Examples 3.3 and 3.4 that there are lattices consisting of functions that are independent of a certain direction of change. In these examples the chosen direction of change is specified by the variables $\mathbf{x}_0 = (x_2, x_3, x_4)$ of the vectorial derivative in (3.17). Generally, this independence is not restricted to one direction of change.

The number of directions of change in which a Boolean function is independent has a strong influence to both the percentage of functions of n variables that satisfy these restrictions and their complexity. Due to the pairs of equal function values only $2^{2^{n-1}}$ of all 2^{2^n} Boolean functions are independent of one selected direction of change. Functions that are independent with regard to two directions of change consist of quadruples of equal function values so that only $2^{2^{n-2}}$ of all 2^{2^n} functions satisfy this property. Hence, the number of directions of change, in which all functions of a lattice are independent, is an important information to characterize a lattice. This information can be stored within an independence matrix.

Definition 3.5 Independence Matrix. The **independence matrix** $\mathrm{IDM}(f)$ of a Boolean function $f(x_1, x_2, \ldots, x_n)$ is a Boolean matrix of n rows and n columns. The columns of the independence matrix are associated with the n variables of the Boolean space in the fixed order $(x_1, x_2, \ldots, x_n)$. The independence matrix has the shape of an echelon; all elements below the main diagonal are equal to 0. Values 1 of a row of the independence matrix indicate a set of variables for which the vectorial derivative of the function $f(x_1, x_2, \ldots, x_n)$ is equal to 0. The following rules ensure the uniqueness of the independence matrix.

1. Values 1 can only occur to the right of a value 1 in the main diagonal of the independence matrix.
2. All values above a value 1 in the main diagonal of the independence matrix are equal to 0.

A Boolean function of n variables has $2^n - 1$ directions of change, but at most n of them are independent from each other. Figure 3.3 depicts this property for four subspaces. All functions $f(\mathbf{x}) = f(x_1, x_2, x_3, x_4)$ with

$$\frac{\partial f(x_1, x_2, x_3, x_4)}{\partial (x_2, x_3, x_4)} \vee \frac{\partial f(x_1, x_2, x_3, x_4)}{\partial (x_1, x_2, x_3)} = 0$$

have equal function values in each of the four groups of nodes indicated by light blue backgrounds in Figure 3.3. It can be seen in Figure 3.3 that all these functions are also independent of the simultaneous change of (x_1, x_4):

$$\frac{\partial f(x_1, x_2, x_3, x_4)}{\partial (x_1, x_4)} = 0 \,.$$

Hence, only two of these three directions of change are independent from each other.

The actual number of independent directions of change is implicitly specified by the independence matrix IDM.

Definition 3.6 Rank. The **rank** of an independence matrix $\mathrm{IDM}(f)$ describes the number of independent directions of change of the Boolean function $f(x_1, x_2, \ldots, x_n)$. The $\mathbf{rank}(\mathrm{IDM}(f))$ is equal to the number of elements 1 in the main diagonal of the unique echelon shape of $\mathrm{IDM}(f)$.

A lattice of Boolean functions can be restricted to functions which do not change their values in the case of the simultaneous change of several variables or even of a set of such directions

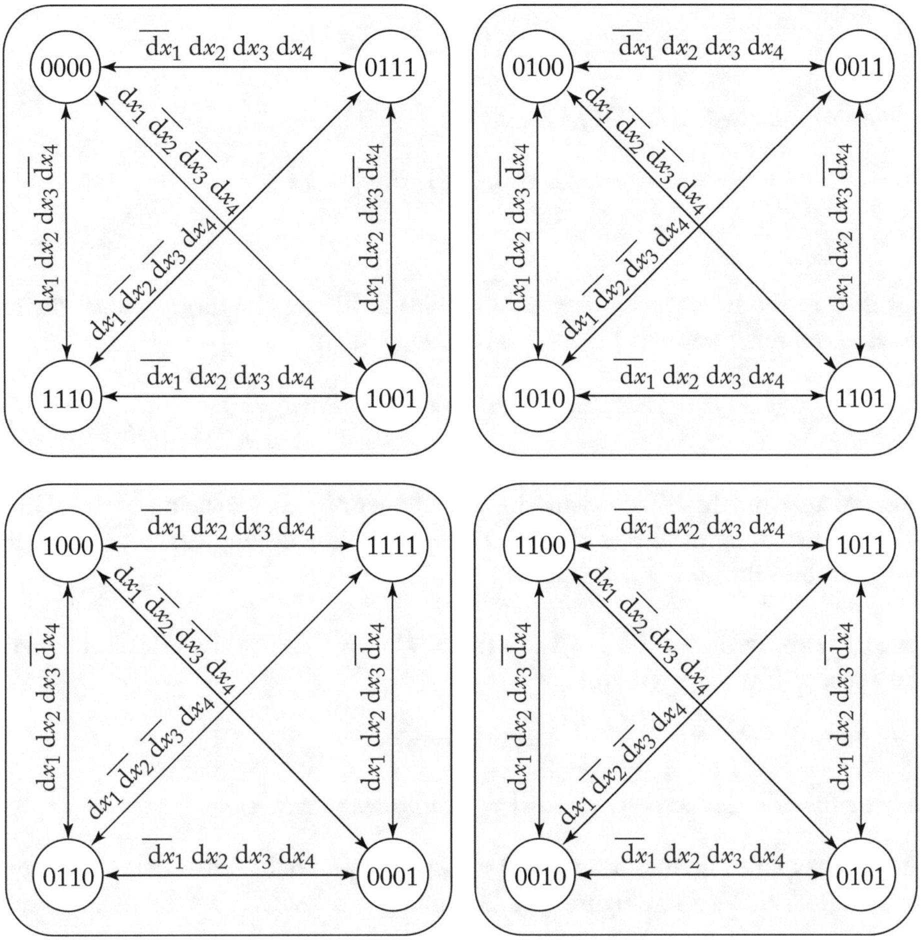

Figure 3.3: All nodes of $\mathbb{B}^4$ ordered into four groups by selected directions of change.

of change. These directions of change can be expressed by a disjunction of appropriate vectorial derivatives which are uniquely indicated in the independence matrix. For a short notation we define the *independence function* $f^{id}(\mathbf{x})$.

Definition 3.7 Independence Function. The **independence function** $f^{id}(\mathbf{x})$ of a Boolean function corresponds to the independence matrix $\mathrm{IDM}(f)$ such that

$$f^{id}(\mathbf{x}) = \bigvee_{i=1}^{n} \frac{\partial f(\mathbf{x})}{\partial \mathbf{x}_{0i}} \,, \tag{3.21}$$

where

$$\frac{\partial f(\mathbf{x})}{\partial \mathbf{x}_{0i}} = 0 \tag{3.22}$$

if all elements of the row i in IDM(f) are equal to 0, and

$$x_j \in \mathbf{x}_{0i} \text{ if } \mathrm{IDM}(f)[i, j] = 1 \; . \tag{3.23}$$

In this way, the Boolean Differential Calculus facilitates the more general definition of lattices of Boolean functions $f_i(\mathbf{x}) \in L\langle f_q(\mathbf{x}), f_r(\mathbf{x}), f^{id}(\mathbf{x})\rangle$:

$$f_q(\mathbf{x}) \wedge \overline{f_i(\mathbf{x})} \vee f_i(\mathbf{x}) \wedge f_r(\mathbf{x}) \vee f^{id}(\mathbf{x}) = 0 \; . \tag{3.24}$$

Definition 3.5 requires a unique specification of the independent directions of change. Before we give an algorithm that satisfies the requirement for an extension of the IDM(f) we explain the used rules by means of Figures 3.3 and 3.4. For dealing easily with directions of change we introduce the binary vector s.

Definition 3.8 Binary Vector (BV). Let $f(\mathbf{x}) = f(x_1, x_2, \ldots, x_n)$ be a Boolean function and $\mathbf{x}_0 \subseteq \mathbf{x}$ be a subset of variables; then,

$$\mathbf{s}_0 = BV(\mathbf{x}_0) \tag{3.25}$$

is a **binary vector** of n elements where $\mathbf{s}_0[i] = 1$ indicates that $x_i \in \mathbf{x}_0$.

The independence matrix of a function IDM(f) is initially generated as a zero matrix, shown as the upper IDM(f) in Figure 3.4a. Assuming it is known that the vectorial derivative with regard to $\mathbf{x}_0 = (x_2, x_3, x_4)$ is equal to 0, i.e., the horizontally neighbored nodes in light blue ranges of Figure 3.3 have equal function values, this direction of change must be included in the IDM(f). The associated vector $\mathbf{s}_0 = (0111)$ is uniquely specified because the IDM(f) is a zero matrix; hence, $\mathbf{s}_{min} = \mathbf{s}_0 = (0111)$ and this binary vector is included into the second row of the IDM(f) so that the most significant bit of $\mathbf{s}_{min}$ is located in the main diagonal as shown in the lower IDM(f) of Figure 3.4a.

Next, we assume that the vectorial derivative with regard to $\mathbf{x}_0 = (x_1, x_2, x_3)$ is equal to 0, i.e., the vertically neighbored nodes in light blue ranges of Figure 3.3 have equal function values. The associated vector $\mathbf{s}_0 = (1110)$ is shown in Figure 3.4b. As can be seen in Figure 3.3, equal functions values in vertically neighbored nodes in light blue ranges also cause that vectorial derivative with regard to $\mathbf{x}_0 = (x_1, x_4)$ is equal to 0. Hence, there is an alternative vector $\mathbf{s}'_0 = (1001)$. Definition 3.5 stipulates a unique specification of each IDM(f). Which of these two vectors $\mathbf{s}_0 = (1110)$ or $\mathbf{s}'_0 = (1001)$ should be selected as a unique representative? We solved this problem such that we interpret these vectors as binary number having the values $1 \cdot 2^3 + 1 \cdot 2^2 +$

Upper row:

$i \backslash j$	1	2	3	4
1	0	0	0	0
2	0	0	0	0
3	0	0	0	0
4	0	0	0	0

$\mathbf{s}_0 = (0\ 1\ 1\ 1)$

$\mathbf{s}_{min} = (0\ 1\ 1\ 1)$

$i \backslash j$	1	2	3	4
1	0	0	0	0
2	0	1	1	1
3	0	0	0	0
4	0	0	0	0

$\mathbf{s}_0 = (1\ 1\ 1\ 0)$

$\mathbf{s}_{min} = (1\ 0\ 0\ 1)$

$i \backslash j$	1	2	3	4
1	1	0	0	1
2	0	1	1	1
3	0	0	0	0
4	0	0	0	0

$\mathbf{s}_0 = (1\ 1\ 1\ 0)$

$\mathbf{s}_{min} = (0\ 0\ 0\ 0)$

$i \backslash j$	1	2	3	4
1	1	0	0	1
2	0	1	1	1
3	0	0	0	0
4	0	0	0	0

$\mathbf{s}_0 = (0\ 0\ 1\ 1)$

$\mathbf{s}_{min} = (0\ 0\ 1\ 1)$

Lower row:

(a)

$i \backslash j$	1	2	3	4
1	0	0	0	0
2	0	1	1	1
3	0	0	0	0
4	0	0	0	0

(b)

$i \backslash j$	1	2	3	4
1	1	0	0	1
2	0	1	1	1
3	0	0	0	0
4	0	0	0	0

(c)

$i \backslash j$	1	2	3	4
1	1	0	0	1
2	0	1	1	1
3	0	0	0	0
4	0	0	0	0

(d)

$i \backslash j$	1	2	3	4
1	1	0	0	1
2	0	1	0	0
3	0	0	1	1
4	0	0	0	0

Figure 3.4: Adding several directions of change to the independence matrix IDM(f).

$1 \cdot 2^1 + 0 \cdot 2^0 = 14$ for (1110) and $1 \cdot 2^3 + 0 \cdot 2^2 + 0 \cdot 2^1 + 1 \cdot 2^0 = 9$ for (1001) and select the binary vector $\mathbf{s}_{min} = (1001)$ with the smallest decimal equivalent. Figure 3.4b shows that this vector is included in the first row of the lower IDM(f) so that again the most significant bit is located in the main diagonal.

The vector $\mathbf{s}_{min}$ is calculated with regard to a given IDM(f) such that it not only specifies a unique direction of change for a new vector $\mathbf{s}_0$, but also indicates by $\mathbf{s}_{min} = \mathbf{0}$ that this direction of change is already covered by the used IDM(f). Figure 3.4c shows such an example. The vector $\mathbf{s}_0 = (1110)$ is not directly visible in the upper IDM(f), but

$$\mathbf{s}_{min} = (1110) \oplus (1001) \oplus (0111) = (0000)$$

detects that $\mathbf{s}_0 = (1110)$ has been previously included into the used IDM(f). This sequence of $\oplus$-operations calculates $\mathbf{s}_{min}$ using $\mathbf{s}_0$ and all the rows of the used IDM(f) in which the bit 1 in the main diagonal appears as a bit 1 in the same position of $\mathbf{s}_0$.

Figure 3.4d shows as last example that a new vector $\mathbf{s}_{min}$ also adjusts other rows of the used IDM(f) to get a unique representation of this matrix. We take the upper IDM(f) of Figure 3.4d as basis and assume that the vectorial derivative with regard to $\mathbf{x}_0 = (x_3, x_4)$ is equal to 0. Hence, we have $\mathbf{s}_0 = (0011)$ and get $\mathbf{s}_{min} = \mathbf{s}_0 = (0011)$ because the elements of the main diagonal in the rows 3 and 4 of the used IDM(f) are equal to 0. Due to this new (third) independent direction of change the Boolean space $\mathbb{B}^4$ is divided into $2^{4-3} = 2$ groups of $2^3 = 8$ nodes that carry the same function values. Due to the eight nodes in these two groups $2^3 - 1 = 7$ vectorial derivatives are equal to 0; ordered by the most significant bit and the decimal equivalent the

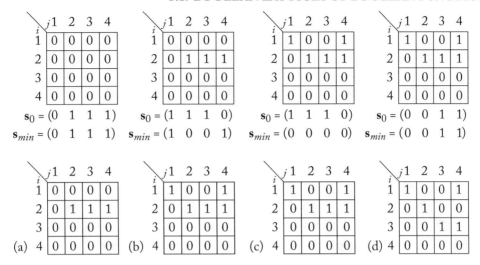

seven possible vectors $\mathbf{s}_0$ are:

$$\begin{vmatrix} (1001) & (0100) & (0011) \\ (1010) & (0111) & \\ (1101) & & \\ (1110) & & \end{vmatrix} ,$$

where the vectors in the upper row have the minimal value of the decimal equivalent and must therefore be used in the rebuilt $\mathrm{IDM}(f)$. The lower $\mathrm{IDM}(f)$ of Figure 3.4d shows the welcome property that all elements located above a value 1 in the main diagonal are equal to 0. This property is used to calculate a unique $\mathrm{IDM}(f)$: the row $\mathrm{IDM}(f)[i]$ is replaced by $\mathrm{IDM}(f)[i] \oplus \mathbf{s}_{min}$ if the most significant bit of $\mathbf{s}_{min}$ appears in this row. The comparison of the upper and lower $\mathrm{IDM}(f)$ of Figure 3.4d shows this effect.

Algorithm 3.7 $\mathbf{s}_{min} = \mathrm{MIDC}(\mathrm{IDM}(f), \mathbf{x}_0)$: Minimal Independent Direction of Change

Input : $\mathbf{x}_0 \subseteq \mathbf{x}$: evaluated subset of variables,
Input : $\mathrm{IDM}(f)$: unique independence matrix of n rows and n columns of $f(\mathbf{x})$
Output : $\mathbf{s}_{min}$: minimal direction of change

1: $j \leftarrow 1$
2: $\mathbf{s}_{min} \leftarrow BV(\mathbf{x}_0)$
3: **while** $j \leq n$ **do**
4:    **if** $(\mathbf{s}_{min}[j] = 1) \wedge (\mathrm{IDM}(f)[j, j] = 1)$ **then**
5:       $\mathbf{s}_{min} \leftarrow \mathbf{s}_{min} \oplus \mathrm{IDM}(f)[j]$
6:    **end if**
7:    $j \leftarrow j + 1$
8: **end while**

Algorithm 3.7 calculates the vector $\mathbf{s}_{min}$ that indicates the minimal direction of change not covered by the given independence matrix $\mathrm{IDM}(f)$ and the set of variables $\mathbf{x}_0$. The function $BV(\mathbf{x}_0)$ in line 2 maps the variables into a *binary vector*, where the bit $\mathbf{s}_{min}[j]$ indicates that the variable x_j belongs to $\mathbf{x}_0$. The initial vector $\mathbf{s}_{min}$ is modified in line 5 if both the bit $\mathbf{s}_{min}[j]$ and the bit in the main diagonal $\mathrm{IDM}(f)[j]$ are equal to 1. The result of Algorithm 3.7 is a uniquely specified vector $\mathbf{s}_{min}$ that must be included into the independence matrix $\mathrm{IDM}(f)$ such that the most significant bit belongs to the main diagonal. A result vector $\mathbf{s}_{min} = \mathbf{0}$ of Algorithm 3.7 indicates that all functions $f(\mathbf{x})$ of the used $\mathrm{IDM}(f)$ satisfy:

$$\frac{\partial f(\mathbf{x})}{\partial \mathbf{x}_0} = 0 . \tag{3.26}$$

Algorithm 3.8 $\mathrm{IDM}(g) = \mathrm{UM}(\mathrm{IDM}(f), \mathbf{x}_0)$: Unique Merge

Input : $\mathbf{x}_0 \subseteq \mathbf{x}$: subset of variables that satisfy (3.26) to merge with $\mathrm{IDM}(f)$,
Input : $\mathrm{IDM}(f)$: unique independence matrix of n rows and n columns of $f(\mathbf{x})$
Output : $\mathrm{IDM}(g)$: unique independence matrix of the same size of $g(\mathbf{x})$ with $\frac{\partial g(\mathbf{x})}{\partial \mathbf{x}_0} = 0$

```
1: IDM(g) ← IDM(f)
2: s_min = MIDC(IDM(f), x_0)
3: if s_min > 0 then
4:    j ← IndexOfMostSignificantBit(s_min)
5:    i ← 1
6:    while i < j do
7:       if IDM(g)[i, j] = 1 then
8:          IDM(g)[i] ← IDM(g)[i] ⊕ s_min
9:       end if
10:      i ← i + 1
11:   end while
12:   IDM(g)[j] ← s_min
13: end if
```

Algorithm 3.8 realizes the unique merge of a give direction of change specified by $\mathbf{x}_0$ into the independence matrix $\mathrm{IDM}(f)$. Initial steps copy $\mathrm{IDM}(f)$ to $\mathrm{IDM}(g)$ and calculate the unique vector $\mathbf{s}_{min}$ for the given $\mathbf{x}_0$ using Algorithm 3.7. The copied independence matrix $\mathrm{IDM}(g)$ must be changed only if $\mathbf{s}_{min} > 0$. The index of the most significant bit j of $\mathbf{s}_{min}$ indicates within $\mathrm{IDM}(g)$ both the column which must be evaluated and the row where $\mathbf{s}_{min}$ must be stored. All rows of $\mathrm{IDM}(g)$ must be equal to 0 in the column j except in the main diagonal. The operations within the while-loop in lines 6–11 perform the needed changes by conditional $\oplus$-operations. The new vector $\mathbf{s}_{min}$ is included into the independence matrix $\mathrm{IDM}(g)$ in line 12.

All derivative operations for a given lattice of Boolean functions

$$f_i(\mathbf{x}) \in L\left\langle f_q(\mathbf{x}), f_r(\mathbf{x}), f^{id}(\mathbf{x})\right\rangle$$

result in a lattice of Boolean functions

$$g_i(\mathbf{x}) \in L\left\langle g_q(\mathbf{x}), g_r(\mathbf{x}), g^{id}(\mathbf{x})\right\rangle .$$

If at least one of the applied directions of change is not included into the independence matrix $\mathrm{IMD}(f)$, then the result lattice is *simpler* than the given lattice, because

$$\mathbf{rank}(\mathrm{IDM}(g)) > \mathbf{rank}(\mathrm{IDM}(f)) .$$

The following three sections provide the theorems that describe the rules to calculate the mark functions for all derivative operations for a given lattice of Boolean functions. The formal proofs of these theorems are given in Steinbach and Posthoff [2013b]. Instead of the technical details of these proofs we explain the rules to calculate the mark functions of the result lattices of all derivative operations so that the reader gets a comprehensive notion.

3.2 VECTORIAL DERIVATIVE OPERATIONS

As mentioned above, all vectorial derivative operations of a given lattice of Boolean functions $f_i(\mathbf{x}) \in L\langle f_q(\mathbf{x}), f_r(\mathbf{x}), f^{id}(\mathbf{x})\rangle$ with regard to $\mathbf{x}_0$ result again in a lattice of Boolean functions where all functions are independent of the simultaneous change of $\mathbf{x}_0$. Based on the background knowledge explored in this section, the associated mark functions $g_q(\mathbf{x})$, $g_r(\mathbf{x})$, and $g^{id}(\mathbf{x})$ can be calculated using Theorem 3.9.

Theorem 3.9 Vectorial Derivative Operations of a Lattice of Boolean Functions. *Let $f(\mathbf{x}) = f(\mathbf{x}_0, \mathbf{x}_1) = f(x_1, x_2, \ldots, x_n)$ be a Boolean function of n variables that belongs to the lattice $L\langle f_q(\mathbf{x}), f_r(\mathbf{x}), f^{id}(\mathbf{x})\rangle$ defined by Equation (3.24) where $f_q(\mathbf{x}) = f_q(\mathbf{x}_0, \mathbf{x}_1)$ and $f_r(\mathbf{x}) = f_r(\mathbf{x}_0, \mathbf{x}_1)$ satisfy (3.5), and $f(\mathbf{x})$ depends on the simultaneous change of the values of all variables of $\mathbf{x}_0$:*

$$MIDC(IDM(f), \mathbf{x}_0) > 0\,. \tag{3.27}$$

*Then all **vectorial derivatives** of $f(\mathbf{x})$ with regard to $\mathbf{x}_0$*

$$g_1(\mathbf{x}) = \frac{\partial f(\mathbf{x}_0, \mathbf{x}_1)}{\partial \mathbf{x}_0} \tag{3.28}$$

belong to a Boolean lattice defined by

$$f_q^{\partial \mathbf{x}_0}(\mathbf{x}_0, \mathbf{x}_1) \wedge \overline{g_1(\mathbf{x})} \vee g_1(\mathbf{x}) \wedge f_r^{\partial \mathbf{x}_0}(\mathbf{x}_0, \mathbf{x}_1) \vee g_1^{id}(\mathbf{x}) = 0 \tag{3.29}$$

with the mark functions of the vectorial derivative of the lattice with regard to $\mathbf{x}_0$

$$f_q^{\partial \mathbf{x}_0}(\mathbf{x}_0, \mathbf{x}_1) = \max_{\mathbf{x}_0} f_q(\mathbf{x}_0, \mathbf{x}_1) \wedge \max_{\mathbf{x}_0} f_r(\mathbf{x}_0, \mathbf{x}_1)\,, \tag{3.30}$$

$$f_r^{\partial \mathbf{x}_0}(\mathbf{x}_0, \mathbf{x}_1) = \min_{\mathbf{x}_0} f_q(\mathbf{x}_0, \mathbf{x}_1) \vee \min_{\mathbf{x}_0} f_r(\mathbf{x}_0, \mathbf{x}_1)\,, \tag{3.31}$$

and the independence function $g_1^{id}(\mathbf{x})$ associated with

$$IDM(g_1) = UM(IDM(f), \mathbf{x}_0)\,; \tag{3.32}$$

*all **vectorial minima** of $f(\mathbf{x})$ with regard to $\mathbf{x}_0$*

$$g_2(\mathbf{x}) = \min_{\mathbf{x}_0} f(\mathbf{x}) \tag{3.33}$$

belong to a Boolean lattice defined by

$$f_q^{\min_{\mathbf{x}_0}}(\mathbf{x}_0, \mathbf{x}_1) \wedge \overline{g_2(\mathbf{x})} \vee g_2(\mathbf{x}) \wedge f_r^{\min_{\mathbf{x}_0}}(\mathbf{x}_0, \mathbf{x}_1) \vee g_2^{id}(\mathbf{x}) = 0 \tag{3.34}$$

with the mark functions of the vectorial minimum of the lattice with regard to $\mathbf{x}_0$

$$f_q^{\min_{\mathbf{x}_0}}(\mathbf{x}_0, \mathbf{x}_1) = \min_{\mathbf{x}_0} f_q(\mathbf{x}_0, \mathbf{x}_1) \, , \tag{3.35}$$

$$f_r^{\min_{\mathbf{x}_0}}(\mathbf{x}_0, \mathbf{x}_1) = \max_{\mathbf{x}_0} f_r(\mathbf{x}_0, \mathbf{x}_1) \, , \tag{3.36}$$

and the independence function $g_2^{id}(\mathbf{x})$ *associated with*

$$IDM(g_2) = UM(IDM(f), \mathbf{x}_0) \, ; \tag{3.37}$$

and all ***vectorial maxima*** *of* $f(\mathbf{x})$ *with regard to* $\mathbf{x}_0$

$$g_3(\mathbf{x}) = \max_{\mathbf{x}_0} f(\mathbf{x}) \tag{3.38}$$

belong to a Boolean lattice defined by

$$f_q^{\max_{\mathbf{x}_0}}(\mathbf{x}_0, \mathbf{x}_1) \wedge \overline{g_3(\mathbf{x})} \vee g_3(\mathbf{x}) \wedge f_r^{\max_{\mathbf{x}_0}}(\mathbf{x}_0, \mathbf{x}_1) \vee g_3^{id}(\mathbf{x}) = 0 \tag{3.39}$$

with the mark functions of the vectorial maximum of the lattice with regard to $\mathbf{x}_0$

$$f_q^{\max_{\mathbf{x}_0}}(\mathbf{x}_0, \mathbf{x}_1) = \max_{\mathbf{x}_0} f_q(\mathbf{x}_0, \mathbf{x}_1) \, , \tag{3.40}$$

$$f_r^{\max_{\mathbf{x}_0}}(\mathbf{x}_0, \mathbf{x}_1) = \min_{\mathbf{x}_0} f_r(\mathbf{x}_0, \mathbf{x}_1) \, , \tag{3.41}$$

and the independence function $g_3^{id}(\mathbf{x})$ *associated with*

$$IDM(g_3) = UM(IDM(f), \mathbf{x}_0) \, . \tag{3.42}$$

The three independence functions are equal to each other:

$$g_1^{id}(\mathbf{x}) = g_2^{id}(\mathbf{x}) = g_3^{id}(\mathbf{x}) = g^{id}(\mathbf{x}) \, , \tag{3.43}$$

with

$$IDM(g) = UM(IDM(f), \mathbf{x}_0) \tag{3.44}$$

and

$$\mathbf{rank}(IDM(g)) = \mathbf{rank}(IDM(f)) + 1 \, . \tag{3.45}$$

The vectorial derivative operations with regard to $\mathbf{x}_0$ compare the function values of pairs of nodes which differ in the assigned values of $\mathbf{x}_0$.

- The *vectorial derivative is equal to* 1 if both a function value 0 and 1 occur in such a pair of nodes. Hence, $f_q^{\partial \mathbf{x}_0}(\mathbf{x}_0, \mathbf{x}_1)$ is equal to 1 if both the vectorial maximum of the ON-set function $f_q(\mathbf{x}_0, \mathbf{x}_1)$ and the OFF-set function $f_r(\mathbf{x}_0, \mathbf{x}_1)$ are equal to 1.

- The *vectorial derivative is equal to* 0 if the two values of such a pair are either equal to 0 or equal to 1. Hence, $f_r^{\partial \mathbf{x}_0}(\mathbf{x}_0, \mathbf{x}_1)$ is equal to 1 if the vectorial minimum of either the ON-set function $f_q(\mathbf{x}_0, \mathbf{x}_1)$ or the OFF-set function $f_r(\mathbf{x}_0, \mathbf{x}_1)$ are equal to 1.

- The *vectorial minimum is equal to* 1 if both function values of such a pair of nodes are equal to 1. Hence, $f_q^{\min_{\mathbf{x}_0}}(\mathbf{x}_0, \mathbf{x}_1)$ is equal to 1 if the vectorial minimum of the ON-set function $f_q(\mathbf{x}_0, \mathbf{x}_1)$ is equal to 1.

- The *vectorial minimum is equal to* 0 if at least one function value of such a pair of nodes is equal to 0. Hence, $f_r^{\min_{\mathbf{x}_0}}(\mathbf{x}_0, \mathbf{x}_1)$ is equal to 1 if the vectorial maximum of the OFF-set function $f_r(\mathbf{x}_0, \mathbf{x}_1)$ is equal to 1.

- The *vectorial maximum is equal to* 1 if at least one function value of such a pair of nodes is equal to 1. Hence, $f_q^{\max_{\mathbf{x}_0}}(\mathbf{x}_0, \mathbf{x}_1)$ is equal to 1 if the vectorial maximum of the ON-set function $f_q(\mathbf{x}_0, \mathbf{x}_1)$ is equal to 1.

- The *vectorial maximum is equal to* 0 if both function values of such a pair of nodes are equal to 0. Hence, $f_r^{\max_{\mathbf{x}_0}}(\mathbf{x}_0, \mathbf{x}_1)$ is equal to 1 if the vectorial minimum of the OFF-set function $f_r(\mathbf{x}_0, \mathbf{x}_1)$ is equal to 1.

If Condition (3.27) of Theorem 3.9 is not satisfied, we have

$$\mathbf{s}_{min} = \mathrm{MIDC}(\mathrm{IDM}(f), \mathbf{x}_0) = \mathbf{0} \tag{3.46}$$

and all functions of the given lattice do not depend on the simultaneous change of $\mathbf{x}_0$. In this case the mark functions of the vectorial derivative operations of the given lattice $L\langle f_q(\mathbf{x}), f_r(\mathbf{x}), f^{id}(\mathbf{x})\rangle$ are:

$$f_q^{\partial \mathbf{x}_0}(\mathbf{x}_0, \mathbf{x}_1) = 0\,, \qquad f_r^{\partial \mathbf{x}_0}(\mathbf{x}_0, \mathbf{x}_1) = 1\,, \qquad \mathrm{IDM}(g_1) = I_n\,, \tag{3.47}$$

$$f_q^{\min_{\mathbf{x}_0}}(\mathbf{x}_0, \mathbf{x}_1) = f_q(\mathbf{x}_0, \mathbf{x}_1)\,, \quad f_r^{\min_{\mathbf{x}_0}}(\mathbf{x}_0, \mathbf{x}_1) = f_r(\mathbf{x}_0, \mathbf{x}_1)\,, \quad \mathrm{IDM}(g_2) = \mathrm{IDM}(f)\,, \tag{3.48}$$

$$f_q^{\max_{\mathbf{x}_0}}(\mathbf{x}_0, \mathbf{x}_1) = f_q(\mathbf{x}_0, \mathbf{x}_1)\,, \quad f_r^{\max_{\mathbf{x}_0}}(\mathbf{x}_0, \mathbf{x}_1) = f_r(\mathbf{x}_0, \mathbf{x}_1)\,, \quad \mathrm{IDM}(g_3) = \mathrm{IDM}(f)\,, \tag{3.49}$$

where I_n is the identity matrix of the size n. From (3.47) follows that the vectorial derivatives with regard to $\mathbf{x}_0$ of all functions $f(\mathbf{x})$ of the given lattice $L\langle f_q(\mathbf{x}), f_r(\mathbf{x}), f^{id}(\mathbf{x})\rangle$, which do not depend on the simultaneous change of $\mathbf{x}_0$, i.e., $\mathbf{s}_{min} = \mathbf{0}$, are equal to the constant function $f(\mathbf{x}) = \mathbf{0}(\mathbf{x})$.

Example 3.10 Vectorial Derivative of a Lattice of Boolean Functions.
The lattice of Boolean functions of Figure 3.1 consists of 16 function of four variables. The mark

functions are:

$$f_q(\mathbf{x}) = \overline{x}_1\overline{x}_2x_4 \vee x_2x_3\overline{x}_4 \vee x_1\overline{x}_3\overline{x}_4 \;, \tag{3.50}$$

$$f_r(\mathbf{x}) = \overline{x}_1\overline{x}_3\overline{x}_4 \vee x_2\overline{x}_3x_4 \vee x_1\overline{x}_2x_3 \;, \tag{3.51}$$

$$f^{id}(\mathbf{x}) = 0 \;. \tag{3.52}$$

Figure 3.5a shows the Karnaugh-maps of $f_q(\mathbf{x})$ and $f_r(\mathbf{x})$ as well as the independence matrix IDM(f) associated with $f^{id}(\mathbf{x})$. The aim is the calculation of the vectorial derivatives of all functions of this lattice with regard to $\mathbf{x}_0 = (x_2, x_3, x_4)$.

As intermediate functions the vectorial minimum and maximum of $f_q(\mathbf{x})$ are shown in Figure 3.5b, and the also needed vectorial minimum and maximum of $f_r(\mathbf{x})$ with regard to the same variables (x_2, x_3, x_4) are shown in Figure 3.5c.

Applying the rules (3.30) and (3.31), the conjunction of the functions in the left Karnaugh-maps of Figures 3.5b and c results in the ON-set function $g_q(\mathbf{x})$ of the wanted lattice, and the disjunction of the functions in the right Karnaugh-maps of Figures 3.5b and c generates the associated OFF-set function $g_r(\mathbf{x})$. Using Algorithm 3.8, the independence matrix IDM(g) of Figure 3.5d is built. The lattice of the calculated vectorial derivatives $L\langle g_q(\mathbf{x}), g_r(\mathbf{x}), g^{id}(\mathbf{x})\rangle$ is specified by the mark functions:

$$g_q(\mathbf{x}) = x_2\overline{x}_3\overline{x}_4 \vee \overline{x}_2x_3x_4 \;, \tag{3.53}$$

$$g_r(\mathbf{x}) = (x_1 \oplus x_2)x_3\overline{x}_4 \vee (x_1 \odot x_2)\overline{x}_3x_4 \;, \tag{3.54}$$

$$g^{id}(\mathbf{x}) = \frac{\partial g(\mathbf{x})}{\partial(x_2, x_3, x_4)} \;. \tag{3.55}$$

Due to the ON-set function $g_q(\mathbf{x})$ and the OFF-set function $g_r(\mathbf{x})$ 8 of the 16 function values are fixed. The **rank**(IDM(g)) $= 1$ specifies that groups of $2^1 = 2$ nodes must carry the same function value so that only half of the 8 remaining function values can be arbitrarily set to the value 0 or 1. Hence, the calculated lattice consists of 16 functions of four variables. These 16 functions are shown in Figure 3.2b as example to explain the properties of a general lattice of Boolean function. The comparison of Figures 3.1b and 3.2b confirms that $g_i(\mathbf{x}), i = 1, \ldots, 16$ are the vectorial derivatives of $f_j(\mathbf{x}), j = 1, \ldots, 16$ with regard to $\mathbf{x}_0 = (x_2, x_3, x_4)$:

$$g_1(\mathbf{x}) = \frac{\partial f_6(\mathbf{x})}{\partial \mathbf{x}_0}\,, \qquad g_2(\mathbf{x}) = \frac{\partial f_3(\mathbf{x})}{\partial \mathbf{x}_0}\,, \qquad g_3(\mathbf{x}) = \frac{\partial f_2(\mathbf{x})}{\partial \mathbf{x}_0}\,, \qquad g_4(\mathbf{x}) = \frac{\partial f_{12}(\mathbf{x})}{\partial \mathbf{x}_0}\,,$$

$$g_5(\mathbf{x}) = \frac{\partial f_{13}(\mathbf{x})}{\partial \mathbf{x}_0}\,, \qquad g_6(\mathbf{x}) = \frac{\partial f_1(\mathbf{x})}{\partial \mathbf{x}_0}\,, \qquad g_7(\mathbf{x}) = \frac{\partial f_9(\mathbf{x})}{\partial \mathbf{x}_0}\,, \qquad g_8(\mathbf{x}) = \frac{\partial f_{10}(\mathbf{x})}{\partial \mathbf{x}_0}\,,$$

$$g_9(\mathbf{x}) = \frac{\partial f_7(\mathbf{x})}{\partial \mathbf{x}_0}\,, \qquad g_{10}(\mathbf{x}) = \frac{\partial f_8(\mathbf{x})}{\partial \mathbf{x}_0}\,, \qquad g_{11}(\mathbf{x}) = \frac{\partial f_{16}(\mathbf{x})}{\partial \mathbf{x}_0}\,, \qquad g_{12}(\mathbf{x}) = \frac{\partial f_4(\mathbf{x})}{\partial \mathbf{x}_0}\,,$$

$$g_{13}(\mathbf{x}) = \frac{\partial f_5(\mathbf{x})}{\partial \mathbf{x}_0}\,, \qquad g_{14}(\mathbf{x}) = \frac{\partial f_{15}(\mathbf{x})}{\partial \mathbf{x}_0}\,, \qquad g_{15}(\mathbf{x}) = \frac{\partial f_{14}(\mathbf{x})}{\partial \mathbf{x}_0}\,, \qquad g_{16}(\mathbf{x}) = \frac{\partial f_{11}(\mathbf{x})}{\partial \mathbf{x}_0}\,.$$

(a)

$f_q(\mathbf{x})$

$x_3\ x_4$					
0 0	0	0	1	1	
0 1	1	0	0	0	
1 1	1	0	0	0	
1 0	0	1	1	0	
	0	1	1	0	x_2
	0	0	1	1	x_1

$f_r(\mathbf{x})$

$x_3\ x_4$					
0 0	1	1	0	0	
0 1	0	1	1	0	
1 1	0	0	0	1	
1 0	0	0	0	1	
	0	1	1	0	x_2
	0	0	1	1	x_1

IDM(f)

$i \backslash j$	1	2	3	4
1	0	0	0	0
2	0	0	0	0
3	0	0	0	0
4	0	0	0	0

(b)

$\max_{\mathbf{x}_0} f_q(\mathbf{x})$

$x_3\ x_4$					
0 0	0	1	1	1	
0 1	1	0	0	1	
1 1	1	0	1	1	
1 0	0	1	1	0	
	0	1	1	0	x_2
	0	0	1	1	x_1

$\min_{\mathbf{x}_0} f_q(\mathbf{x})$

$x_3\ x_4$					
0 0	0	0	0	0	
0 1	1	0	0	0	
1 1	0	0	0	0	
1 0	0	1	0	0	
	0	1	1	0	x_2
	0	0	1	1	x_1

(c)

$\max_{\mathbf{x}_0} f_r(\mathbf{x})$

$x_3\ x_4$					
0 0	1	1	1	0	
0 1	0	1	1	0	
1 1	1	1	0	1	
1 0	1	0	0	1	
	0	1	1	0	x_2
	0	0	1	1	x_1

$\min_{\mathbf{x}_0} f_r(\mathbf{x})$

$x_3\ x_4$					
0 0	0	0	0	0	
0 1	0	0	1	0	
1 1	0	0	0	0	
1 0	0	0	0	1	
	0	1	1	0	x_2
	0	0	1	1	x_1

(d)

$g_q(\mathbf{x})$

$x_3\ x_4$					
0 0	0	1	1	0	
0 1	0	0	0	0	
1 1	1	0	0	1	
1 0	0	0	0	0	
	0	1	1	0	x_2
	0	0	1	1	x_1

$g_r(\mathbf{x})$

$x_3\ x_4$					
0 0	0	0	0	0	
0 1	1	0	1	0	
1 1	0	0	0	0	
1 0	0	1	0	1	
	0	1	1	0	x_2
	0	0	1	1	x_1

IDM(g)

$i \backslash j$	1	2	3	4
1	0	0	0	0
2	0	1	1	1
3	0	0	0	0
4	0	0	0	0

Figure 3.5: Calculation of the vectorial derivatives of a lattice of Boolean functions: (a) the given lattice $L\langle f_q(\mathbf{x}), f_r(\mathbf{x}), f^{id}(\mathbf{x})\rangle$; (b) vectorial minimum and maximum of $f_q(\mathbf{x})$; (c) vectorial minimum and maximum of $f_r(\mathbf{x})$; and (d) the lattice $L\langle g_q(\mathbf{x}), g_r(\mathbf{x}), g^{id}(\mathbf{x})\rangle$ of all vectorial derivatives of $L\langle f_q(\mathbf{x}), f_r(\mathbf{x}), f^{id}(\mathbf{x})\rangle$ with regard to $\mathbf{x}_0 = (x_2, x_3, x_4)$.

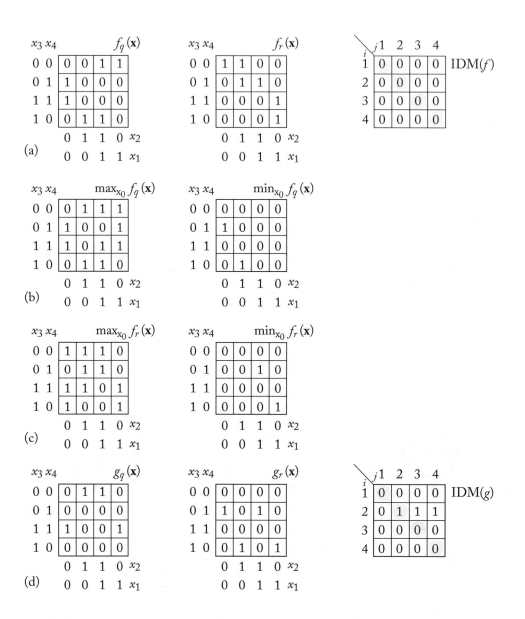

3.3 SINGLE DERIVATIVE OPERATIONS

The restriction of the set of variables $\mathbf{x}_0$ to the single variables x_i leads to the adapted Theorem 3.11 for derivative operations with regard to a single variable for all functions $f(\mathbf{x})$ of the given lattice $L\langle f_q(\mathbf{x}), f_r(\mathbf{x}), f^{id}(\mathbf{x})\rangle$.

Theorem 3.11 Single Derivative Operations of a Lattice of Boolean Functions. *Let $f(\mathbf{x}) = f(x_i, \mathbf{x}_1) = f(x_1, x_2, \ldots, x_n)$ be a Boolean function of n variables that belongs to the lattice defined by (3.24) where $f_q(\mathbf{x}) = f_q(x_i, \mathbf{x}_1)$ and $f_r(\mathbf{x}) = f_r(x_i, \mathbf{x}_1)$ satisfy (3.5), and $f(\mathbf{x})$ depends on x_i:*

$$MIDC(IDM(f), x_i) > 0 \,. \tag{3.56}$$

*Then all **derivatives of $f(\mathbf{x})$ with regard to** x_i*

$$g_1(\mathbf{x}) = \frac{\partial f(x_i, \mathbf{x}_1)}{\partial x_i} \tag{3.57}$$

belong to a Boolean lattice defined by

$$f_q^{\partial x_i}(\mathbf{x}_1) \wedge \overline{g_1(\mathbf{x})} \vee g_1(\mathbf{x}) \wedge f_r^{\partial x_i}(\mathbf{x}_1) \vee g_1^{id}(\mathbf{x}) = 0 \tag{3.58}$$

with the mark functions of the derivative of the lattice with regard to x_i

$$f_q^{\partial x_i}(\mathbf{x}_1) = \max_{x_i} f_q(x_i, \mathbf{x}_1) \wedge \max_{x_i} f_r(x_i, \mathbf{x}_1) \,, \tag{3.59}$$

$$f_r^{\partial x_i}(\mathbf{x}_1) = \min_{x_i} f_q(x_i, \mathbf{x}_1) \vee \min_{x_i} f_r(x_i, \mathbf{x}_1) \,, \tag{3.60}$$

and the independence function $g_1^{id}(\mathbf{x})$ associated with

$$IDM(g_1) = UM(IDM(f), x_i) \,; \tag{3.61}$$

*all **minima of $f(\mathbf{x})$ with regard to** x_i*

$$g_2(\mathbf{x}) = \min_{x_i} f(x_i, \mathbf{x}_1) \tag{3.62}$$

belong to a Boolean lattice defined by

$$f_q^{\min_{x_i}}(\mathbf{x}_1) \wedge \overline{g_2(\mathbf{x})} \vee g_2(\mathbf{x}) \wedge f_r^{\min_{x_i}}(\mathbf{x}_1) \vee g_2^{id}(\mathbf{x}) = 0 \tag{3.63}$$

with the mark functions of the minimum of the lattice with regard to x_i

$$f_q^{\min_{x_i}}(\mathbf{x}_1) = \min_{x_i} f_q(x_i, \mathbf{x}_1) \,, \tag{3.64}$$

$$f_r^{\min_{x_i}}(\mathbf{x}_1) = \max_{x_i} f_r(x_i, \mathbf{x}_1) \,, \tag{3.65}$$

and the independence function $g_2^{id}(\mathbf{x})$ *associated with*

$$IDM(g_2) = UM(IDM(f), x_i) \; ; \tag{3.66}$$

and all ***maxima of*** $f(\mathbf{x})$ ***with regard to*** x_i

$$g_3(\mathbf{x}) = \max_{x_i} f(x_i, \mathbf{x}_1) \tag{3.67}$$

belong to a Boolean lattice defined by

$$f_q^{\max x_i}(\mathbf{x}_1) \wedge \overline{g_3(\mathbf{x})} \vee g_3(\mathbf{x}) \wedge f_r^{\max x_i}(\mathbf{x}_1) \vee g_3^{id}(\mathbf{x}) = 0 \tag{3.68}$$

with the mark functions of the maximum of the lattice with regard to x_i

$$f_q^{\max x_i}(\mathbf{x}_1) = \max_{x_i} f_q(x_i, \mathbf{x}_1) \; , \tag{3.69}$$

$$f_r^{\max x_i}(\mathbf{x}_1) = \min_{x_i} f_r(x_i, \mathbf{x}_1) \; , \tag{3.70}$$

and the independence function $g_3^{id}(\mathbf{x})$ *associated with*

$$IDM(g_3) = UM(IDM(f), x_i) \; . \tag{3.71}$$

The three independence functions are equal to each other:

$$g_1^{id}(\mathbf{x}) = g_2^{id}(\mathbf{x}) = g_3^{id}(\mathbf{x}) = g^{id}(\mathbf{x}) \; , \tag{3.72}$$

with

$$IDM(g) = UM(IDM(f), x_i) \tag{3.73}$$

and

$$\mathbf{rank}(IDM(g)) = \mathbf{rank}(IDM(f)) + 1 \; . \tag{3.74}$$

The derivative operations with regard to a single variable restrict the simultaneous change of all values of $\mathbf{x}_0$ to the change of the value of the selected variable x_i. However, besides of the other directions of change, the derivative operations with regard to one variable also compare the function values of pairs of nodes which differ in the assigned values of x_i. Hence, in the rules to calculate the mark functions of derivative operations with regard to one variable of a lattice of Boolean functions $f_i(\mathbf{x}) \in L\langle f_q(\mathbf{x}), f_r(\mathbf{x}), f^{id}(\mathbf{x})\rangle$ use instead of the vectorial derivative operations with regard to $\mathbf{x}_0$ the associated derivative operations with regard to x_i.

If Condition (3.56) of Theorem 3.11 is not satisfied, we have

$$\mathbf{s}_{min} = \mathrm{MIDC}(\mathrm{IDM}(f), x_i) = \mathbf{0} \tag{3.75}$$

and all functions of the given lattice do not depend on the change of x_i. In this case the mark functions of the derivative operations with regard to x_i of the given lattice $L\langle f_q(\mathbf{x}), f_r(\mathbf{x}), f^{id}(\mathbf{x})\rangle$ are:

$$f_q^{\partial x_i}(\mathbf{x}_1) = 0\,, \qquad f_r^{\partial x_i}(\mathbf{x}_1) = 1\,, \qquad \mathrm{IDM}(g_1) = I_n\,, \tag{3.76}$$

$$f_q^{\min x_i}(\mathbf{x}_1) = f_q(\mathbf{x}_1)\,, \qquad f_r^{\min x_i}(\mathbf{x}_1) = f_r(\mathbf{x}_1)\,, \qquad \mathrm{IDM}(g_2) = \mathrm{IDM}(f)\,, \tag{3.77}$$

$$f_q^{\max x_i}(\mathbf{x}_1) = f_q(\mathbf{x}_1)\,, \qquad f_r^{\max x_i}(\mathbf{x}_1) = f_r(\mathbf{x}_1)\,, \qquad \mathrm{IDM}(g_3) = \mathrm{IDM}(f)\,, \tag{3.78}$$

where I_n is the identity matrix of the size n. From (3.76) follows that the derivative with regard to x_i of all functions $f(\mathbf{x})$ of the given lattice $L\langle f_q(\mathbf{x}), f_r(\mathbf{x}), f^{id}(\mathbf{x})\rangle$, which do not depend on the change of x_i, i.e., $\mathbf{s}_{min} = \mathbf{0}$, are equal to the constant function $f(\mathbf{x}) = \mathbf{0}(\mathbf{x})$.

3.4 m-FOLD DERIVATIVE OPERATIONS

Repeated derivative operations of the same type with regard to different variables are summarized to m-fold derivative operations. It is a consequence of Theorem 3.11 that each m-fold derivative operation of a given lattice $L\langle f_q(\mathbf{x}), f_r(\mathbf{x}), f^{id}(\mathbf{x})\rangle$ results again in a lattice of Boolean functions.

Theorem 3.12 ***m*-fold Derivative Operations of a Lattice of Boolean Functions.** *Let $f(\mathbf{x}) = f(\mathbf{x}_0, \mathbf{x}_1) = f(x_1, x_2, \ldots, x_n)$ be a Boolean function of n variables that belongs to the lattice defined by (3.24) where $f_q(\mathbf{x}) = f_q(\mathbf{x}_0, \mathbf{x}_1)$ and $f_r(\mathbf{x}) = f_r(\mathbf{x}_0, \mathbf{x}_1)$ satisfy (3.5), and there is at least one $x_{0i} \in \mathbf{x}_0$ with*

$$\frac{\partial f(\mathbf{x})}{\partial x_{0i}} \neq 0\,. \tag{3.79}$$

*Then all m-**fold derivatives** of $f(\mathbf{x}) = f(\mathbf{x}_0, \mathbf{x}_1)$ with regard to $\mathbf{x}_0 = (x_{01}, x_{02}, \ldots, x_{0m})$*

$$g_1(\mathbf{x}_1) = \frac{\partial^m f(\mathbf{x}_0, \mathbf{x}_1)}{\partial x_{01} \partial x_{02} \ldots \partial x_{0m}} \tag{3.80}$$

belong to a Boolean lattice defined by

$$f_q^{\partial x_1 \partial x_2 \ldots \partial x_m}(\mathbf{x}_1) \wedge \overline{g_1(\mathbf{x}_1)} \vee g_1(\mathbf{x}_1) \wedge f_r^{\partial x_1, \partial x_2, \ldots \partial x_m}(\mathbf{x}_1) \vee g_1^{id}(\mathbf{x}) = 0 \tag{3.81}$$

with the mark functions of the m-fold derivatives with regard to $\mathbf{x}_0$

$$f_q^{\partial x_1 \partial x_2 \ldots \partial x_m}(\mathbf{x}_1) = \frac{\partial^m f_q(\mathbf{x}_0, \mathbf{x}_1)}{\partial x_1 \partial x_2 \ldots \partial x_m} \wedge \min_{\mathbf{x}_0}{}^m (f_q(\mathbf{x}_0, \mathbf{x}_1) \vee f_r(\mathbf{x}_0, \mathbf{x}_1))\,, \tag{3.82}$$

$$f_r^{\partial x_1, \partial x_2, \ldots \partial x_m}(\mathbf{x}_1) = \overline{\frac{\partial^m f_q(\mathbf{x}_0, \mathbf{x}_1)}{\partial x_1 \partial x_2 \ldots \partial x_m}} \wedge \min_{\mathbf{x}_0}{}^m (f_q(\mathbf{x}_0, \mathbf{x}_1) \vee f_r(\mathbf{x}_0, \mathbf{x}_1)\,, \tag{3.83}$$

and the independence function $g_1^{id}(\mathbf{x})$ associated with $IDM(g_1)$ satisfies

$$\forall x_{0i} \in \mathbf{x}_0 : \qquad MIDC(IDM(g_1), x_{0i}) = 0 \, ; \tag{3.84}$$

*all m-**fold minima** of $f(\mathbf{x}) = f(\mathbf{x}_0, \mathbf{x}_1)$ with regard to $\mathbf{x}_0 = (x_{01}, x_{02}, \ldots, x_{0m})$*

$$g_2(\mathbf{x}_1) = \min_{\mathbf{x}_0}{}^m f(\mathbf{x}_0, \mathbf{x}_1) \tag{3.85}$$

belong to a Boolean lattice defined by

$$f_q^{\min_{\mathbf{x}_0}^m}(\mathbf{x}_1) \wedge \overline{g_2(\mathbf{x}_1)} \vee g_2(\mathbf{x}_1) \wedge f_r^{\min_{\mathbf{x}_0}^m}(\mathbf{x}_1) \vee g_2^{id}(\mathbf{x}) = 0 \tag{3.86}$$

with the mark functions of the m-fold minima with regard to $\mathbf{x}_0$

$$f_q^{\min_{\mathbf{x}_0}^m}(\mathbf{x}_1) = \min_{\mathbf{x}_0}{}^m f_q(\mathbf{x}_0, \mathbf{x}_1) \, , \tag{3.87}$$

$$f_r^{\min_{\mathbf{x}_0}^m}(\mathbf{x}_1) = \max_{\mathbf{x}_0}{}^m f_r(\mathbf{x}_0, \mathbf{x}_1) \, , \tag{3.88}$$

and the independence function $g_2^{id}(\mathbf{x})$ associated with $IDM(g_2)$ satisfies

$$\forall x_{0i} \in \mathbf{x}_0 : \qquad MIDC(IDM(g_2), x_{0i}) = 0 \, ; \tag{3.89}$$

*all m-**fold maxima** of $f(\mathbf{x}) = f(\mathbf{x}_0, \mathbf{x}_1)$ with regard to $\mathbf{x}_0 = (x_{01}, x_{02}, \ldots, x_{0m})$*

$$g_3(\mathbf{x}_1) = \max_{\mathbf{x}_0}{}^m f(\mathbf{x}_0, \mathbf{x}_1) \tag{3.90}$$

belong to a Boolean lattice defined by

$$f_q^{\max_{\mathbf{x}_0}^m}(\mathbf{x}_1) \wedge \overline{g_3(\mathbf{x}_1)} \vee g_3(\mathbf{x}_1) \wedge f_r^{\max_{\mathbf{x}_0}^m}(\mathbf{x}_1) \vee g_3^{id}(\mathbf{x}) = 0 \tag{3.91}$$

with the mark functions of the m-fold maxima with regard to $\mathbf{x}_0$

$$f_q^{\max_{\mathbf{x}_0}^m}(\mathbf{x}_1) = \max_{\mathbf{x}_0}{}^m f_q(\mathbf{x}_0, \mathbf{x}_1) \, , \tag{3.92}$$

$$f_r^{\max_{\mathbf{x}_0}^m}(\mathbf{x}_1) = \min_{\mathbf{x}_0}{}^m f_r(\mathbf{x}_0, \mathbf{x}_1) \, , \tag{3.93}$$

and the independence function $g_3^{id}(\mathbf{x})$ associated with $IDM(g_3)$ satisfies

$$\forall x_{0i} \in \mathbf{x}_0 : \qquad MIDC(IDM(g_3), x_{0i}) = 0 \, ; \tag{3.94}$$

*and all Δ-**operations** of $f(\mathbf{x}) = f(\mathbf{x}_0, \mathbf{x}_1)$ with regard to $\mathbf{x}_0 = (x_{01}, x_{02}, \ldots, x_{0m})$*

$$g_4(\mathbf{x}_1) = \Delta_{\mathbf{x}_0} f(\mathbf{x}_0, \mathbf{x}_1) \tag{3.95}$$

belong to a Boolean lattice defined by

$$f_q^{\Delta_{\mathbf{x}_0}}(\mathbf{x}_1) \wedge \overline{g_4(\mathbf{x}_1)} \vee g_4(\mathbf{x}_1) \wedge f_r^{\Delta_{\mathbf{x}_0}}(\mathbf{x}_1) \vee g_4^{id}(\mathbf{x}) = 0 \tag{3.96}$$

with the mark functions of the Δ*-operations with regard to* $\mathbf{x}_0$

$$f_q^{\Delta_{\mathbf{x}_0}}(\mathbf{x}_1) = \max_{\mathbf{x}_0}{}^m f_q(\mathbf{x}_0, \mathbf{x}_1) \wedge \max_{\mathbf{x}_0}{}^m f_r(\mathbf{x}_0, \mathbf{x}_1) , \tag{3.97}$$

$$f_r^{\Delta_{\mathbf{x}_0}}(\mathbf{x}_1) = \min_{\mathbf{x}_0}{}^m f_q(\mathbf{x}_0, \mathbf{x}_1) \vee \min_{\mathbf{x}_0}{}^m f_r(\mathbf{x}_0, \mathbf{x}_1) , \tag{3.98}$$

and the independence function $g_4^{id}(\mathbf{x})$ *associated with* $IDM(g_4)$ *satisfies*

$$\forall x_{0i} \in \mathbf{x}_0 : \quad MIDC(IDM(g_4), x_{0i}) = 0 . \tag{3.99}$$

The four independence functions are equal to each other:

$$g_1^{id}(\mathbf{x}) = g_2^{id}(\mathbf{x}) = g_3^{id}(\mathbf{x}) = g_4^{id}(\mathbf{x}) = g^{id}(\mathbf{x}) , \tag{3.100}$$

with

$$IDM(g) = UM(\ldots UM(IDM(f), x_{01}), \ldots, x_{0m}) \tag{3.101}$$

and

$$\mathbf{rank}(IDM(f)) + 1 \leq \mathbf{rank}(IDM(g)) \leq \mathbf{rank}(IDM(f)) + m . \tag{3.102}$$

The m-fold derivative operations with regard to $\mathbf{x}_0$ evaluate the function values of subspaces of 2^m nodes which differ at least in one value of $\mathbf{x}_0$.

- A unique decision whether an odd or an even number of function values occur in a subspace is only possible if all function values belong either to the ON-set or to the OFF-set of the evaluated subspace. This property is satisfied if the the m-fold minimum of the disjunction of the ON-set function $f_q(\mathbf{x}_0, \mathbf{x}_1)$ and the OFF-set function $f_r(\mathbf{x}_0, \mathbf{x}_1)$ is equal to 1.
- The *m-fold derivative is equal to* 1 if an odd number of function values 1 occurs in such a subspace. Hence, $f_q^{\partial x_1 \partial x_2 \ldots \partial x_m}(\mathbf{x}_1)$ is equal to 1 if in addition to the condition of the first item the m-fold derivative of the ON-set function $f_q(\mathbf{x}_0, \mathbf{x}_1)$ equal to 1.
- *The m-fold derivative is equal to* 0 if an even number of function values 1 occurs in such a subspace. Hence, $f_r^{\partial x_1 \partial x_2 \ldots \partial x_m}(\mathbf{x}_1)$ is equal to 1 if additionally to the condition of the first item the m-fold derivative of the ON-set function $f_q(\mathbf{x}_0, \mathbf{x}_1)$ is equal to 0.
- The *m-fold minimum is equal to* 1 if all function values in such a subspace are equal to 1. Hence, $f_q^{\min_{\mathbf{x}_0}^m}(\mathbf{x}_1)$ is equal to 1 if the m-fold minimum of the ON-set function $f_q(\mathbf{x}_0, \mathbf{x}_1)$ is equal to 1.

- The *m-fold minimum is equal to* 0 if at least one function value in such a subspace is equal to 0. Hence, $f_r^{\min^m_{\mathbf{x}_0}}(\mathbf{x}_1)$ is equal to 1 if the m-fold maximum of the OFF-set function $f_r(\mathbf{x}_0, \mathbf{x}_1)$ is equal to 1.

- The *m-fold maximum is equal to* 1 if at least one function value in such a subspace is equal to 1. Hence, $f_q^{\max^m_{\mathbf{x}_0}}(\mathbf{x}_1)$ is equal to 1 if the m-fold maximum of the ON-set function $f_q(\mathbf{x}_0, \mathbf{x}_1)$ is equal to 1.

- The *m-fold maximum is equal to* 0 if all function values in such a subspace are equal to 0. Hence, $f_r^{\max^m_{\mathbf{x}_0}}(\mathbf{x}_1)$ is equal to 1 if the m-fold minimum of the OFF-set function $f_r(\mathbf{x}_0, \mathbf{x}_1)$ is equal to 1.

- The *Δ-operation is equal to* 1 if both at least one function value 1 and one function value 0 occur in in such a subspace. Hence, $f_q^{\Delta \mathbf{x}_0}(\mathbf{x}_1)$ is equal to 1 if both the m-fold maximum of the ON-set function $f_q(\mathbf{x}_0, \mathbf{x}_1)$ and the m-fold maximum of the OFF-set function $f_r(\mathbf{x}_0, \mathbf{x}_1)$ are equal to 1.

- The *Δ-operation is equal to* 0 if all function values in such a subspace are either equal to 0 or equal to 1. Hence, $f_r^{\Delta \mathbf{x}_0}(\mathbf{x}_1)$ is equal to 1 if either the m-fold minimum of the ON-set function $f_q(\mathbf{x}_0, \mathbf{x}_1)$ or the m-fold minimum of the OFF-set function $f_r(\mathbf{x}_0, \mathbf{x}_1)$ is equal to 1.

If Condition (3.79) of Theorem 3.12 is not satisfied, we have

$$\forall x_{0i} \in \mathbf{x}_0 : \qquad \mathbf{s}_{min} = \text{MIDC}(\text{IDM}(f), x_{0i}) = \mathbf{0} \tag{3.103}$$

and all functions of the given lattice do not depend on the change of all $x_{0i} \in \mathbf{x}_0$. In this case the mark functions of the m-fold derivative operations with regard to $\mathbf{x}_0$ of the given lattice $L\langle f_q(\mathbf{x}), f_r(\mathbf{x}), f^{id}(\mathbf{x})\rangle$ are:

$$f_q^{\partial x_1 \ldots \partial x_m}(\mathbf{x}_1) = 0\,, \qquad f_r^{\partial x_1 \ldots \partial x_m}(\mathbf{x}_1) = 1\,, \qquad \text{IDM}(g_1) = I_n\,, \tag{3.104}$$

$$f_q^{\min^m_{\mathbf{x}_0}}(\mathbf{x}_1)) = f_q(\mathbf{x}_1)\,, \qquad f_r^{\min^m_{\mathbf{x}_0}}(\mathbf{x}_1) = f_r(\mathbf{x}_1)\,, \qquad \text{IDM}(g_2) = \text{IDM}(f)\,, \tag{3.105}$$

$$f_q^{\max^m_{\mathbf{x}_0}}(\mathbf{x}_1) = f_q(\mathbf{x}_1)\,, \qquad f_r^{\max^m_{\mathbf{x}_0}}(\mathbf{x}_1) = f_r(\mathbf{x}_1)\,, \qquad \text{IDM}(g_3) = \text{IDM}(f)\,, \tag{3.106}$$

$$f_q^{\Delta \mathbf{x}_0}(\mathbf{x}_1) = 0\,, \qquad f_q^{\Delta \mathbf{x}_0}(\mathbf{x}_1) = 1\,, \qquad \text{IDM}(g_4) = I_n\,, \tag{3.107}$$

where I_n is the identity matrix of the size n. From (3.104) and (3.107) follows that both the m-fold derivative and the Δ-operation with regard to $\mathbf{x}_0$ of all functions $f(\mathbf{x})$ of the given lattice $L\langle f_q(\mathbf{x}), f_r(\mathbf{x}), f^{id}(\mathbf{x})\rangle$, which do not depend on the change of all $x_{0i} \in \mathbf{x}_0$, i.e., $\mathbf{s}_{min} = \mathbf{0}$ for all these x_{0i}, are equal to the constant function $f(\mathbf{x}) = \mathbf{0}(\mathbf{x})$.

SUMMARY

A Boolean lattice of Boolean functions is a set of Boolean functions that satisfy the axioms of a Boolean Algebra. The calculation of each derivative operation for all Boolean functions of such a lattice results again in a Boolean lattice of Boolean functions, which is in some cases a more general Boolean lattice. It is not necessary to calculate a derivative operation for all functions of such a lattice separately, but calculated mark functions characterize the derived lattice. A derivative operations simplifies a lattice such that no function of the result lattice depends on the used direction of change. There are only n independent directions of change for functions of n variables. The independence matrix uniquely specifies all the directions of change all the functions of a lattice are not depending on.

EXERCISES

3.1 The functions of the ON-set

$$f_q(x_1, x_2, x_3, x_4, x_5) = \overline{x}_1(x_4 \oplus x_5) \vee x_1 x_2(x_3 \oplus x_4) \vee \overline{x}_2(x_3 \oplus x_5) \tag{3.108}$$

and the OFF-set

$$f_r(x_1, x_2, x_3, x_4, x_5) = x_1\overline{x}_2(\overline{x}_3 \oplus x_5) \vee \overline{x}_2(\overline{x}_3\overline{x}_4\overline{x}_5 \vee x_3 x_4 x_5) \tag{3.109}$$

are given.

(a) Do these mark functions satisfy the restriction (3.5) of a Boolean lattice?

(b) How many functions $f_i(\mathbf{x})$ belong to a Boolean lattice that is defined by the lattice Equation (3.9) and the mark functions f_q (3.108) and f_r (3.109)?

(c) Does this lattice contain at least one function that does not depend on the simultaneous change of the variables (x_3, x_4, x_5)?

(d) How many functions $g_i(\mathbf{x})$ belong to the restricted Boolean lattice

$$g_q(\mathbf{x}) \wedge \overline{g_i(\mathbf{x})} \vee g_i(\mathbf{x}) \wedge g_r(\mathbf{x}) \vee \frac{\partial g(\mathbf{x})}{\partial(x_3, x_4, x_5)} = 0 \tag{3.110}$$

with the mark functions $g_q = f_q$ (3.108) and $g_r = f_r$ (3.109)?

3.2 The vectorial derivatives of all functions $f_i(\mathbf{x})$ of the lattice specified by the mark functions f_q (3.108) and f_r (3.109) and

$$f_q(\mathbf{x}) \wedge \overline{f_i(\mathbf{x})} \vee f_i(\mathbf{x}) \wedge f_r(\mathbf{x}) \vee \frac{\partial f(\mathbf{x})}{\partial(x_3, x_4, x_5)} = 0 \tag{3.111}$$

with regard to $\mathbf{x}_0 = (x_1, x_3, x_4)$ have to be calculated.

(a) Which property have all result functions $g_i(\mathbf{x})$?

(b) Calculate the mark functions $f_q^{\partial \mathbf{x}_0}(\mathbf{x}_0, \mathbf{x}_1)$ and $f_r^{\partial \mathbf{x}_0}(\mathbf{x}_0, \mathbf{x}_1)$.

(c) How many functions $f_i(\mathbf{x})$ belong to a Boolean lattice of the calculated vectorial derivatives?

(d) What does the independence matrix IDM(f) look like? Apply Algorithms 3.7 and 3.8 to compute the independence matrix

$$\mathrm{IDM}(g) = \mathrm{UM}(\mathrm{IDM}(f), (x_1, x_3, x_4)) \,.$$

All functions of the calculated lattice are independent of the simultaneous change of either (x_3, x_4, x_5) or (x_1, x_3, x_4). For which additional directions of change are all these functions independent as well?

(e) Determine **rank**(IDM(f)) and **rank**(IDM(g)) of the given and the resulting independence matrices.

CHAPTER 4

Differentials and Differential Operations

4.1 DIFFERENTIAL OF A BOOLEAN VARIABLE

All derivative operations describe the change behavior for a certain direction of change, but the used direction of change is not explicitly visible in the result functions. However, there are applications where the direction of change is needed. The differentials of Boolean variables bridge this gap.

Definition 4.1 Differential of a Boolean Variable. Let x_i be a Boolean variable then

$$\mathrm{d}x_i = \begin{cases} 1\,, & \text{if } x_i \text{ changes its value} \\ 0\,, & \text{if } x_i \text{ does not change its value} \end{cases} \tag{4.1}$$

is the differential of x_i. The *change of the value* means the transition from $x_i = 0$ to $x_i = 1$ or vice versa.

There are two important aspects in the relation between the differential $\mathrm{d}x_i$ and the associated Boolean variable x_i.

1. The differential $\mathrm{d}x_i$ is an independent Boolean variable. However, the differential $\mathrm{d}x_i$ is associated with the Boolean variable x_i because the differential $\mathrm{d}x_i$ describes the change of the value of x_i. Hence, the space for modeling is extended by the differential $\mathrm{d}x_i$.

2. The differential $\mathrm{d}x_i$ is also a Boolean variable. Hence, all the laws of the Boolean Algebra remain valid for these differentials.

Differentials of Boolean variables can be used to describe a directed graph. The nodes are determined by the values of Boolean variables, and the associated differentials describe the edges. Figure 4.1a depicts the graphical representation of a directed graph $G(\mathbf{x}, \mathbf{dx})$ in which the edge beginning at the node $(0, 0)$ leads to the cycle $(1, 1) - (0, 1) - (1, 0) - (1, 1)$.

The edges of a directed graph can be described by a graph equation:

$$G(\mathbf{x}, \mathbf{dx}) = 1\,. \tag{4.2}$$

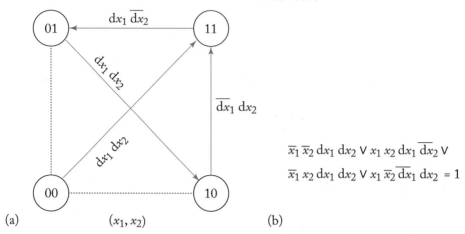

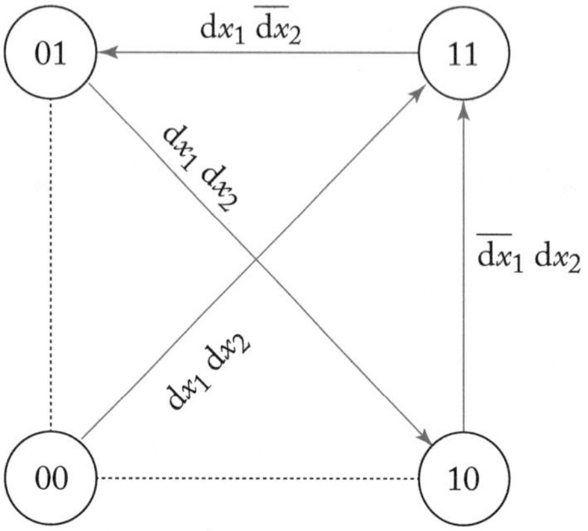

(a) (x_1, x_2) (b)

$$\overline{x}_1\,\overline{x}_2\,dx_1\,dx_2 \vee x_1\,x_2\,dx_1\,\overline{dx_2} \vee \overline{x}_1\,x_2\,dx_1\,dx_2 \vee x_1\,\overline{x}_2\,\overline{dx_1}\,dx_2 = 1$$

Figure 4.1: Directed graph $G(x_1, x_2, dx_1, dx_2)$: (a) graphical representation and (b) associated graph equation.

Figure 4.1b shows the graph equation that describes the four edges of the graph of Figure 4.1a. Graph Equation (4.2) opens a very wide field of applications in graph theory or control systems. The following example show the control of an engraving machine.

Example 4.2 Movements for the Character 4. A simple engraving machine uses for each character a width of one and a height of two units. The left position of the engraving cutter is encoded by $x = 0$ and the right one by $x = 1$. The positions of the two units of the height are encoded by the Boolean variables y_1 and y_2 such that $(y_1, y_2) = (0, 0)$ indicates the bottom, $(1, 0)$ the medium, and $(1, 1)$ the upper position. The assignment $(x_1, x_2) = (0, 1)$ is not used. The engraving cutter is located above the workpiece for $z = 0$ and engraves it for $z = 1$. A Boolean variable a is used to decide about alternative directions of the movement. A condition for this control unit: only one differential of a variable can be equal to 1. The graph equation

$$\begin{aligned}
&\overline{x}\,\overline{y}_1\,\overline{y}_2\,\overline{z}\,\overline{dx}\,dy_1\,\overline{dy_2}\,\overline{dz} \vee \overline{x}\,y_1\,\overline{y}_2\,\overline{z}\,dx\,\overline{dy_1}\,\overline{dy_2}\,\overline{dz} \vee x\,y_1\,\overline{y}_2\,\overline{z}\,\overline{dx}\,\overline{dy_1}\,\overline{dy_2}\,dz \vee \\
&\overline{a}\,x\,y_1\,\overline{y}_2\,z\,dx\,\overline{dy_1}\,\overline{dy_2}\,\overline{dz} \vee \overline{x}\,y_1\,\overline{y}_2\,z\,\overline{dx}\,\overline{dy_1}\,dy_2\,\overline{dz} \vee \overline{x}\,y_1\,y_2\,z\,\overline{dx}\,\overline{dy_1}\,\overline{dy_2}\,dz \vee \\
&\overline{x}\,y_1\,y_2\,\overline{z}\,dx\,\overline{dy_1}\,\overline{dy_2}\,\overline{dz} \vee x\,y_1\,y_2\,\overline{z}\,\overline{dx}\,\overline{dy_1}\,\overline{dy_2}\,dz \vee x\,y_1\,y_2\,z\,\overline{dx}\,\overline{dy_1}\,dy_2\,\overline{dz} \vee \\
&a\,x\,y_1\,\overline{y}_2\,z\,\overline{dx}\,dy_1\,\overline{dy_2}\,\overline{dz} \vee x\,\overline{y}_1\,\overline{y}_2\,z\,\overline{dx}\,\overline{dy_1}\,\overline{dy_2}\,dz \vee x\,\overline{y}_1\,\overline{y}_2\,\overline{z}\,dx\,\overline{dy_1}\,\overline{dy_2}\,\overline{dz} = 1
\end{aligned} \tag{4.3}$$

describes the 12 movements used to engrave the character 4.

Each solution vector of (4.3) is represented by an edge in the graph of Figure 4.2. Starting from the node (0000), non-negated differentials dx, dy_1, dy_2, and dz indicate the direction of change of each single step. The sequence (0000) – (0100) – (1100) – (1101) brings the engraving cutter in the start position. The node (1101) is visited twice in the sequence of edges.

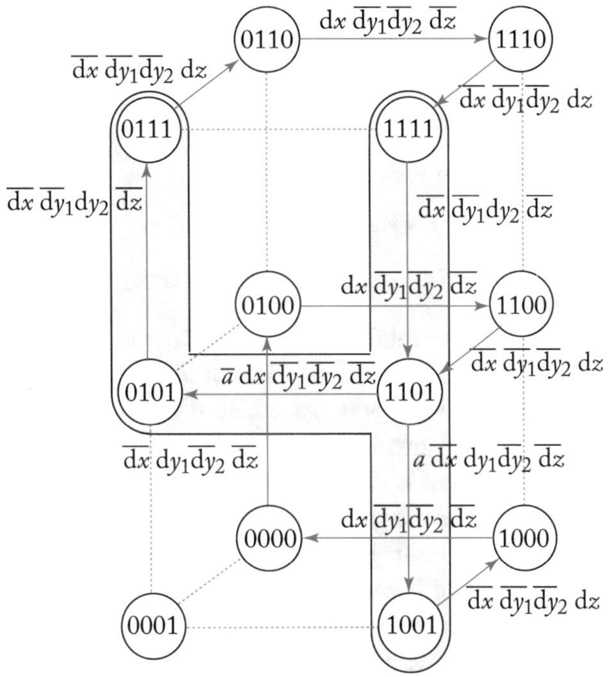

Figure 4.2: Graph to control an engraving machine for the character 4.

The value 0 of decision variable a selects the upper cycle (1101) – (0101) – (0111) – (0110) – (1110) – (1111) – (1101), and the value 1 starts the sequence (1101) – (1001) – (1000) – (0000) back to the initial node. The light blue background emphasizes that the movements with $z = 1$ controls the engraving machine such that the character 4 appears.

4.2 TOTAL DIFFERENTIAL OPERATIONS

Differential operations aim to the common description of the change behavior of Boolean functions.

The differential of Boolean function explores the changes of this function in all directions of change of their variables **x**. The direction of change is specified by the differentials $\mathrm{d}x_i$ in the following way:

$$x_i \oplus \mathrm{d}x_i = \begin{cases} x_i \,, & \text{if } x_i \text{ did not change its value, i.e., } \mathrm{d}x_i = 0 \\ \overline{x}_i \,, & \text{if } x_i \text{ changed its value, i.e., } \mathrm{d}x_i = 1 \,. \end{cases} \tag{4.4}$$

The substitution of this expression for all variables x_i in the function $f(\mathbf{x})$ leads to the function

$$f(\mathbf{x} \oplus \mathbf{dx}) = f(x_1 \oplus \mathrm{d}x_1, x_2 \oplus \mathrm{d}x_2, \ldots, x_n \oplus \mathrm{d}x_n) \tag{4.5}$$

that determines the function value by both the selection of a basic point in the Boolean space $\mathbf{x} = \mathbf{c}$ and the direction of change $\mathrm{d}\mathbf{x} = \mathrm{d}\mathbf{c}$. Using this function the complete change behavior of a Boolean function can be explored.

Definition 4.3 Total Differential.
Let $f(\mathbf{x})$ be a Boolean function of n variables. Then

$$\mathrm{d}_{\mathbf{x}} f(\mathbf{x}) = f(\mathbf{x}) \oplus f(\mathbf{x} \oplus \mathbf{dx}) \tag{4.6}$$

is the (**total**) **differential** of $f(\mathbf{x})$ with regard to all variables $\mathbf{x}$.

Definition (4.6) shows that the total differential of $f(\mathbf{x})$ is a Boolean function in $\mathbb{B}^{2n}$ that depends on both the variables $\mathbf{x}$ and the differentials of all variables $\mathbf{dx}$. The total differential describes edges in a graph which connects nodes where the function $f(\mathbf{x})$ has different values. All 2^n directions of change are taken into account. The term *total* emphasizes that all variables x_i are included into the calculation. This is the general case; hence, the term *total* can be avoided.

Figure 4.3a shows as example the graph of a differential of a function of three variables. The color *blue* has been taken for the five nodes with $f(\mathbf{x}) = 1$, and the *white* color indicates the three nodes with $f(\mathbf{x}) = 0$. The function value changes between these two sets of nodes. Each white node is connected by five edges with the blue nodes and vice versa, each blue node is connected by three edges with the white nodes. There are no loop edges on the nodes because the function value does not change for $\overline{\mathrm{d}x_1}\,\overline{\mathrm{d}x_2}\,\overline{\mathrm{d}x_3}$.

The differential of $f(\mathbf{x})$ is equal to 1 if the pair of nodes, selected by the assignment of $2 * n$ constant values to $(\mathbf{x}, \mathbf{dx})$, carry different function values. Therefore, there is a direct relationship between the differential of $f(\mathbf{x})$ and all derivatives that compare function values of pairs of nodes, i.e., all vectorial derivatives.

The application of the Shannon decomposition to Definition (4.6) reveals this relationship:

$$\begin{aligned}
\mathrm{d}_{(x_1,x_2)} f(x_1, x_2) &= f(x_1, x_2) \oplus f(x_1 \oplus \mathrm{d}x_1, x_2 \oplus \mathrm{d}x_2) \\
&= (f(x_1, x_2) \oplus f(x_1, x_2 \oplus \mathrm{d}x_2))\,\overline{\mathrm{d}x_1} \oplus (f(x_1, x_2) \oplus f(\overline{x}_1, x_2 \oplus \mathrm{d}x_2))\,\mathrm{d}x_1 \\
&= (f(x_1, x_2) \oplus f(x_1, x_2))\,\overline{\mathrm{d}x_1}\,\overline{\mathrm{d}x_2} \oplus (f(x_1, x_2) \oplus f(\overline{x}_1, x_2))\,\mathrm{d}x_1\overline{\mathrm{d}x_2} \oplus \\
&\quad (f(x_1, x_2) \oplus f(x_1, \overline{x}_2))\,\overline{\mathrm{d}x_1}\mathrm{d}x_2 \oplus (f(x_1, x_2) \oplus f(\overline{x}_1, \overline{x}_2))\,\mathrm{d}x_1\mathrm{d}x_2 \\
&= \frac{\partial f(x_1, x_2)}{\partial x_1}\mathrm{d}x_1\overline{\mathrm{d}x_2} \oplus \frac{\partial f(x_1, x_2)}{\partial x_2}\overline{\mathrm{d}x_1}\mathrm{d}x_2 \oplus \frac{\partial f(x_1, x_2)}{\partial (x_1, x_2)}\mathrm{d}x_1\mathrm{d}x_2 \,.
\end{aligned}$$

By generalizing to a function of n variables, the total differential of $f(\mathbf{x})$ is constructed by all vectorial derivatives:

$$\begin{aligned}
\mathrm{d}_{\mathbf{x}} f(\mathbf{x}) = {} & \frac{\partial f(\mathbf{x})}{\partial x_1}\,\mathrm{d}x_1\,\overline{\mathrm{d}x_2}\ldots\overline{\mathrm{d}x_n} \oplus \cdots \oplus \frac{\partial f(\mathbf{x})}{\partial x_n}\,\overline{\mathrm{d}x_1}\,\overline{\mathrm{d}x_2}\ldots\mathrm{d}x_n \oplus \\
& \frac{\partial f(\mathbf{x})}{\partial (x_1, x_2)}\,\mathrm{d}x_1\,\mathrm{d}x_2\ldots\overline{\mathrm{d}x_n} \oplus \cdots \oplus \frac{\partial f(\mathbf{x})}{\partial \mathbf{x}}\,\mathrm{d}x_1\,\mathrm{d}x_2\ldots\mathrm{d}x_n \,.
\end{aligned} \tag{4.7}$$

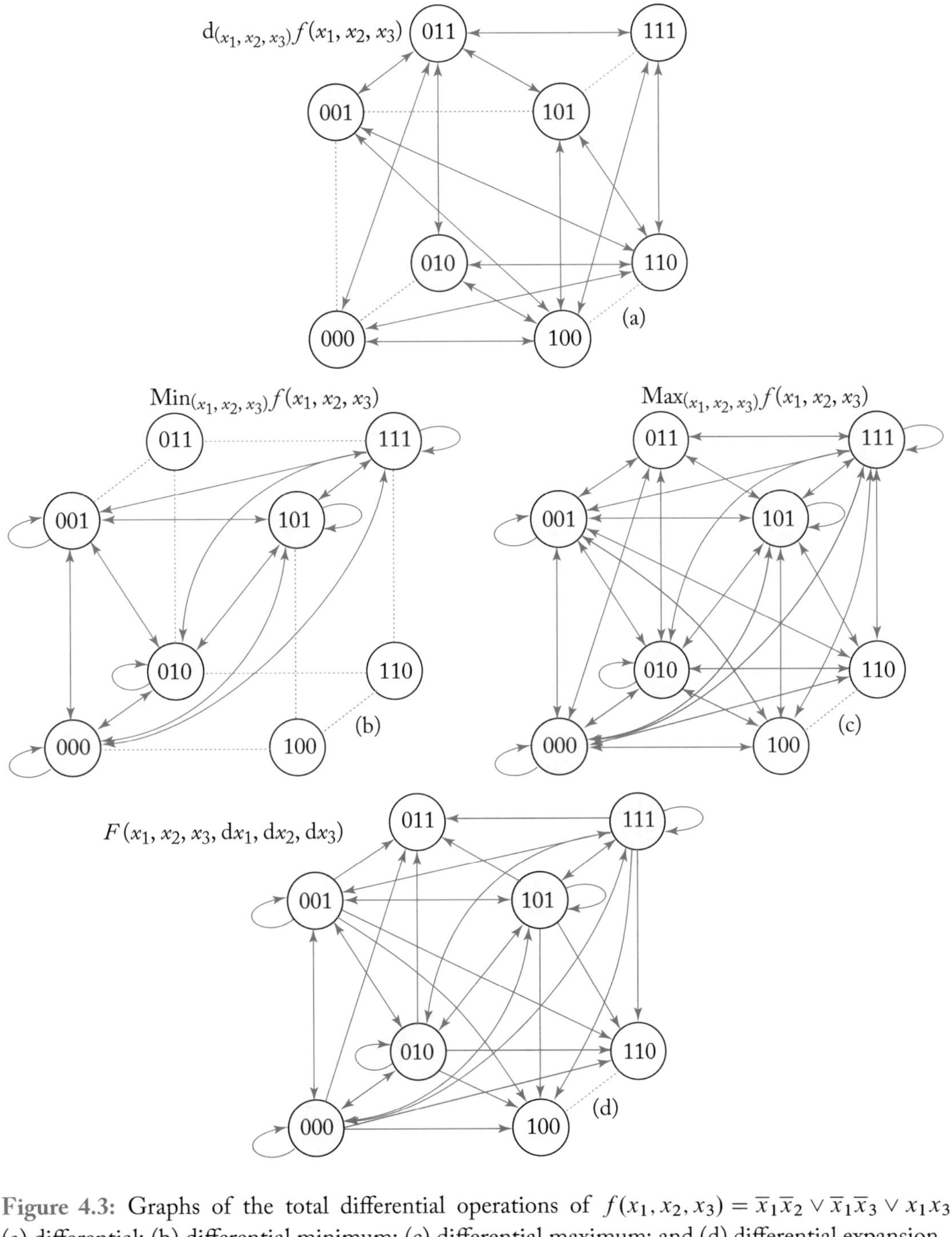

Figure 4.3: Graphs of the total differential operations of $f(x_1, x_2, x_3) = \overline{x}_1\overline{x}_2 \vee \overline{x}_1\overline{x}_3 \vee x_1x_3$: (a) differential; (b) differential minimum; (c) differential maximum; and (d) differential expansion.

Vice versa, all vectorial derivatives can be derived from the total differential:

$$\frac{\partial f(x_i, \mathbf{x}_1)}{\partial x_i} = \mathrm{d}_{\mathbf{x}} f(x_i, \mathbf{x}_1) \mid_{\mathrm{d}x_i=1, \mathbf{dx}_1=\mathbf{0}} , \tag{4.8}$$

$$\frac{\partial f(\mathbf{x}_0, \mathbf{x}_1)}{\partial \mathbf{x}_0} = \mathrm{d}_{\mathbf{x}} f(\mathbf{x}_0, \mathbf{x}_1) \mid_{\mathbf{dx}_0=\mathbf{1}, \mathbf{dx}_1=\mathbf{0}} . \tag{4.9}$$

Similarly, the total differential minimum and maximum summarize the results of all associated vectorial derivative operations.

Definition 4.4 Total Differential Operations.
Let $f(\mathbf{x})$ be a Boolean function of n variables. Then

$$\mathrm{Min}_{\mathbf{x}} f(\mathbf{x}) = f(\mathbf{x}) \wedge f(\mathbf{x} \oplus \mathbf{dx}) \tag{4.10}$$

is the (**total**) **differential minimum** of $f(\mathbf{x})$ with regard to $\mathbf{x}$,

$$\mathrm{Max}_{\mathbf{x}} f(\mathbf{x}) = f(\mathbf{x}) \vee f(\mathbf{x} \oplus \mathbf{dx}) \tag{4.11}$$

is the (**total**) **differential maximum** of $f(\mathbf{x})$ with regard to $\mathbf{x}$, and

$$F(\mathbf{x}, \mathbf{dx}) = f(\mathbf{x}) \wedge \bigwedge_{i=1}^{n} (\mathrm{d}x_i \vee \overline{\mathrm{d}x_i}) \tag{4.12}$$

is the (**total**) **differential expansion** of $f(\mathbf{x})$ with regard to $\mathbf{x}$.

Definition 4.4 shows that the total differential minimum and the total differential maximum evaluate the relationship between the function $f(\mathbf{x})$ and the same function under consideration of certain changes $\mathbf{dx}$. Due to the used operation,

- the result of the total differential minimum describes edges in a graph which connect nodes where the function $f(\mathbf{x})$ is equal to 1, and
- the result of the total differential maximum describes edges in a graph which connect nodes where the function $f(\mathbf{x})$ carries at least once the value 1.

The total differential expansion is an embedding of the Boolean function $f(\mathbf{x})$ into the Boolean space $\mathbb{B}^{2n}$ of the variables $\mathbf{x}$ and $\mathbf{dx}$. This expansion can be used for comparisons between the total differential operations. The term *total* again emphasizes that all variables x_i are included into the calculation of the differential operations, and because this is the general case, the term *total* can be omitted. Figure 4.3 depicts the graphs of all total differential operations of a function $f(x_1, x_2, x_3)$. Blue nodes indicate the function values 1. While the edges of the differential in Figure 4.3a connect nodes of different function values, the edges of the differential minimum connect only nodes with $f(x_1, x_2, x_3) = 1$; see Figure 4.3b. The edges of the differential maximum in Figure 4.3c connect all pairs of nodes where at least one node is equal to 1. The comparison of these upper three graphs shows that each edge appears in two of these graphs. This is not a special property of the chosen example, but generally, it holds that

$$\mathrm{d}_{\mathbf{x}} f(\mathbf{x}) \oplus \mathrm{Min}_{\mathbf{x}} f(\mathbf{x}) \oplus \mathrm{Max}_{\mathbf{x}} f(\mathbf{x}) = 0 \,. \tag{4.13}$$

The graph of the differential expansion in Figure 4.3d contains only edges beginning at the nodes with $f(x_1, x_2, x_3) = 1$, but these edges reach each node of the Boolean space. The comparison of the lower three graphs in Figure 4.3 shows that all edges of the differential minimum also appear in the differential expansion, and the differential maximum contains in comparison with the differential expansion additional edges from $f(x_1, x_2, x_3) = 0$ to $f(x_1, x_2, x_3) = 1$. Again, this is not a special property of the chosen example but, generally, it holds that

$$\mathrm{Min}_{\mathbf{x}} f(\mathbf{x}) \le F(\mathbf{x}, \mathbf{dx}) \le \mathrm{Max}_{\mathbf{x}} f(\mathbf{x}) = 0 \,. \tag{4.14}$$

Straightforward calculations reveal that the differential minimum comprises all vectorial minima, and the differential maximum shows an analog property:

$$\begin{aligned}
\mathrm{Min}_{\mathbf{x}} f(\mathbf{x}) = {} & f(\mathbf{x})\, \overline{\mathrm{d}x_1}\, \overline{\mathrm{d}x_2} \ldots \overline{\mathrm{d}x_n} \vee \min_{x_1} f(\mathbf{x})\, \mathrm{d}x_1\, \overline{\mathrm{d}x_2} \ldots \overline{\mathrm{d}x_n} \vee \ldots \vee \\
& \min_{x_n} f(\mathbf{x})\, \overline{\mathrm{d}x_1}\, \overline{\mathrm{d}x_2} \ldots \mathrm{d}x_n \vee \min_{(x_1,x_2)} f(\mathbf{x})\, \mathrm{d}x_1\, \mathrm{d}x_2 \ldots \overline{\mathrm{d}x_n} \vee \ldots \vee \\
& \min_{\mathbf{x}} f(\mathbf{x})\, \mathrm{d}x_1\, \mathrm{d}x_2 \ldots \mathrm{d}x_n \,,
\end{aligned} \tag{4.15}$$

$$\begin{aligned}
\mathrm{Max}_{\mathbf{x}} f(\mathbf{x}) = {} & f(\mathbf{x})\, \overline{\mathrm{d}x_1}\, \overline{\mathrm{d}x_2} \ldots \overline{\mathrm{d}x_n} \vee \max_{x_1} f(\mathbf{x})\, \mathrm{d}x_1\, \overline{\mathrm{d}x_2} \ldots \overline{\mathrm{d}x_n} \vee \ldots \vee \\
& \max_{x_n} f(\mathbf{x})\, \overline{\mathrm{d}x_1}\, \overline{\mathrm{d}x_2} \ldots \mathrm{d}x_n \vee \max_{(x_1,x_2)} f(\mathbf{x})\, \mathrm{d}x_1\, \mathrm{d}x_2 \ldots \overline{\mathrm{d}x_n} \vee \ldots \vee \\
& \max_{\mathbf{x}} f(\mathbf{x})\, \mathrm{d}x_1\, \mathrm{d}x_2 \ldots \mathrm{d}x_n \,.
\end{aligned} \tag{4.16}$$

Hence, alternatively to Definition 4.4, both the differential minimum and maximum can be composed of all correspondent derivative operations. Vice versa, the minimum and maximum of $f(\mathbf{x}) = f(x_i, \mathbf{x}_1)$ with regard to x_i can be selected from the correspondent differential operation:

$$\min_{x_i} f(x_i, \mathbf{x}_1) = \mathrm{Min}_{\mathbf{x}} f(x_i, \mathbf{x}_1) \,|_{\mathrm{d}x_i=1, \mathbf{dx}_1=0} \,, \tag{4.17}$$

$$\max_{x_i} f(x_i, \mathbf{x}_1) = \mathrm{Max}_{\mathbf{x}} f(x_i, \mathbf{x}_1) \,|_{\mathrm{d}x_i=1, \mathbf{dx}_1=0} \,, \tag{4.18}$$

and the assignment of the value 1 to several differentials $\mathrm{d}x_i \in \mathbf{dx}_0$ leads to the correspondent vectorial derivative operations:

$$\min_{\mathbf{x}_0} f(\mathbf{x}_0, \mathbf{x}_1) = \mathrm{Min}_{\mathbf{x}} f(\mathbf{x}_0, \mathbf{x}_1) \,|_{\mathbf{dx}_0=1, \mathbf{dx}_1=0} \,, \tag{4.19}$$

$$\max_{\mathbf{x}_0} f(\mathbf{x}_0, \mathbf{x}_1) = \mathrm{Max}_{\mathbf{x}} f(\mathbf{x}_0, \mathbf{x}_1) \,|_{\mathbf{dx}_0=1, \mathbf{dx}_1=0} \,. \tag{4.20}$$

It is an advantage of the total differential operations that these operations provide comprehensive information about the change behavior of the explored function with regard to all directions of change. The drawback, however, is that the number of variables of each total differential operation is twice the number of the given function. Hence, the total differential operations are suitable for compact representations of comprehensive change behaviors.

4.3 PARTIAL DIFFERENTIAL OPERATIONS

The effort to calculate and store total differential operations can be reduced when the change behavior is only needed for a subset of directions of change.

Definition 4.5 Partial Differential Operations.
Let $\mathbf{x}_0 = (x_1, x_2, \dots, x_m)$, $\mathbf{x}_1 = (x_{m+1}, \dots, x_n)$ be two disjoint sets of Boolean variables, and $f(\mathbf{x}_0, \mathbf{x}_1) = f(x_1, x_2 \dots, x_n) = f(\mathbf{x})$ be a Boolean function of n variables; then,

$$\mathrm{d}_{\mathbf{x}_0} f(\mathbf{x}_0, \mathbf{x}_1) = f(\mathbf{x}_0, \mathbf{x}_1) \oplus f(\mathbf{x}_0 \oplus \mathbf{dx}_0, \mathbf{x}_1) \tag{4.21}$$

is the **partial differential** of $f(\mathbf{x})$ with regard to $\mathbf{x}_0$,

$$\mathrm{Min}_{\mathbf{x}_0} f(\mathbf{x}_0, \mathbf{x}_1) = f(\mathbf{x}_0, \mathbf{x}_1) \wedge f(\mathbf{x}_0 \oplus \mathbf{dx}_0, \mathbf{x}_1) \tag{4.22}$$

is the **partial differential minimum** of $f(\mathbf{x})$ with regard to $\mathbf{x}_0$,

$$\mathrm{Max}_{\mathbf{x}_0} f(\mathbf{x}_0, \mathbf{x}_1) = f(\mathbf{x}_0, \mathbf{x}_1) \vee f(\mathbf{x}_0 \oplus \mathbf{dx}_0, \mathbf{x}_1) \tag{4.23}$$

is the **partial differential maximum** of $f(\mathbf{x})$ with regard to $\mathbf{x}_0$, and

$$F(\mathbf{x}_0, \mathbf{x}_1, \mathbf{dx}_0) = f(\mathbf{x}_0, \mathbf{x}_1) \wedge \bigwedge_{i=1}^{m} (\mathrm{d}x_i \vee \overline{\mathrm{d}x_i}) \tag{4.24}$$

is the **partial differential expansion** of $f(\mathbf{x})$ with regard to $\mathbf{x}_0$.

The partial differential operations describe the same change behavior as the total differential operations but take only into account changes in the directions covered by $\mathbf{dx}_0$. The following formulas show for $\mathbf{x}_0 = (x_1, x_2)$ how the vectorial derivative operations are summarized within the partial differential operations:

$$
\begin{aligned}
\mathrm{d}_{(x_1,x_2)} f(x_1,x_2,\mathbf{x}_1) = {} & \frac{\partial f(x_1,x_2,\mathbf{x}_1)}{\partial x_1}\, \mathrm{d}x_1\, \overline{\mathrm{d}x_2} \,\oplus\, \frac{\partial f(x_1,x_2,\mathbf{x}_1)}{\partial x_2}\, \overline{\mathrm{d}x_1}\, \mathrm{d}x_2 \oplus \\
& \frac{\partial f(x_1,x_2,\mathbf{x}_1)}{\partial (x_1,x_2)}\, \mathrm{d}x_1\, \mathrm{d}x_2 \,, && (4.25) \\
\mathrm{Min}_{(x_1,x_2)} f(x_1,x_2,\mathbf{x}_1) = {} & f(x_1,x_2,\mathbf{x}_1)\, \overline{\mathrm{d}x_1}\, \overline{\mathrm{d}x_2} \,\vee \min_{x_1} f(x_1,x_2,\mathbf{x}_1)\, \mathrm{d}x_1\, \overline{\mathrm{d}x_2} \vee \\
& \min_{x_2} f(x_1,x_2,\mathbf{x}_1)\, \overline{\mathrm{d}x_1}\, \mathrm{d}x_2 \vee \min_{(x_1,x_2)} f(x_1,x_2,\mathbf{x}_1)\, \mathrm{d}x_1\, \mathrm{d}x_2 \,, && (4.26) \\
\mathrm{Max}_{(x_1,x_2)} f(x_1,x_2,\mathbf{x}_1) = {} & f(x_1,x_2,\mathbf{x}_1)\, \overline{\mathrm{d}x_1}\, \overline{\mathrm{d}x_2} \,\vee \max_{x_1} f(x_1,x_2,\mathbf{x}_1)\, \mathrm{d}x_1\, \overline{\mathrm{d}x_2} \vee \\
& \max_{x_2} f(x_1,x_2,\mathbf{x}_1)\, \overline{\mathrm{d}x_1}\, \mathrm{d}x_2 \vee \max_{(x_1,x_2)} f(x_1,x_2,\mathbf{x}_1)\, \mathrm{d}x_1\, \mathrm{d}x_2 \,. && (4.27)
\end{aligned}
$$

Figure 4.4 depicts the graphs of all partial differential operations with regard to (x_1, x_2) of the same function $f(x_1, x_2, x_3)$ as used in Figure 4.3. It can be seen that the edges are restricted to the subspaces $x_3 = 0$ and $x_3 = 1$. Within these subspaces the $\oplus$-operation of the upper three graphs is again equal to 0 and the partial differential minimum, the partial differential expansion, and the partial differential maximum satisfy a chain of inequalities. The adopted general relationships between the partial differential operations are:

$$
\begin{aligned}
\mathrm{d}_{\mathbf{x}_0} f(\mathbf{x}_0,\mathbf{x}_1) \oplus \mathrm{Min}_{\mathbf{x}_0} f(\mathbf{x}_0,\mathbf{x}_1) \oplus \mathrm{Max}_{\mathbf{x}_0} f(\mathbf{x}_0,\mathbf{x}_1) &= 0 \,, && (4.28) \\
\mathrm{Min}_{\mathbf{x}_0} f(\mathbf{x}_0,\mathbf{x}_1) \le F(\mathbf{x}, \mathbf{dx}_0) \le \mathrm{Max}_{\mathbf{x}_0} f(\mathbf{x}_0,\mathbf{x}_1) &= 0 \,. && (4.29)
\end{aligned}
$$

The partial differential operation of $f(\mathbf{x}_0, \mathbf{x}_1) = f(x_1, x_2, \ldots, x_m, x_{m+1}, \ldots, x_n)$ with regard to $\mathbf{x}_0 = (x_1, x_2, \ldots, x_m)$ only comprise vectorial derivative operations with regard to subsets of $\mathbf{x}_0$:

$$
\begin{aligned}
\mathrm{d}_{\mathbf{x}_0} f(\mathbf{x}_0,\mathbf{x}_1) = {} & \frac{\partial f(\mathbf{x}_0,\mathbf{x}_1)}{\partial x_1}\, \mathrm{d}x_1\, \overline{\mathrm{d}x_2} \ldots \overline{\mathrm{d}x_m} \oplus \cdots \oplus \frac{\partial f(\mathbf{x}_0,\mathbf{x}_1)}{\partial x_m}\, \overline{\mathrm{d}x_1}\, \overline{\mathrm{d}x_2} \ldots \mathrm{d}x_m \oplus \\
& \frac{\partial f(\mathbf{x}_0,\mathbf{x}_1)}{\partial (x_1,x_2)}\, \mathrm{d}x_1\, \mathrm{d}x_2 \ldots \overline{\mathrm{d}x_m} \oplus \cdots \oplus \frac{\partial f(\mathbf{x}_0,\mathbf{x}_1)}{\partial \mathbf{x}_0}\, \mathrm{d}x_1\, \mathrm{d}x_2 \ldots \mathrm{d}x_m \,, && (4.30) \\
\mathrm{Min}_{\mathbf{x}_0} f(\mathbf{x}_0,\mathbf{x}_1) = {} & f(\mathbf{x}_0,\mathbf{x}_1)\, \overline{\mathrm{d}x_1}\, \overline{\mathrm{d}x_2} \ldots \overline{\mathrm{d}x_m} \,\vee \min_{x_1} f(\mathbf{x}_0,\mathbf{x}_1)\, \mathrm{d}x_1\, \overline{\mathrm{d}x_2} \ldots \overline{\mathrm{d}x_m} \vee \cdots \vee \\
& \min_{x_m} f(\mathbf{x}_0,\mathbf{x}_1)\, \overline{\mathrm{d}x_1}\, \overline{\mathrm{d}x_2} \ldots \mathrm{d}x_m \vee \min_{(x_1,x_2)} f(\mathbf{x}_0,\mathbf{x}_1)\, \mathrm{d}x_1\, \mathrm{d}x_2 \ldots \overline{\mathrm{d}x_m} \vee \cdots \vee \\
& \min_{\mathbf{x}} f(\mathbf{x}_0,\mathbf{x}_1)\, \mathrm{d}x_1\, \mathrm{d}x_2 \ldots \mathrm{d}x_m \,, && (4.31) \\
\mathrm{Max}_{\mathbf{x}_0} f(\mathbf{x}_0,\mathbf{x}_1) = {} & f(\mathbf{x})\, \overline{\mathrm{d}x_1}\, \overline{\mathrm{d}x_2} \ldots \overline{\mathrm{d}x_m} \,\vee \max_{x_1} f(\mathbf{x}_0,\mathbf{x}_1)\, \mathrm{d}x_1\, \overline{\mathrm{d}x_2} \ldots \overline{\mathrm{d}x_m} \vee \cdots \vee \\
& \max_{x_m} f(\mathbf{x}_0,\mathbf{x}_1)\, \overline{\mathrm{d}x_1}\, \overline{\mathrm{d}x_2} \ldots \mathrm{d}x_m \vee \max_{(x_1,x_2)} f(\mathbf{x}_0,\mathbf{x}_1)\, \mathrm{d}x_1\, \mathrm{d}x_2 \ldots \overline{\mathrm{d}x_m} \vee \ldots \vee \\
& \max_{\mathbf{x}} f(\mathbf{x}_0,\mathbf{x}_1)\, \mathrm{d}x_1\, \mathrm{d}x_2 \ldots \mathrm{d}x_m \,, && (4.32)
\end{aligned}
$$

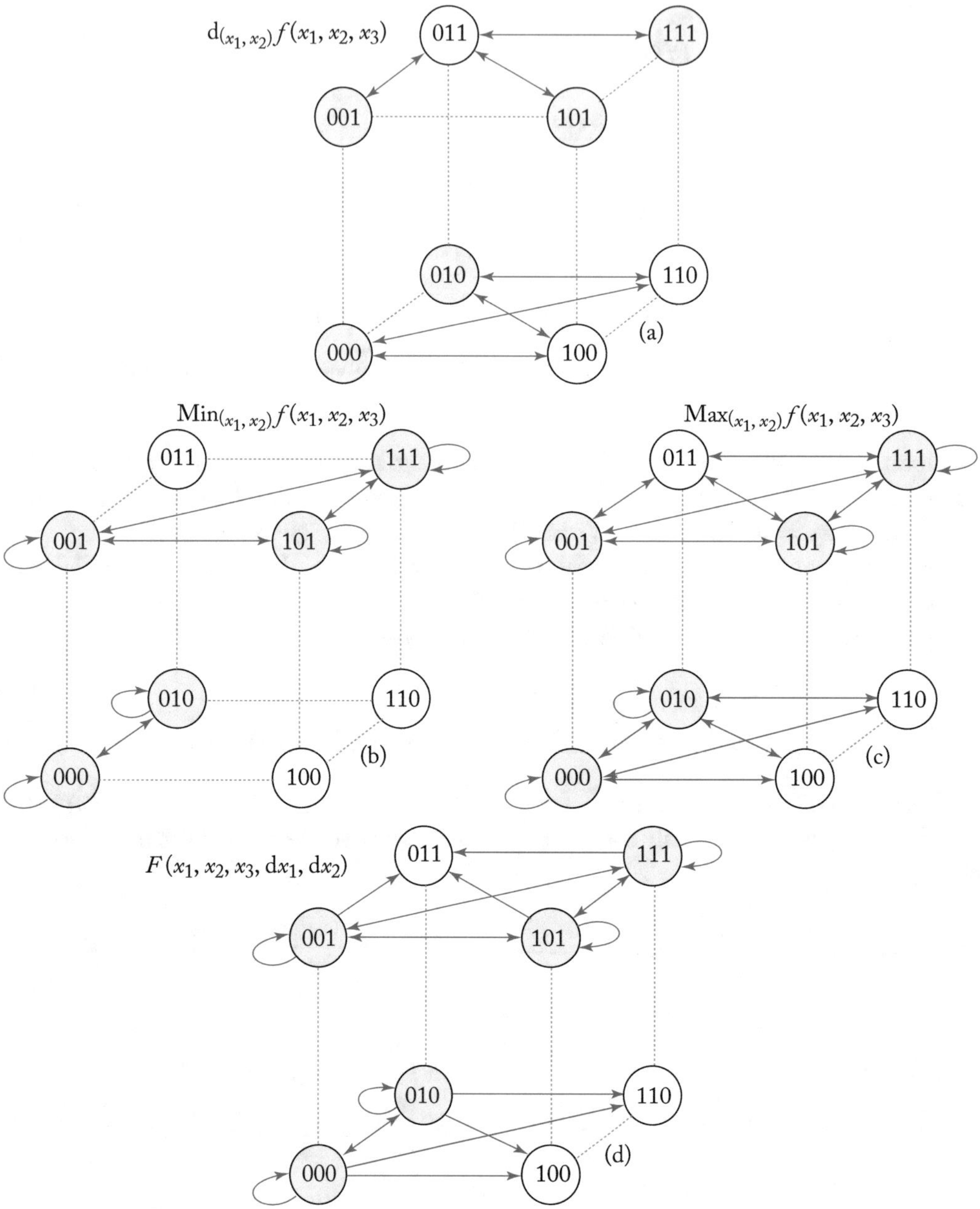

Figure 4.4: Graphs of the partial differential operations of $f(x_1, x_2, x_3) = \overline{x}_1\overline{x}_2 \vee \overline{x}_1\overline{x}_3 \vee x_1x_3$ with regard to (x_1, x_2): (a) partial differential; (b) partial differential minimum; (c) partial differential maximum; and (d) partial differential expansion.

and therefore only these derivative operations can be derived from the partial differential operations, where $x_i \in \mathbf{x}_0$ and $\mathbf{x}_s \subseteq \mathbf{x}_0$:

$$\frac{\partial f(\mathbf{x}_0, \mathbf{x}_1)}{\partial x_i} = \mathrm{d}_{\mathbf{x}_0} f(\mathbf{x}_0, \mathbf{x}_1) \,|_{\mathrm{d}x_i=1,(\mathbf{dx}_0 \backslash \mathrm{d}x_i)=\mathbf{0}} \;, \tag{4.33}$$

$$\min_{x_i} f(\mathbf{x}_0, \mathbf{x}_1) = \mathrm{Min}_{\mathbf{x}_0} f(\mathbf{x}_0, \mathbf{x}_1) \,|_{\mathrm{d}x_i=1,(\mathbf{dx}_0 \backslash \mathrm{d}x_i)=\mathbf{0}} \;, \tag{4.34}$$

$$\max_{x_i} f(\mathbf{x}_0, \mathbf{x}_1) = \mathrm{Max}_{\mathbf{x}_0} f(\mathbf{x}_0, \mathbf{x}_1) \,|_{\mathrm{d}x_i=1,(\mathbf{dx}_0 \backslash \mathrm{d}x_i)=\mathbf{0}} \;, \tag{4.35}$$

$$\frac{\partial f(\mathbf{x}_0, \mathbf{x}_1)}{\partial \mathbf{x}_s} = \mathrm{d}_{\mathbf{x}_0} f(\mathbf{x}_0, \mathbf{x}_1) \,|_{\mathbf{dx}_s=\mathbf{1},(\mathbf{dx}_0 \backslash \mathbf{dx}_s)=\mathbf{0}} \;, \tag{4.36}$$

$$\min_{\mathbf{x}_s} f(\mathbf{x}_0, \mathbf{x}_1) = \mathrm{Min}_{\mathbf{x}_0} f(\mathbf{x}_0, \mathbf{x}_1) \,|_{\mathbf{dx}_s=\mathbf{1},(\mathbf{dx}_0 \backslash \mathbf{dx}_s)=\mathbf{0}} \;, \tag{4.37}$$

$$\max_{\mathbf{x}_s} f(\mathbf{x}_0, \mathbf{x}_1) = \mathrm{Max}_{\mathbf{x}_0} f(\mathbf{x}_0, \mathbf{x}_1) \,|_{\mathbf{dx}_s=\mathbf{1},(\mathbf{dx}_0 \backslash \mathbf{dx}_s)=\mathbf{0}} \;. \tag{4.38}$$

4.4 m-FOLD DIFFERENTIAL OPERATIONS

The set of variables $\mathbf{x}_0$ that specify the partial differential operations can be restricted to a single variable x_i. The result of such partial differential operations of $f(x_1, x_2, \ldots, x_n)$ is a Boolean function that depends on all variables $x_j, j = 1, \ldots, n$, and the differential $\mathrm{d}x_i$. Subsequent calculations of the same type of partial differential operations with regard to other variables lead to m-fold differential operations.

Definition 4.6 ***m*-fold Differential Operations.**
Let $f(\mathbf{x}) = f(\mathbf{x}_0, \mathbf{x}_1)$ be a Boolean function of n variables and $\mathbf{x}_0 = (x_1, x_2, \ldots, x_m)$, then

$$\mathrm{d}^m_{\mathbf{x}_0} f(\mathbf{x}_0, \mathbf{x}_1) = \mathrm{d}_{x_m}(\ldots(\mathrm{d}_{x_2}(\mathrm{d}_{x_1} f(\mathbf{x}_0, \mathbf{x}_1)))\ldots) \tag{4.39}$$

is the m-**fold differential** of $f(\mathbf{x})$ with regard to $\mathbf{x}_0$,

$$\mathrm{Min}^m_{\mathbf{x}_0} f(\mathbf{x}_0, \mathbf{x}_1) = \mathrm{Min}_{x_m}(\ldots(\mathrm{Min}_{x_2}(\mathrm{Min}_{x_1} f(\mathbf{x}_0, \mathbf{x}_1)))\ldots) \tag{4.40}$$

the m-**fold differential minimum** of $f(\mathbf{x})$ with regard to $\mathbf{x}_0$,

$$\mathrm{Max}^m_{\mathbf{x}_0} f(\mathbf{x}_0, \mathbf{x}_1) = \mathrm{Max}_{x_m}(\ldots(\mathrm{Max}_{x_2}(\mathrm{Max}_{x_1} f(\mathbf{x}_0, \mathbf{x}_1)))\ldots) \tag{4.41}$$

the m-**fold differential maximum** of $f(\mathbf{x})$ with regard to $\mathbf{x}_0$, and

$$\vartheta_{\mathbf{x}_0} f(\mathbf{x}_0, \mathbf{x}_1) = \mathrm{Min}^m_{\mathbf{x}_0} f(\mathbf{x}_0, \mathbf{x}_1) \oplus \mathrm{Max}^m_{\mathbf{x}_0} f(\mathbf{x}_0, \mathbf{x}_1) \tag{4.42}$$

the ϑ-**operation** of $f(\mathbf{x})$ with regard to $\mathbf{x}_0$.

A detailed analysis leads to a surprising result of the m-fold differential:

$$\mathrm{d}^m_{\mathbf{x}_0} f(\mathbf{x}_0, \mathbf{x}_1) = \frac{\partial^m f(\mathbf{x}_0, \mathbf{x}_1)}{\partial x_1 \partial x_2 \ldots \partial x_m} \wedge \mathrm{d}x_1 \mathrm{d}x_2 \ldots \mathrm{d}x_m \;. \tag{4.43}$$

The other m-fold differential operations cover similarly to the partial differential operations all associated m-fold derivative operations with regard to all subsets of $\mathbf{x}_0$:

$$\begin{aligned}
\mathrm{Min}^m_{\mathbf{x}_0} f(\mathbf{x}_0, \mathbf{x}_1) = & f(\mathbf{x})\, \overline{\mathrm{d}x_1}\, \overline{\mathrm{d}x_2} \dots \overline{\mathrm{d}x_m} \vee \min_{x_1} f(\mathbf{x})\, \mathrm{d}x_1\, \overline{\mathrm{d}x_2} \dots \overline{\mathrm{d}x_m} \vee \dots \vee \\
& \min_{x_m} f(\mathbf{x})\, \overline{\mathrm{d}x_1}\, \overline{\mathrm{d}x_2} \dots \mathrm{d}x_m \vee \min_{(x_1,x_2)}{}^2 f(\mathbf{x})\, \mathrm{d}x_1\, \mathrm{d}x_2 \dots \overline{\mathrm{d}x_m} \vee \dots \vee \\
& \min_{\mathbf{x}_0}{}^m f(\mathbf{x})\, \mathrm{d}x_1\, \mathrm{d}x_2 \dots \mathrm{d}x_m \,, \qquad (4.44)
\end{aligned}$$

$$\begin{aligned}
\mathrm{Max}^m_{\mathbf{x}_0} f(\mathbf{x}_0, \mathbf{x}_1) = & f(\mathbf{x})\, \overline{\mathrm{d}x_1}\, \overline{\mathrm{d}x_2} \dots \overline{\mathrm{d}x_m} \vee \max_{x_1} f(\mathbf{x})\, \mathrm{d}x_1\, \overline{\mathrm{d}x_2} \dots \overline{\mathrm{d}x_m} \vee \dots \vee \\
& \max_{x_m} f(\mathbf{x})\, \overline{\mathrm{d}x_1}\, \overline{\mathrm{d}x_2} \dots \mathrm{d}x_m \vee \max_{(x_1,x_2)}{}^2 f(\mathbf{x})\, \mathrm{d}x_1\, \mathrm{d}x_2 \dots \overline{\mathrm{d}x_m} \vee \dots \vee \\
& \max_{\mathbf{x}_0}{}^m f(\mathbf{x})\, \mathrm{d}x_1\, \mathrm{d}x_2 \dots \mathrm{d}x_m \,, \qquad (4.45)
\end{aligned}$$

$$\begin{aligned}
\vartheta_{\mathbf{x}_0} f(\mathbf{x}_0, \mathbf{x}_1) = & \Delta_{x_1} f(\mathbf{x})\, \mathrm{d}x_1\, \overline{\mathrm{d}x_2} \dots \overline{\mathrm{d}x_m} \oplus \dots \oplus \Delta_{x_m} f(\mathbf{x})\, \overline{\mathrm{d}x_1}\, \overline{\mathrm{d}x_2} \dots \mathrm{d}x_m \oplus \\
& \Delta_{(x_1,x_2)} f(\mathbf{x})\, \mathrm{d}x_1\, \mathrm{d}x_2 \dots \overline{\mathrm{d}x_m} \oplus \dots \oplus \Delta_{\mathbf{x}_0} f(\mathbf{x})\, \mathrm{d}x_1\, \mathrm{d}x_2 \dots \mathrm{d}x_m \,. \qquad (4.46)
\end{aligned}$$

Vice versa, the m-fold derivative operations can be extracted from the m-fold differential operations, where $x_i \in \mathbf{x}_0$ and $\mathbf{x}_s \subseteq \mathbf{x}_0$, as follows:

$$\frac{\partial^m f(\mathbf{x}_0, \mathbf{x}_1)}{\partial x_1 \partial x_2 \dots \partial x_m} = \mathrm{d}^m_{\mathbf{x}_0} f(\mathbf{x}_0, \mathbf{x}_1) \mid_{\mathbf{d}x_0 = \mathbf{1}} \,, \qquad (4.47)$$

$$\frac{\partial f(\mathbf{x}_0, \mathbf{x}_1)}{\partial x_i} = \vartheta_{\mathbf{x}_0} f(\mathbf{x}_0, \mathbf{x}_1) \mid_{\mathrm{d}x_i = 1, (\mathbf{dx}_0 \backslash \mathrm{d}x_i) = \mathbf{0}} \,, \qquad (4.48)$$

$$\min_{x_i} f(\mathbf{x}_0, \mathbf{x}_1) = \mathrm{Min}^m_{\mathbf{x}_0} f(\mathbf{x}_0, \mathbf{x}_1) \mid_{\mathrm{d}x_i = 1, (\mathbf{dx}_0 \backslash \mathrm{d}x_i) = \mathbf{0}} \,, \qquad (4.49)$$

$$\max_{x_i} f(\mathbf{x}_0, \mathbf{x}_1) = \mathrm{Max}^m_{\mathbf{x}_0} f(\mathbf{x}_0, \mathbf{x}_1) \mid_{\mathrm{d}x_i = 1, (\mathbf{dx}_0 \backslash \mathrm{d}x_i) = \mathbf{0}} \,, \qquad (4.50)$$

$$\Delta_{\mathbf{x}_s} f(\mathbf{x}_0, \mathbf{x}_1) = \vartheta_{\mathbf{x}_0} f(\mathbf{x}_0, \mathbf{x}_1) \mid_{\mathbf{dx}_s = \mathbf{1}, (\mathbf{dx}_0 \backslash \mathbf{dx}_s) = \mathbf{0}} \,, \qquad (4.51)$$

$$\min_{\mathbf{x}_s}{}^{|\mathbf{x}_s|} f(\mathbf{x}_0, \mathbf{x}_1) = \mathrm{Min}^m_{\mathbf{x}_0} f(\mathbf{x}_0, \mathbf{x}_1) \mid_{\mathbf{dx}_s = \mathbf{1}, (\mathbf{dx}_0 \backslash \mathbf{dx}_s) = \mathbf{0}} \,, \qquad (4.52)$$

$$\max_{\mathbf{x}_s}{}^{|\mathbf{x}_s|} f(\mathbf{x}_0, \mathbf{x}_1) = \mathrm{Max}^m_{\mathbf{x}_0} f(\mathbf{x}_0, \mathbf{x}_1) \mid_{\mathbf{dx}_s = \mathbf{1}, (\mathbf{dx}_0 \backslash \mathbf{dx}_s) = \mathbf{0}} \,. \qquad (4.53)$$

The meaning of the m-fold differential is not obviously recognizable from their definition. The relationships (4.43)–(4.53) are not only a possibility for transformations between the m-fold differential operations and the m-fold derivative operations, but also reveal the meaning of m-fold differential operations. Figure 4.5 supports the understanding of the m-fold differential operations using the same given function as evaluated for total and partial differential operations.

Blue nodes in the four graphs of Figure 4.5 again indicate the function values 1. Edges in graphs of m-fold differential operations require a special interpretation. Differently from the total and partial differential operations, the m-fold differential operations do not evaluate pairs of function values, but all values of a subspace specified by an edge. We take as an example the

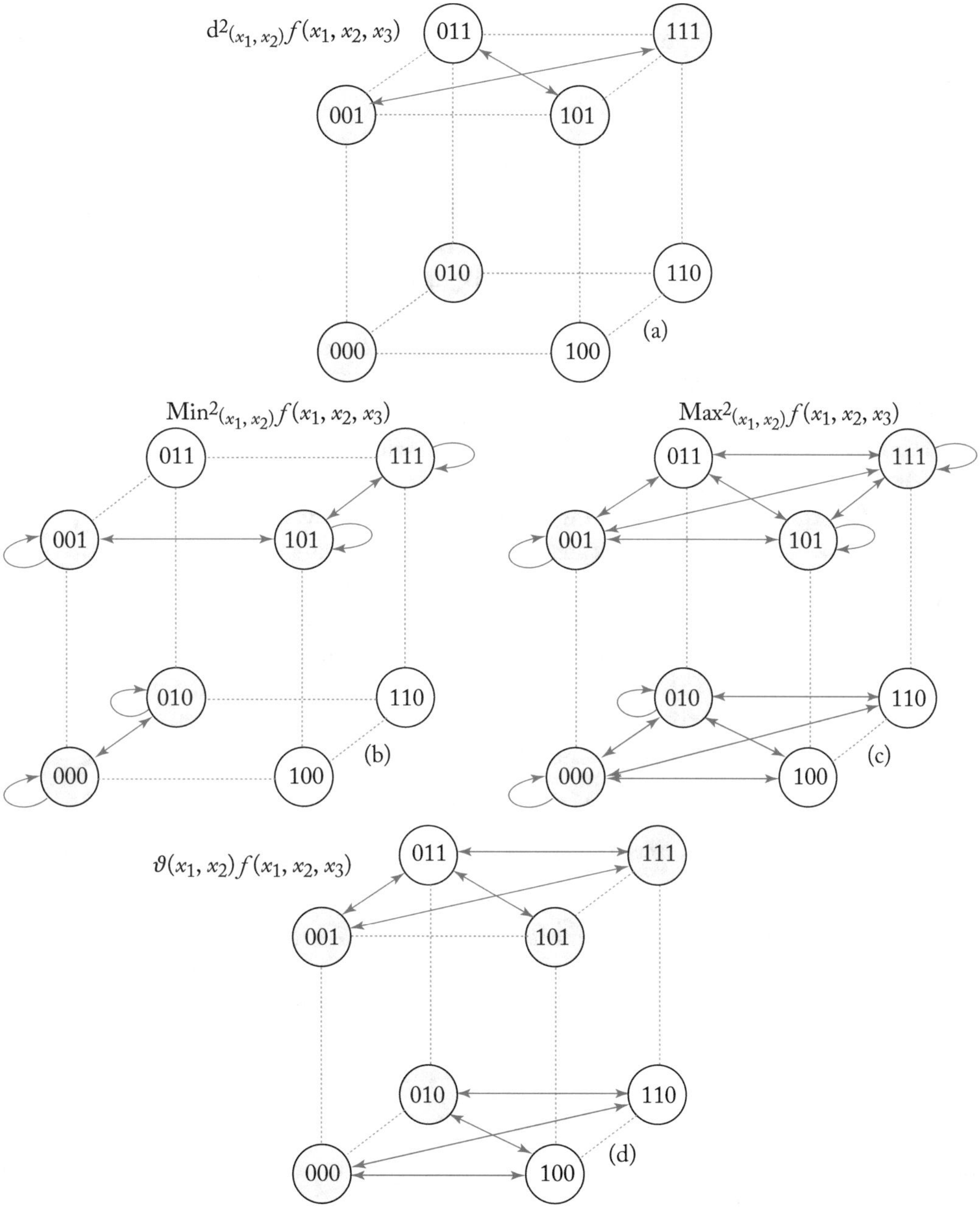

Figure 4.5: Graphs of the 2-fold differential operations of $f(x_1, x_2, x_3) = \overline{x}_1\overline{x}_2 \vee \overline{x}_1\overline{x}_3 \vee x_1x_3$ with regard to (x_1, x_2): (a) 2-fold differential; (b) 2-fold differential minimum; (c) 2-fold differential maximum; and (d) ϑ-operation.

edge of Figure 4.5a between the nodes $(x_1, x_2, x_3) = (001)$ and $(x_1, x_2, x_3) = (111)$. These two nodes differ in the values of x_1 and x_2; hence, the evaluated subspace is defined by the subspace of the four nodes (001), (011), (101), and (111) characterized by all possible assignments to these variables x_1 and x_2. The other edge of Figure 4.5a between the nodes $(x_1, x_2, x_3) = (011)$ and $(x_1, x_2, x_3) = (101)$ indicates the same subspace.

With this background knowledge it is easy to interpret the 2-fold differential operations of Figure 4.5. The edges of the 2-fold differential of Figure 4.5a indicate that the subspace $x_3 = 1$ contains an odd number of function values 1. The subspace $x_3 = 0$ contains an even number of function values 1; hence, the conjunction $\overline{x}_3 \, dx_1 \, dx_2$ does not belong to the 2-fold differential of the evaluated function. As emphasized in (4.43), the m-fold differential only evaluates subspaces $\mathbb{B}^m$. Hence, the 2-fold differential can only contain edges between nodes where the values of all variables of $\mathbf{x}_0$, in this example x_1 and x_2, change.

Figure 4.5b depicts the 2-fold differential minimum with regard to (x_1, x_2). Edges of a 2-fold minimum indicate subspaces $\mathbb{B}^0$, $\mathbb{B}^1$, and $\mathbb{B}^2$ in which all function values are equal to 1. The maximal subspace that only contains function values 1 is in this example $\mathbb{B}^1$.

The 2-fold differential maximum with regard to (x_1, x_2) evaluates subspaces $\mathbb{B}^0$, $\mathbb{B}^1$, and $\mathbb{B}^2$. An edge of a 2-fold maximum indicates a subspace that contains at least one function value 1. Subspaces of all possible sizes satisfy this condition in the example of Figure 4.5c.

Figure 4.5d shows the ϑ-operation with regard to (x_1, x_2). An edge of the ϑ-operation indicates a subspace that contains different function values. Hence, no graph of a ϑ-operation can have a loop on a node. The ϑ-operation in the example of Figure 4.5d contains subspaces $\mathbb{B}^1$ and $\mathbb{B}^2$ with different function values.

Due to Definition (4.42) the m-fold differential operations satisfy the equation:

$$\operatorname{Min}^m_{\mathbf{x}_0} f(\mathbf{x}_0, \mathbf{x}_1) \oplus \operatorname{Max}^m_{\mathbf{x}_0} f(\mathbf{x}_0, \mathbf{x}_1) \oplus \vartheta_{\mathbf{x}_0} f(\mathbf{x}_0, \mathbf{x}_1) = 0 \; . \tag{4.54}$$

The following example shows the expressiveness of graph equations with differential operations.

Example 4.7 All Static Function Hazards. A glitch is the actual occurrence of a spurious signal of a circuit. Figure 4.6 shows that such a glitch can occur when the assignment to the input pattern changes from $\mathbf{x} = \mathbf{c}_0$ to $\mathbf{x} = \mathbf{c}_1$ where $f(\mathbf{x} = \mathbf{c}_0) = f(\mathbf{x} = \mathbf{c}_1)$. A hazard is the possibility that a glitch can occur (see Steinbach and Posthoff [2010b]). A *static function hazard* describes a change between two input patterns that satisfies two properties:

1. the function values of the two input patterns are the same and
2. at least one opposite function value belongs to an input pattern that can be reached when the different input bits are separately toggled in an arbitrary order.

Figure 4.6 shows the two types of static function hazards. All static function hazards are described by the solution vectors $(\mathbf{x}, \mathbf{dx})$ of

$$\vartheta_{\mathbf{x}} f(\mathbf{x}) \wedge \overline{\mathrm{d}_{\mathbf{x}} f(\mathbf{x})} = 1 \; . \tag{4.55}$$

The complement of the total differential of the explored function $f(\mathbf{x})$ is equal to 1 for all pairs of input patterns that hold the first condition for all directions of change. The second condition is compactly expressed for each associated subspace by the ϑ-operation.

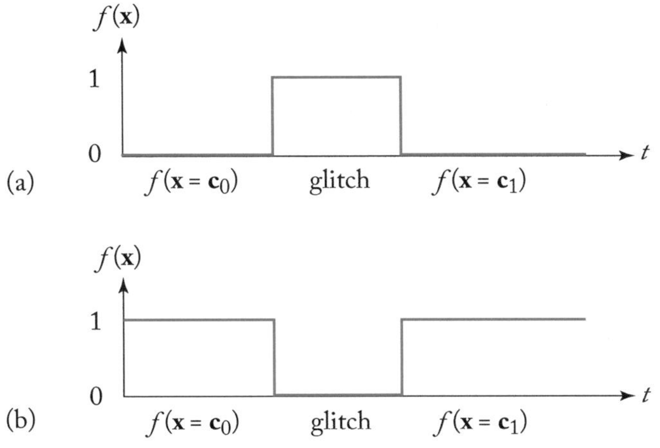

Figure 4.6: Types of static function hazards: (a) static-0 hazard and (b) static-1 hazard.

The two types of static functions hazards can also be calculated separately. All static-0 function hazards are described by the solution vectors $(\mathbf{x}, \mathbf{dx})$ of

$$\overline{f(\mathbf{x})} \wedge \vartheta_{\mathbf{x}} f(\mathbf{x}) \wedge \overline{\mathrm{d}_{\mathbf{x}} f(\mathbf{x})} = 1 \, , \tag{4.56}$$

and all static-1 function hazards are the solution of

$$f(\mathbf{x}) \wedge \vartheta_{\mathbf{x}} f(\mathbf{x}) \wedge \overline{\mathrm{d}_{\mathbf{x}} f(\mathbf{x})} = 1 \, . \tag{4.57}$$

SUMMARY

The differential of a Boolean variable is itself a Boolean variable that determines whether or not the basic variable changes its value. While the direction of change is fixed for all kinds of derivative operations, several directions of change are taken into account for differential operations. The total differential operations evaluate the change behavior between all pairs of point in the Boolean space. The partial differential operations restrict these evaluation to pairs of point which differ only in the subset of variables $\mathbf{x}_0 \subseteq \mathbf{x}$. The m-fold differential operations evaluate all function values of a subspace which are determined by the non-negated differentials of variables.

EXERCISES

4.1 The character 2 has to be top down engraved using the engraving machine, introduced in Example 4.2.

(a) Draw the graph of a complete cycle of the movements starting from the node $(x, y_1, y_2, z) = (0000)$.

(b) Which graph equation describes these movements?

4.2 Verify (4.43) for the 2-fold differential of $f(x_1, x_2)$ with regard to (x_1, x_2).

CHAPTER 5

Applications

5.1 PROPERTIES OF BOOLEAN FUNCTIONS

Commonly with the introduction of the derivative operations we mentioned already how these operations can be used to detect certain properties of the used Boolean function. Table 5.1 summarizes and extends these rules.

Table 5.1: Properties of Boolean functions

Property	Rule
$f(\mathbf{x})$ is independent of x_i	$\frac{\partial f(\mathbf{x})}{\partial x_i} = 0$
$f(\mathbf{x})$ is linear in x_i	$\frac{\partial f(\mathbf{x})}{\partial x_i} = 1$
$f(\mathbf{x})$ is monotonously rising with regard to x_i	$\overline{x}_i\, f(\mathbf{x})\, \frac{\partial f(\mathbf{x})}{\partial x_i} = 0$
$f(\mathbf{x})$ is monotonously falling with regard to x_i	$x_i\, f(\mathbf{x})\, \frac{\partial f(\mathbf{x})}{\partial x_i} = 0$
$f(\mathbf{x})$ is strong monotonously rising with regard to x_i	$\overline{x}_i\, f(\mathbf{x}) \vee \overline{\frac{\partial f(\mathbf{x})}{\partial x_i}} = 0$
$f(\mathbf{x})$ is strong monotonously falling with regard to x_i	$x_i\, f(\mathbf{x}) \vee \overline{\frac{\partial f(\mathbf{x})}{\partial x_i}} = 0$
$f(\mathbf{x})$ is selfdual	$\frac{\partial f(\mathbf{x})}{\partial \mathbf{x}} = 1$
$f(\mathbf{x})$ is symmetric with regard to (x_i, x_j)	$(x_i \oplus x_j)\, \overline{\frac{\partial f(\mathbf{x})}{\partial (x_i, x_j)}} = 0$
$f(\mathbf{x})$ is symmetric with regard to x_i, $i = 1,\ldots, k$	$\bigvee_{i=1}^{k} (x_1 \oplus x_i)\, \overline{\frac{\partial f(\mathbf{x})}{\partial (x_1, x_i)}} = 0$
$f(\mathbf{x}) = f(\mathbf{x}_0, \mathbf{x}_1)$ is constant in the subspaces $\mathbf{x}_0 = \mathbf{c}$	$\mathbf{x}_0\, \Delta_{\mathbf{x}_1} f(\mathbf{x}) = 0$
$f(\mathbf{x}) = f(\mathbf{x}_0, \mathbf{x}_1)$ is equal to 1 in the subspaces $\mathbf{x}_0 = \mathbf{c}$	$\mathbf{x}_0\, \overline{\min_{\mathbf{x}_1} f(\mathbf{x})} = 0$

Table 5.1 does not completely enumerate the properties of Boolean functions. Further basic properties as well as combinations of them can be specified. All derivative operations are also Boolean functions. Table 5.2 summarizes the independence properties of functions created by derivative operations. These rules confirm that all derivative operations simplify the given Boolean function.

Table 5.2: Independence properties of derivative operations

Property	Rule
$\frac{\partial f(\mathbf{x})}{\partial x_i}$ is independent of x_i	$\frac{\partial}{\partial x_i}\left(\frac{\partial f(\mathbf{x})}{\partial x_i}\right) = 0$
$\min_{x_i} f(\mathbf{x})$ is independent of x_i	$\frac{\partial}{\partial x_i}(\min_{x_i} f(\mathbf{x})) = 0$
$\max_{x_i} f(\mathbf{x})$ is independent of x_i	$\frac{\partial}{\partial x_i}(\max_{x_i} f(\mathbf{x})) = 0$
$\frac{\partial^m f(\mathbf{x}_0, \mathbf{x}_1)}{\partial x_1 \ldots \partial x_m}$ is independent of $x_i \in \mathbf{x}_0 = \{x_1, \ldots, x_m\}$	$\bigvee_{i=1}^{m} \frac{\partial}{\partial x_i}\left(\frac{\partial^m f(\mathbf{x})}{\partial x_1 \ldots \partial x_m}\right) = 0$
$\min_{\mathbf{x}_0}{}^m f(\mathbf{x})$ is independent of $x_i \in \mathbf{x}_0 = \{x_1, \ldots, x_m\}$	$\bigvee_{i=1}^{m} \frac{\partial}{\partial x_i}(\min_{\mathbf{x}_0}{}^m f(\mathbf{x})) = 0$
$\max_{\mathbf{x}_0}{}^m f(\mathbf{x})$ is independent of $x_i \in \mathbf{x}_0 = \{x_1, \ldots, x_m\}$	$\bigvee_{i=1}^{m} \frac{\partial}{\partial x_i}(\max_{\mathbf{x}_0}{}^m f(\mathbf{x})) = 0$
$\Delta_{\mathbf{x}_0} f(\mathbf{x}_0, \mathbf{x}_1)$ is independent of $x_i \in \mathbf{x}_0 = \{x_1, \ldots, x_m\}$	$\bigvee_{i=1}^{m} \frac{\partial}{\partial x_i}(\Delta_{\mathbf{x}_0} f(\mathbf{x}_0, \mathbf{x}_1)) = 0$
$\frac{\partial f(\mathbf{x}_0, \mathbf{x}_1)}{\partial \mathbf{x}_0}$ is independent of the simultaneous change of all variables of $\mathbf{x}_0$	$\frac{\partial}{\partial \mathbf{x}_0}\left(\frac{\partial f(\mathbf{x})}{\partial \mathbf{x}_0}\right) = 0$
$\min_{\mathbf{x}_0} f(\mathbf{x}_0, \mathbf{x}_1)$ is independent of the simultaneous change of all variables of $\mathbf{x}_0$	$\frac{\partial}{\partial \mathbf{x}_0}(\min_{\mathbf{x}_0} f(\mathbf{x}_0, \mathbf{x}_1)) = 0$
$\max_{\mathbf{x}_0} f(\mathbf{x}_0, \mathbf{x}_1)$ is independent of the simultaneous change of all variables of $\mathbf{x}_0$	$\frac{\partial}{\partial \mathbf{x}_0}(\max_{\mathbf{x}_0} f(\mathbf{x}_0, \mathbf{x}_1)) = 0$

A general property of all derivative operations is that they are closely related to other derivative operations. All vectorial derivative operations with regard to an arbitrarily selected direction of change satisfy, e.g., the rule that the EXOR-operation of two of them results in the third vectorial derivative operation with regard to the same direction of change. Such relations are not restricted to a fixed direction of change. Knowing, for example, the vectorial derivatives of the same function with regard to two different directions of change, then a third vectorial derivative with regard to the symmetric difference of the known directions of change is the result of an EXOR-operation between the given derivatives.

5.2 SOLUTION OF A BOOLEAN EQUATION WITH REGARD TO VARIABLES

The solution of a Boolean equation with regard to variables is a very powerful concept. It has the same aim as the well-known solution of a real-valued equation with regard to one of the involved variables. Let us take as example the equation:

$$3x + 2y = 4x - 2y \ . \tag{5.1}$$

Solving this equation with regard to the real-valued variable y requires transformation steps that do not change the solution set of this equation, but bring the equation into a form where only the variable y appears on the left-hand side and an expression of the remaining variables (here only x) on the right-hand side. This expression determines a function $f(x)$ that specifies for each value of x the associated function value y.

The solution of an equation remains unchanged when the same expression is added, subtracted, multiplied, or divided (except by 0). Adding $2y$ on both sides separates the variable y to the left-hand side and the subtraction of $3x$ on both sides removes the variable x from the left-hand side. The division by 4 on both sides lead to the wanted form where the left-hand side of the equation is equal to y:

$$\begin{aligned} 3x + 4y &= 4x \\ 4y &= x \\ y &= \frac{x}{4} \ . \end{aligned} \tag{5.2}$$

The substitution of the function $f(x) = \frac{x}{4}$ for the variable y in all places of y in the given Equation (5.1) leads to the same expression on both sides of the equation:

$$\begin{aligned} 3x + 2 \cdot \frac{x}{4} &= 4x - 2 \cdot \frac{x}{4} \\ \frac{12x + 2x}{4} &= \frac{16x - 2x}{4} \\ \frac{14x}{4} &= \frac{14x}{4} \ . \end{aligned} \tag{5.3}$$

Linear equations, e.g., (5.1), are very often solved with regard to y. However, basically such equations can be solved with regard to each of the involved variables. As solution of the Equation (5.1) with regard to x we get:

$$\begin{aligned} -x + 2y &= -2y \\ -x &= -4y \\ x &= 4y \ . \end{aligned} \tag{5.4}$$

The substitution of the function $f(y) = 4y$ in all places of x in the given Equation (5.1) lead again to an identity:

$$\begin{aligned} 3 \cdot 4y + 2y &= 4 \cdot 4y - 2y \\ 12y + 2y &= 16y - 2y \\ 14y &= 14y \; . \end{aligned} \tag{5.5}$$

Now we are going to transfer this principle to Boolean equations. First we want to solve an arbitrary Boolean equation with regard to one selected variable. From Section 1.4 we know that each Boolean equation can be transformed into a homogeneous characteristic Equation (1.64). Hence, without loss of generality we restrict ourselves to the solution of a characteristic equation

$$F(\mathbf{x}, y) = 1 \tag{5.6}$$

with regard to the selected variable y. That means we try to find a Boolean function

$$y = f(\mathbf{x}) \tag{5.7}$$

such that

$$F(\mathbf{x}, f(\mathbf{x})) = 1 \tag{5.8}$$

is a tautology.

We have the same aim as demonstrated for the real-valued equation, but the nonlinearity of the AND- and OR-operations cumber the application of the same methods. Hence, we are faced with the question whether for each Boolean Equation (5.6) a solution with regard to y exists. A simple consideration leads to the conclusion that there are functions $F(\mathbf{x}, y)$ for which no solution of Equation (5.6) with regard to y exists. No solution exists if $F(\mathbf{x}, y)$ determines for an arbitrary $\mathbf{x} = \mathbf{c}_0$ that $F(\mathbf{x} = \mathbf{c}_0, y = 0) = F(\mathbf{x} = \mathbf{c}_0, y = 1) = 0$, because in this case the substitution of the function $f(\mathbf{x})$ (5.7) cannot modify the function $F(\mathbf{x}, y)$ into a function that is constant equal to 1. Hence, a condition for the solution of Equation (5.6) with regard to y is

$$\max_{y} F(\mathbf{x}, y) = 1 \; . \tag{5.9}$$

Example 5.1 Equation $F_0(\mathbf{x}, y) = 1$ that is not Solvable with Regard to y.

The solution of the equation

$$F_0(\mathbf{x}, y) = \overline{x}_1 \oplus x_2 \oplus \overline{x}_1 y = 1 \tag{5.10}$$

consists of four solution vectors (x_1, x_2, y) which are $\{(000), (011), (110), (111)\}$. These four solution vectors are indicated by values 1 in the Karnaugh-map of $F_0(\mathbf{x}, y)$ (see Figure 5.1a). However, the function $F_0(\mathbf{x}, y)$ is equal to 0 for both $(x_1, x_2, y) = (100)$ and $(x_1, x_2, y) = (101)$. Hence, a function $y = f(\mathbf{x})$ such that

$$\forall \mathbf{x} : \quad F_0(\mathbf{x}, y) = 1 \tag{5.11}$$

does not exist. This property is detected by Condition (5.9) that is not satisfied for Equation (5.10) (see Figure 5.10b).

(a) $F_0(\mathbf{x}, y)$

y					
0	1	0	1	0	
1	0	1	1	0	
	0	1	1	0	x_2
	0	0	1	1	x_1

(b) $\max\limits_{y} F_0(\mathbf{x}, y)$

1	1	1	0	
0	1	1	0	x_2
0	0	1	1	x_1

Figure 5.1: Function $F_0(\mathbf{x}, y)$ (a) and its maximum with regard to y (b).

Next, we explore Equation (5.6) that is uniquely solvable with regard to y. We derive the condition for such equations by the construction of $F(\mathbf{x}, y)$ for an arbitrary function

$$y = f(\mathbf{x}) \; . \tag{5.12}$$

The solution set of this function remains unchanged when on both sides an $\odot$-operation with $f(\mathbf{x})$ is executed:

$$\begin{aligned} y \odot f(\mathbf{x}) &= f(\mathbf{x}) \odot f(\mathbf{x}) \\ F(\mathbf{x}, y) &= 1 \; . \end{aligned} \tag{5.13}$$

The linear relation between the variable y and the arbitrary function $f(\mathbf{x})$ can be verified by the derivative of $F(\mathbf{x}, y)$ with regard to y:

$$\begin{aligned} \frac{\partial F(\mathbf{x}, y)}{\partial y} &= \frac{\partial (y \odot f(\mathbf{x}))}{\partial y} \\ &= (0 \odot f(\mathbf{x})) \oplus (1 \odot f(\mathbf{x})) \\ &= \overline{f(\mathbf{x})} \oplus f(\mathbf{x}) \\ &= 1 \oplus f(\mathbf{x}) \oplus f(\mathbf{x}) \\ &= 1 \; . \end{aligned} \tag{5.14}$$

Hence, we have the following theorem.

Theorem 5.2 Uniquely Solvable Boolean Equation. *The Boolean equation* $F(\mathbf{x}, y) = 1$ (5.6) *is uniquely solvable with regard to y if and only if*

$$\frac{\partial F(\mathbf{x}, y)}{\partial y} = 1 \; . \tag{5.15}$$

The solution function $f(\mathbf{x})$ is determined by:

$$f(\mathbf{x}) = \max_y \left(y \wedge F(\mathbf{x}, y)\right) . \tag{5.16}$$

Example 5.3 Equation $F_1(\mathbf{x}, y) = 1$ that is Uniquely Solvable with Regard to y.
The solution of the equation

$$F_1(\mathbf{x}, y) = x_2(x_1 \oplus y) \vee \overline{x}_2\overline{y} = 1 \tag{5.17}$$

consists of four solution vectors (x_1, x_2, y) which are $\{(000), (011), (110), (100)\}$. These four solution vectors are indicated by values 1 in the Karnaugh-map of $F_1(\mathbf{x}, y)$ (see Figure 5.2a). Each column of this Karnaugh-map contains both a value 0 and a value 1; hence, the condition

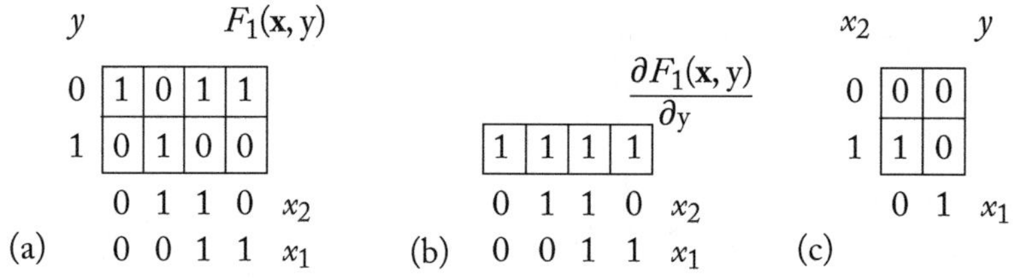

Figure 5.2: Unique solution of Equation (5.17) with regard to y: (a) $F_1(\mathbf{x}, y)$; (b) satisfied condition; and (c) solution function $y = f_1(\mathbf{x})$.

(5.15) is satisfied (see Figure 5.2b) so that Equation (5.17) is uniquely solvable with regard to y. The solution function $y = f_1(\mathbf{x})$ is calculated using (5.16):

$$\begin{aligned} f_1(\mathbf{x}) &= \max_y \left(y \wedge F_1(\mathbf{x}, y)\right) \\ &= \max_y \left(y \wedge (x_2(x_1 \oplus y) \vee \overline{x}_2\overline{y})\right) \\ &= \max_y \left(y \wedge (x_1x_2\overline{y} \vee \overline{x}_1x_2y \vee \overline{x}_2\overline{y})\right) \\ &= \max_y \left(\overline{x}_1x_2y\right) \\ &= \overline{x}_1x_2 . \end{aligned} \tag{5.18}$$

The comparison with the Karnaugh-map of Figure 5.2a shows that the function values of $f_1(\mathbf{x})$ are taken from the row with $y = 1$. The substitution of $f_1(\mathbf{x})$ (5.18) into Equation (5.17) confirms the correctness of this solution:

$$
\begin{aligned}
x_2(x_1 \oplus y) \vee \overline{x}_2\overline{y} &= 1 \\
x_2(x_1 \oplus (\overline{x}_1 x_2)) \vee \overline{x}_2\overline{(\overline{x}_1 x_2)} &= 1 \\
x_2(\overline{x}_1\overline{x}_1 x_2 \vee x_1(x_1 \vee \overline{x}_2)) \vee \overline{x}_2(x_1 \vee \overline{x}_2) &= 1 \\
\overline{x}_1 x_2 \vee x_1 x_2 \vee \overline{x}_2 &= 1 \\
x_2 \vee \overline{x}_2 &= 1 \\
1 &= 1 \; .
\end{aligned}
\tag{5.19}
$$

If Equation (5.6) is solvable with regard to y (confirmed by Condition (5.9)), but not uniquely solvable (Condition (5.15) is not satisfied), more than one function $y = f(\mathbf{x})$ solve this equation. These functions belong to a lattice that can be described by the mark functions $f_q(\mathbf{x})$ and $f_r(\mathbf{x})$.

The Karnaugh-map of Figure 5.1a is helpful to determine these mark functions. Each column consists of two function values. All possible combinations of function values occur.

- The first column determines the mark function $f_r(\mathbf{x})$ because the value 1 of $F_0(\mathbf{x}, y)$ appears for $y = 0$ and the value 0 for $y = 1$.
- The second column determines the mark function $f_q(\mathbf{x})$ because the value 1 of $F_0(\mathbf{x}, y)$ appears for $y = 1$ and the value 0 for $y = 0$.
- The third column determines the mark function $f_\varphi(\mathbf{x})$ because the value 1 of $F_0(\mathbf{x}, y)$ appears for both $y = 0$ and $y = 1$.
- The fourth column determines the situation where the solution with regard to y is forbidden because the value 0 of $F_0(\mathbf{x}, y)$ appears for both $y = 0$ and $y = 1$.

From this observation we can specify formulas of the three mark functions:

$$f_q(\mathbf{x}) = \max_y \left(y \wedge F(\mathbf{x}, y)\right) \wedge \overline{\max_y \left(\overline{y} \wedge F(\mathbf{x}, y)\right)} \; , \tag{5.20}$$

$$f_r(\mathbf{x}) = \max_y \left(\overline{y} \wedge F(\mathbf{x}, y)\right) \wedge \overline{\max_y \left(y \wedge F(\mathbf{x}, y)\right)} \; , \tag{5.21}$$

$$f_\varphi(\mathbf{x}) = \min_y F(\mathbf{x}, y) \; . \tag{5.22}$$

These mark functions of the lattice have to be calculated very often. Therefore, simpler formulas are welcome. We demonstrate the sequence of transformations that simplify $f_q(\mathbf{x})$. First we extend the expression by a zero-function which is built by the conjunction of a second term and its complement:

$$
\begin{aligned}
f_q(\mathbf{x}) &= \max_y (y \wedge F(\mathbf{x}, y)) \wedge \overline{\max_y (\overline{y} \wedge F(\mathbf{x}, y))} \\
&= \left(\max_y (y \wedge F(\mathbf{x}, y)) \wedge \overline{\max_y (\overline{y} \wedge F(\mathbf{x}, y))} \right) \vee \\
&\qquad \left(\max_y (\overline{y} \wedge F(\mathbf{x}, y)) \wedge \overline{\max_y (\overline{y} \wedge F(\mathbf{x}, y))} \right) .
\end{aligned}
\tag{5.23}
$$

We apply the distributive law and thereafter the rule that the disjunction of two maximum expressions results in the maximum of these expressions:

$$
\begin{aligned}
f_q(\mathbf{x}) &= \overline{\max_y (\overline{y} \wedge F(\mathbf{x}, y))} \wedge \left(\max_y (y \wedge F(\mathbf{x}, y)) \vee \max_y (\overline{y} \wedge F(\mathbf{x}, y)) \right) \\
&= \overline{\max_y (\overline{y} \wedge F(\mathbf{x}, y))} \wedge \max_y (y \wedge F(\mathbf{x}, y) \vee \overline{y} \wedge F(\mathbf{x}, y)) .
\end{aligned}
\tag{5.24}
$$

Now, the distributive law can be applied to the expression within the maximum, $(y \vee \overline{y})$ results in the value 1 and can be avoided due to the $\wedge$-operation. The resulting maximum expression is the condition (5.9) that must be equal to 1 in order to solve the given Equation (5.6) with regard to y.

$$
\begin{aligned}
f_q(\mathbf{x}) &= \overline{\max_y (\overline{y} \wedge F(\mathbf{x}, y))} \wedge \left(\max_y ((y \vee \overline{y}) \wedge F(\mathbf{x}, y)) \right) \\
&= \overline{\max_y (\overline{y} \wedge F(\mathbf{x}, y))} \wedge \max_y F(\mathbf{x}, y) \\
&= \overline{\max_y (\overline{y} \wedge F(\mathbf{x}, y))} \wedge 1 \\
&= \overline{\max_y (\overline{y} \wedge F(\mathbf{x}, y))} .
\end{aligned}
\tag{5.25}
$$

Similarly, the mark function $f_r(\mathbf{x})$ can be simplified so that we get the preferred formulas for the calculation of the mark functions of the lattice containing all functions that solve Equation (5.6) under the condition (5.9):

$$
f_q(\mathbf{x}) = \overline{\max_y (\overline{y} \wedge F(\mathbf{x}, y))} , \tag{5.26}
$$

$$
f_r(\mathbf{x}) = \overline{\max_y (y \wedge F(\mathbf{x}, y))} . \tag{5.27}
$$

Equipped with the knowledge to solve a Boolean equation with regard to one variable, we are going to generalize this method for more than one variable.

Definition 5.4 Solution of a Boolean Equation with Regard to Several Variables y. Let $\mathbf{x} = (x_1, \ldots, x_k)$, $\mathbf{y} = (y_1, \ldots, y_m)$, $F(\mathbf{x}, \mathbf{y})$ be a function of $n = k + m$ variables. The equation

$$
F(\mathbf{x}, \mathbf{y}) = 1 \tag{5.28}
$$

can be solved with regard to the variables $y_1, \ldots, y_m$ if there are m functions

$$\begin{aligned} y_1 &= f_1(\mathbf{x}) , \\ &\vdots \\ y_m &= f_m(\mathbf{x}) , \end{aligned} \tag{5.29}$$

with

$$F(\mathbf{x}, f_1(\mathbf{x}), \ldots, f_m(\mathbf{x})) = 1 . \tag{5.30}$$

The functions $f_1(\mathbf{x}), \ldots, f_m(\mathbf{x})$ (5.29) specify function values $y_1, \ldots, y_m$ for each assignment of the independent variables $\mathbf{x}$. Hence, a solution of Equation (5.28) with regard to $\mathbf{y}$ exists only if there is at least one function value 1 of $F(\mathbf{x}, \mathbf{y})$ for each assignment of $\mathbf{x}$. This condition for the solvability of Equation (5.28) with regard to $\mathbf{y}$ can be expressed by:

$$\max_{\mathbf{y}}{}^m F(\mathbf{x}, \mathbf{y}) = 1 . \tag{5.31}$$

The comparison of the conditions for the solvability of a characteristic equation $F(\mathbf{x}, y_0, \mathbf{y}_1) = 1$ with regard to the single variable y_0 and the set of variables $\mathbf{y} = \{y_0\} \cup \mathbf{y}_1$ leads to a surprising effect.

Theorem 5.5 *If $F(\mathbf{x}, y_0, \mathbf{y}_1) = 1$ is solvable with regard to y_0 it is also solvable with regard to each set of variables that contains y_0.*

Proof. This theorem follows directly from the solvability condition of $F(\mathbf{x}, y_0, \mathbf{y}_1) = 1$ with regard to y_0

$$\max_{y_0} F(\mathbf{x}, y_0, \mathbf{y}_1) = 1 \tag{5.32}$$

and the rule (2.118) that an m-fold maximum is larger than or equal to a single maximum:

$$\max_{y_0} F(\mathbf{x}, y_0, \mathbf{y}_1) \leq \max_{(y_0, \mathbf{y}_1)}{}^m F(\mathbf{x}, y_0, \mathbf{y}_1) . \tag{5.33}$$

□

If the solvability condition (5.31) is satisfied, it can separately be verified for each function $f_i(\mathbf{x})$, $i = 1, \ldots, m$, whether this function is uniquely specified by the given Equation (5.28) or can be chosen from a lattice.

Theorem 5.6 *The Boolean equation $F(\mathbf{x}, \mathbf{y}) = 1$ (5.28) is uniquely solvable with regard to y_i, $i = 1, \ldots, m$, if and only if*

$$\frac{\partial \left(\max_{(\mathbf{y} \setminus y_i)}^{(m-1)} F(\mathbf{x}, \mathbf{y}) \right)}{\partial y_i} = 1 . \tag{5.34}$$

The solution function $f_i(\mathbf{x})$ is determined by:

$$f_i(\mathbf{x}) = \max_{\mathbf{y}}{}^m (y_i \wedge F(\mathbf{x}, \mathbf{y})) \ . \tag{5.35}$$

Theorem 5.6 generalizes Theorem 5.2 for a selected function within a solution with regard to several variables. The $(m-1)$-fold maximum in the condition (5.34) eliminates all variables $\mathbf{y}$ except y_i and the subsequent derivative verifies the linearity with regard to y_i. The m-fold maximum in (5.35) eliminates both the variable y_i belonging to the function $f_i(\mathbf{x})$ and all other variables $y_j \in (\mathbf{y} \setminus y_i)$.

If Equation (5.28) is solvable with regard to $\mathbf{y}$, but not uniquely solvable with regard to certain variables $y_i \in \mathbf{y}$ then the associated function $f_i(\mathbf{x})$ can taken from a lattice specified by:

$$f_{qi}(\mathbf{x}) = \overline{\max_{\mathbf{y}}{}^m (\overline{y}_i \wedge F(\mathbf{x}, \mathbf{y}))} \ , \tag{5.36}$$

$$f_{ri}(\mathbf{x}) = \max_{\mathbf{y}}{}^m (y_i \wedge F(\mathbf{x}, \mathbf{y})) \ . \tag{5.37}$$

Unfortunately, it can be that certain combinations of functions of these lattices do not solve the given equation. This drawback can be overcome by the iterative execution of the following steps:

1. the calculation of the mark functions of only one lattice $L_i\langle f_{qi}(\mathbf{x}), f_{ri}(\mathbf{x})\rangle$ using (5.36) and (5.37);
2. the selection of an arbitrary function $f_i(\mathbf{x})$ of this lattice $L_i < f_{qi}(\mathbf{x}), f_{ri}(\mathbf{x}) >$; and
3. the substitution of the function $y_i = f_i(\mathbf{x})$ into $F(\mathbf{x}, \mathbf{y}) = 1$ that excludes the variable y_i from the function $F(\mathbf{x}, \mathbf{y})$.

The simplified function $F(\mathbf{x}, y_1, \dots, y_{i-1}, y_{i+1}, \dots, y_m)$ of step 3 can be calculated by

$$F(\mathbf{x}, y_1, \dots, y_{i-1}, y_{i+1}, \dots, y_m) = \max_{y_i} ((y_i \odot f_i(\mathbf{x})) \wedge F(\mathbf{x}, \mathbf{y})) \ , \tag{5.38}$$

where $f_i(\mathbf{x})$ is the function selected in step 2. The simplified equation

$$F(\mathbf{x}, y_1, \dots, y_{i-1}, y_{i+1}, \dots, y_m) = 1 \tag{5.39}$$

can be solved with regard to the remaining variables $(y_1, \dots, y_{i-1}, y_{i+1}, \dots, y_m)$ without restriction originated by the selected function $f_i(\mathbf{x})$.

Example 5.7 Solution of an Equation $F(\mathbf{x}, \mathbf{y}) = 1$ with Regard to $\mathbf{y}$. The equation

$$\begin{aligned} &\overline{x}_1\overline{x}_2\overline{y}_1(x_3 \oplus y_3) \vee \overline{x}_1 x_2 x_3 y_1 \vee \overline{x}_1 x_2 y_1 y_2 \vee \\ &x_1 x_2 \overline{y}_1 \overline{y}_2 y_3 \vee x_1 x_2 x_3 \overline{y}_1 y_2 \vee x_1 \overline{x}_2 x_3 y_1 y_3 \vee x_1 \overline{x}_2 y_1 \overline{y}_2 \overline{y}_3 = 1 \end{aligned} \tag{5.40}$$

has to be solved with regard to $\mathbf{y} = (y_1, y_2, y_3)$. The expression on the left-hand side of Equation (5.40) describes the function $F(\mathbf{x}, \mathbf{y})$. The condition (5.31) for the solution of (5.40) with regard to $\mathbf{y} = (y_1, y_2, y_3)$ is satisfied:

$$\begin{aligned} \max_{\mathbf{y}}{}^m F(\mathbf{x}, \mathbf{y}) &= \overline{x}_1\overline{x}_2 \vee \overline{x}_1 x_2 x_3 \vee \overline{x}_1 x_2 \vee x_1 x_2 \vee x_1 x_2 x_3 \vee x_1 \overline{x}_2 x_3 \vee x_1 \overline{x}_2 \\ &= 1 \ . \end{aligned} \tag{5.41}$$

The check whether Equation (5.40) is uniquely solvable with regard to one or more variables $\mathbf{y}$ by means of the condition (5.34) leads to the results:

$$\frac{\partial \left(\max^2_{(y_2, y_3)} F(\mathbf{x}, \mathbf{y}) \right)}{\partial y_1} = 1 \ , \tag{5.42}$$

$$\frac{\partial \left(\max^2_{(y_1, y_3)} F(\mathbf{x}, \mathbf{y}) \right)}{\partial y_2} = (x_1 \vee x_2)\overline{x}_3 \ , \tag{5.43}$$

$$\frac{\partial \left(\max^2_{(y_1, y_2)} F(\mathbf{x}, \mathbf{y}) \right)}{\partial y_3} = \overline{x}_1\overline{x}_2 \vee x_1\overline{x}_3 \ . \tag{5.44}$$

Due to Theorem 5.6, Equation (5.40) is uniquely solvable with regard to y_1 and the solution function of y_2 and y_3 can be chosen from lattices. Using (5.35) we get for the function $y_1 = f_1(\mathbf{x})$:

$$\begin{aligned} f_1(\mathbf{x}) &= \max_{\mathbf{y}}{}^3 \left(y_1 \wedge F(\mathbf{x}, \mathbf{y}) \right) \\ &= \max_{\mathbf{y}}{}^3 \left(\overline{x}_1 x_2 x_3 y_1 \vee \overline{x}_1 x_2 y_1 y_2 \vee x_1 \overline{x}_2 x_3 y_1 y_3 \vee x_1 \overline{x}_2 y_1 \overline{y}_2 \overline{y}_3 \right) \\ &= \overline{x}_1 x_2 \vee x_1 \overline{x}_2 \\ &= x_1 \oplus x_2 \ . \end{aligned} \tag{5.45}$$

The mark function of the lattice for $y_2 = f_2(\mathbf{x})$ can be calculated using (5.36) and (5.37) without any restrictions:

$$f_{q2}(\mathbf{x}) = \overline{\max_{\mathbf{y}}{}^3 \left(\overline{y}_2 \wedge F(\mathbf{x}, \mathbf{y}) \right)} = \overline{x}_1 x_2 \overline{x}_3 \ , \tag{5.46}$$

$$f_{r2}(\mathbf{x}) = \overline{\max_{\mathbf{y}}{}^3 \left(y_2 \wedge F(\mathbf{x}, \mathbf{y}) \right)} = x_1 \overline{x}_3 \ . \tag{5.47}$$

This lattice is determined by one value 1 and two values 0 and consists of $2^5 = 32$ functions due to the remaining five don't-cares. We choose the simple function

$$f_2(\mathbf{x}) = \overline{x}_1 \tag{5.48}$$

as solution function for y_2 from the calculated lattice $L_2 \langle f_{q2}(\mathbf{x}), f_{r2}(\mathbf{x}) \rangle$.

In order to avoid conflicts between the chosen functions $f_2(\mathbf{x})$ and $f_3(\mathbf{x})$ from the lattices L_2 and L_3 we restrict the function $F(\mathbf{x}, y_1, y_2, y_3)$ to $F(\mathbf{x}, y_1, y_3)$ using (5.38):

$$\begin{aligned}
F(\mathbf{x}, y_1, y_3) &= \max_{y_2}\left((y_2 \odot f_2(\mathbf{x})) \wedge F(\mathbf{x}, y_1, y_2, y_3)\right) \\
&= \max_{y_2}\left((y_2 \odot \overline{x}_1) \wedge F(\mathbf{x}, y_1, y_2, y_3)\right) \\
&= \max_{y_2}\left(\overline{x}_1\overline{x}_2\overline{y}_1 y_2(x_3 \oplus y_3) \vee \overline{x}_1 x_2 x_3 y_1 y_2 \vee \overline{x}_1 x_2 y_1 y_2 \vee\right. \\
&\qquad \left. x_1 x_2 \overline{y}_1 \overline{y}_2 y_3 \vee x_1 \overline{x}_2 x_3 y_1 \overline{y}_2 y_3 \vee x_1 \overline{x}_2 y_1 \overline{y}_2 \overline{y}_3\right) \\
&= \overline{x}_1\overline{x}_2\overline{y}_1(x_3 \oplus y_3) \vee \overline{x}_1 x_2 y_1 \vee x_1 x_2 \overline{y}_1 y_3 \vee x_1\overline{x}_2 x_3 y_1 y_3 \vee x_1 \overline{x}_2 y_1 \overline{y}_3 \,. \qquad (5.49)
\end{aligned}$$

This restricted equation

$$F(\mathbf{x}, y_1, y_3) = 1 \qquad (5.50)$$

retains the already verified solvability with regard to y_3. The mark function of the lattice for $y_3 = f_3(\mathbf{x})$ can again be calculated using (5.36) and (5.37):

$$f_{q3}(\mathbf{x}) = \overline{\max_{(y_1,y_3)}{}^2 \left(\overline{y}_3 \wedge F(\mathbf{x}, y_1, y_3)\right)} = x_1 x_2 \vee \overline{x}_1\overline{x}_2\overline{x}_3 \,, \qquad (5.51)$$

$$f_{r3}(\mathbf{x}) = \overline{\max_{(y_1,y_3)}{}^2 \left(y_3 \wedge F(\mathbf{x}, y_1, y_3)\right)} = \overline{x}_1\overline{x}_2 x_3 \vee x_1\overline{x}_2\overline{x}_3 \,. \qquad (5.52)$$

This lattice is determined by three values 1 and two values 0 and consists of $2^3 = 8$ functions due to the remaining three don't-cares. We choose the function

$$f_3(\mathbf{x}) = \overline{x}_1\overline{x}_3 \vee x_2 \qquad (5.53)$$

as solution function for y_3 from the calculated lattice $L_3\langle f_{q3}(\mathbf{x}), f_{r3}(\mathbf{x})\rangle$.

The computed complete solution of Equation (5.40) with regard to $\mathbf{y} = (y_1, y_2, y_3)$ is:

$$\begin{aligned}
y_1 &= x_1 \oplus x_2 \,, \\
y_2 &= \overline{x}_1 \,, \\
y_3 &= \overline{x}_1\overline{x}_3 \vee x_2 \,. \qquad (5.54)
\end{aligned}$$

The substitution of these functions into (5.40) confirms the correctness of this solution. The system of Equations (5.54) can be transformed into the characteristic equation:

$$\begin{aligned}
(y_1 \odot (x_1 \oplus x_2)) \wedge (y_2 \odot (\overline{x}_1)) \wedge (y_3 \odot (\overline{x}_1\overline{x}_3 \vee x_2)) &= 1 \\
F'(\mathbf{x}, \mathbf{y}) &= 1 \,. \qquad (5.55)
\end{aligned}$$

Due to the applied construction, Equation (5.55) is uniquely solvable with regard to $\mathbf{y} = (y_1, y_2, y_3)$. The left-hand side functions $F(\mathbf{x}, \mathbf{y})$ of (5.40) and $F'(\mathbf{x}, \mathbf{y})$ of (5.55) satisfy

$$F'(\mathbf{x}, \mathbf{y}) < F(\mathbf{x}, \mathbf{y}) \,. \qquad (5.56)$$

Values 1 in the Karnaugh-maps of Figure 5.3 specify both the function values 1 of the functions $F(\mathbf{x}, \mathbf{y})$ and $F'(\mathbf{x}, \mathbf{y})$ and the solutions of the associated Equations (5.40) and (5.55). The comparison of these Karnaugh-maps confirms the property (5.56). The unique solvability of (5.55) with regard to $\mathbf{y}$ becomes visible by exactly one value 1 in each column of the Karnaugh-map of $F'(\mathbf{x}, \mathbf{y})$ in Figure 5.3b.

(a) $F(\mathbf{x}, \mathbf{y})$

$y_1\ y_2\ y_3$									
0 0 0	0	1	0	0	0	0	0	0	
0 0 1	1	0	0	0	1	1	0	0	
0 1 1	1	0	0	0	0	1	0	0	
0 1 0	0	1	0	0	0	1	0	0	
1 1 0	0	0	1	1	0	0	0	0	
1 1 1	0	0	1	1	0	0	1	0	
1 0 1	0	0	1	0	0	0	1	0	
1 0 0	0	0	1	0	0	0	1	1	
	0	1	1	0	0	1	1	1	x_3
	0	0	1	1	1	1	0	0	x_2
	0	0	0	0	1	1	1	1	x_1

(b) $F'(\mathbf{x}, \mathbf{y})$

$y_1\ y_2\ y_3$									
0 0 0	0	0	0	0	0	0	0	0	
0 0 1	0	0	0	0	1	1	0	0	
0 1 1	1	0	0	0	0	0	0	0	
0 1 0	0	1	0	0	0	0	0	0	
1 1 0	0	0	0	0	0	0	0	0	
1 1 1	0	0	1	1	0	0	0	0	
1 0 1	0	0	0	0	0	0	0	0	
1 0 0	0	0	0	0	0	0	1	1	
	0	1	1	0	0	1	1	1	x_3
	0	0	1	1	1	1	0	0	x_2
	0	0	0	0	1	1	1	1	x_1

Figure 5.3: Karaugh-maps of the explored functions/equations: (a) $F(\mathbf{x}, \mathbf{y})$ and (b) $F'(\mathbf{x}, \mathbf{y})$.

The Karnaugh-map of $F(\mathbf{x}, \mathbf{y})$ in Figure 5.3a can be used to identify the number of all combinations of solution functions of Equation (5.40): it is the product of the values 1 of the eight columns of Karnaugh-map of $F(\mathbf{x}, \mathbf{y})$:

$$n_{sol}^{F(\mathbf{x},\mathbf{y})} = 2 \cdot 2 \cdot 4 \cdot 2 \cdot 1 \cdot 3 \cdot 3 \cdot 1 = 288 \;. \tag{5.57}$$

5.3 COMPUTATION OF GRAPHS

Based on their structure graphs are very widely applied. They consist of vertices connected by edges. Both the vertices and edges can be labeled. There is an extensive theory providing many relationships and algorithms to calculate graphs with certain properties or to analyze given graphs.

The vertices of graphs can be encoded by Boolean values. In this way Boolean variables and Boolean functions can be used to specify the vertices of a graph. Using pairs of encoded vertices Boolean functions are even able to describe a graph completely.

The Boolean Differential Calculus introduced the differentials of Boolean variables. These differentials describe the change of Boolean variables. Hence, an edge of a graph can be alternatively specified by one encoded vertex and the direction of change determined by differentials of

Boolean variables. This additional possibility is a welcome simplification for many applications. There are many other books about graph theory. We present here only a small example that shows the benefits of differentials of Boolean variables to solve a given problem.

Example 5.8 Ferryman, Wolf, Goat, and Cabbage. This well-known problem can be shortly described as follows. A ferryman, a wolf, a goat, and a cabbage want to cross a river, but there is only a small boat. To reach the other bank the following conditions must be satisfied.

1. The wolf and the goat or the goat and the cabbage cannot stay without ferryman at one bank, otherwise the wolf feeds the goat or the goat feeds the cabbage.
2. Only the ferryman is able to control the boat to cross the river.
3. Due to the small boat the ferryman can transfer only one passenger.
4. The states forbidden by condition 1 cannot be reached from other states.
5. It is not useful that all four travelers remain at one side of the river.

We use the Boolean variables of Table 5.3 to model this problem.

Table 5.3: Position of the travelers modeled by Boolean variables

Traveler	Left Bank	Right Bank
ferryman	f	$\overline{f}$
wolf	w	$\overline{w}$
goat	g	$\overline{g}$
cabbage	c	$\overline{c}$

The first rule can be expressed by a Boolean equation for each bank:

$$\overline{f} \wedge (w\, g \vee g\, c) = 0\,, \tag{5.58}$$
$$f \wedge (\overline{w}\,\overline{g} \vee \overline{g}\,\overline{c}) = 0\,. \tag{5.59}$$

The second rule can be expressed by means of differentials of Boolean variables; without moving of the ferryman $\overline{\mathrm{d}f}$ neither the wolf $\mathrm{d}w$ nor the goat $\mathrm{d}g$ nor the cabbage $\mathrm{d}c$ can move over the river:

$$\overline{\mathrm{d}f} \wedge (\mathrm{d}w \vee \mathrm{d}g \vee \mathrm{d}c) = 0\,. \tag{5.60}$$

The third rule also restricts the movement of the travelers and can be expressed by means of differentials of Boolean variables:

$$\mathrm{d}f\, \mathrm{d}w\, \mathrm{d}g \vee \mathrm{d}f\, \mathrm{d}w\, \mathrm{d}c \vee \mathrm{d}f\, \mathrm{d}g\, \mathrm{d}c = 0\,. \tag{5.61}$$

The fourth rule can be modeled such that each variable of Equations (5.58) and (5.59) is replaced by the $\oplus$-operation of this variable and the associated differential:

$$\overline{(f \oplus \mathrm{d}f)} \wedge ((w \oplus \mathrm{d}w)\,(g \oplus \mathrm{d}g) \vee (g \oplus \mathrm{d}g)\,(c \oplus \mathrm{d}c)) = 0\,, \tag{5.62}$$

$$(f \oplus \mathrm{d}f) \wedge (\overline{(w \oplus \mathrm{d}w)}\,\overline{(g \oplus \mathrm{d}g)} \vee \overline{(g \oplus \mathrm{d}g)}\,\overline{(c \oplus \mathrm{d}c)}) = 0\,. \tag{5.63}$$

The last rule requires that not all differentials are equal to 0:

$$\overline{\mathrm{d}f} \wedge \overline{\mathrm{d}w} \wedge \overline{\mathrm{d}g} \wedge \overline{\mathrm{d}c} = 0\,. \tag{5.64}$$

The solution of the system of Equations (5.58), ..., (5.64) results in the solution set of all edges of the graph of the explored problem. The edges are described by the state (the positions of the traveler on the banks) and the direction of change (the travelers moving across the river). Figure 5.4a shows the solution set, and Figure 5.4b depicts the associated graph.

	f	w	g	c	df	dw	dg	c
	0	1	0	0	1	0	0	1
	0	0	0	1	1	1	0	0
	1	1	1	0	1	1	0	0
	0	0	1	0	1	0	0	1
	1	0	1	1	1	0	0	1
$S =$	0	0	1	0	1	1	0	0
	–	0	1	0	1	0	0	0
	1	1	0	1	1	0	0	1
	0	–	0	–	1	0	1	0
	1	1	0	1	1	1	0	0
	–	1	0	1	1	0	0	0
	1	–	1	–	1	0	1	0

(a)

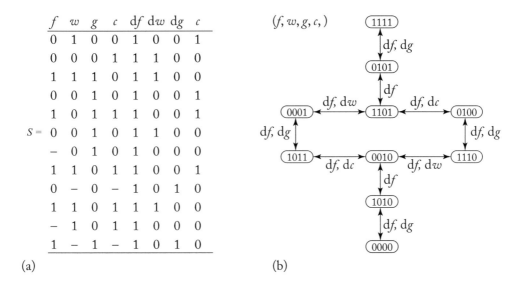

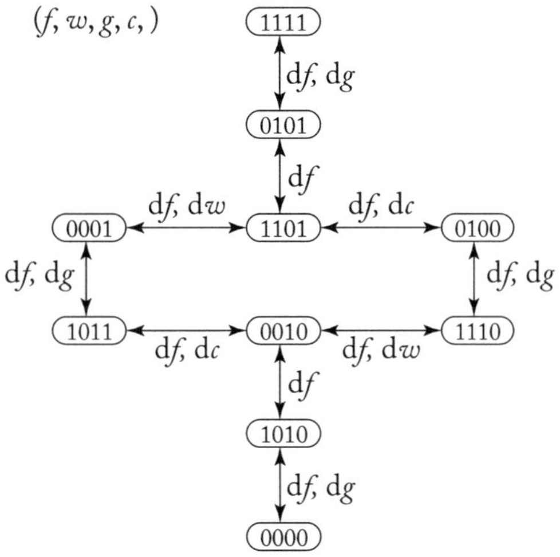

(b)

Figure 5.4: Ferryman, wolf, goat, and cabbage: (a) solution set and (b) graph.

5.4 ANALYSIS OF DIGITAL CIRCUITS

The basic analysis problem for digital circuits is to determine the realized behavior of a given circuit structure. We present a universal approach that can be applied to all circuits that work with Boolean values. This approach needs specifications of the used switching elements as a Boolean equation or the associated solution set and the circuit structure where all connection wires are uniquely labeled.

First, we demonstrate this approach by means of a combinational circuit of logic gates. Table 5.4 summarizes a the description of a selection of logic gates. This list can be extended for other gates or adapted, e.g, for gates with negated inputs or outputs. In the context of gates or circuits we use the term *phase list* (PHL) for the solution set of the equation that specifies the behavior, because each solution vector describes the values that can be observed on the inputs and outputs of the explored item.

Table 5.4: Logic gates: name, symbol, equation, and phase list

Name	Symbol	Equation	Phase List
NOT	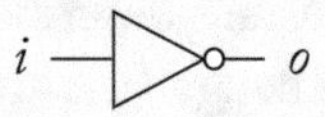	$o = \overline{i}$	i o 0 1 1 0
AND	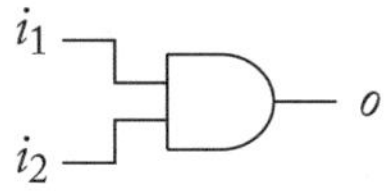	$o = i_1 \wedge i_2$	i_1 i_2 o 0 – 0 1 0 0 1 1 1
NAND	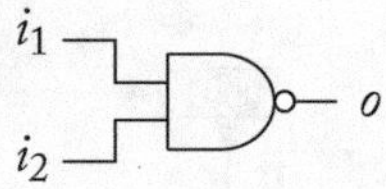	$o = \overline{i_1 \wedge i_2}$	i_1 i_2 o 0 – 1 1 0 1 1 1 0
OR	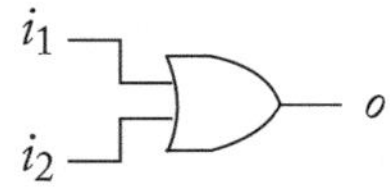	$o = i_1 \vee i_2$	i_1 i_2 o 0 0 0 1 – 1 0 1 1
EXOR	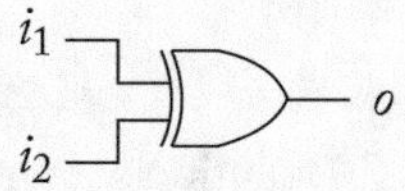	$o = i_1 \oplus i_2$	i_1 i_2 o 0 0 0 0 1 1 1 0 1 1 1 0

As first step of this approach we assign to each connection wire a unique variable. As usual we use indexed variables x_i for the inputs, y_j for the outputs, and g_k for the remaining internal connections of the circuit. Next we prepare a system of equations that contains for each gate the associated equation taken from Table 5.4. Now we can solve this system of equation and get a set of solution vectors $(\mathbf{x}, \mathbf{g}, \mathbf{y})$.

Each vector describes a particular behavior that can be observed on all connection wires of the circuit for the selected input pattern. From this solution set we can build the associated characteristic function $F(\mathbf{x}, \mathbf{g}, \mathbf{y})$ that describes the complete global behavior. This function contains on the one hand the comprehensive behavior information of the analyzed circuit, but is very large due to the variables of the internal wires. Using the m-fold maximum the more compact global system function $F(\mathbf{x}, \mathbf{y})$ of the input-output behavior can be calculated:

$$F(\mathbf{x}, \mathbf{y}) = \max_{\mathbf{g}}{}^m F(\mathbf{x}, \mathbf{g}, \mathbf{y}) \,. \tag{5.65}$$

As an alternative to the system of equations a set of local phase lists can be built. The intersection of these phase lists results in the same solution set as in the case of the before mentioned system of equations.

(a)

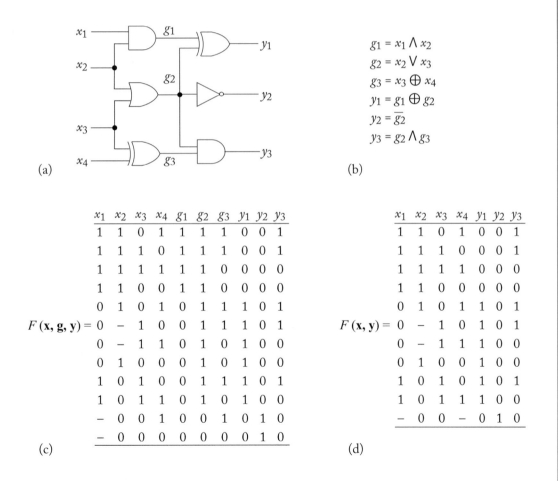

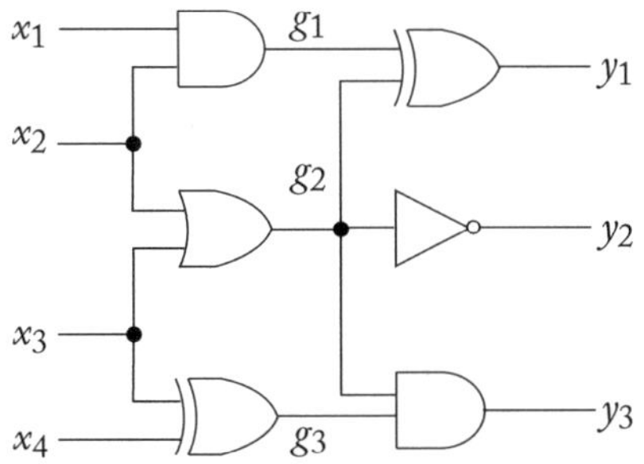

(b)

$$\begin{aligned}
g_1 &= x_1 \wedge x_2 \\
g_2 &= x_2 \vee x_3 \\
g_3 &= x_3 \oplus x_4 \\
y_1 &= g_1 \oplus g_2 \\
y_2 &= \overline{g_2} \\
y_3 &= g_2 \wedge g_3
\end{aligned}$$

(c) $F(\mathbf{x}, \mathbf{g}, \mathbf{y}) =$

x_1	x_2	x_3	x_4	g_1	g_2	g_3	y_1	y_2	y_3
1	1	0	1	1	1	1	0	0	1
1	1	1	0	1	1	1	0	0	1
1	1	1	1	1	1	0	0	0	0
1	1	0	0	1	1	0	0	0	0
0	1	0	1	0	1	1	1	0	1
0	–	1	0	0	1	1	1	0	1
0	–	1	1	0	1	0	1	0	0
0	1	0	0	0	1	0	1	0	0
1	0	1	0	0	1	1	1	0	1
1	0	1	1	0	1	0	1	0	0
–	0	0	1	0	0	1	0	1	0
–	0	0	0	0	0	0	0	1	0

(d) $F(\mathbf{x}, \mathbf{y}) =$

x_1	x_2	x_3	x_4	y_1	y_2	y_3
1	1	0	1	0	0	1
1	1	1	0	0	0	1
1	1	1	1	0	0	0
1	1	0	0	0	0	0
0	1	0	1	1	0	1
0	–	1	0	1	0	1
0	–	1	1	1	0	0
0	1	0	0	1	0	0
1	0	1	0	1	0	1
1	0	1	1	1	0	0
–	0	0	–	0	1	0

Figure 5.5: Analysis of a combinational circuit: (a) circuit structure; (b) system of equations; (c) global system function $F(\mathbf{x}, \mathbf{g}, \mathbf{y})$; and (d) input-output system function $F(\mathbf{x}, \mathbf{y})$.

Example 5.9 Analysis of a Combinational Circuit. Figure 5.5a depicts a combinational circuit. The behavior of this circuit has to be calculated. The Boolean variables $x_1, \dots, x_4$, $g_1, \dots, g_3$, and $y_1, \dots, y_3$ are already assigned in Figure 5.5a. Using these variables the equations of Figure 5.5b can be derived. Figure 5.5c shows the solution of this system of equations. The 12 ternary vectors express the Boolean values on all wires for all 16 input patterns. Using (5.65) the simplified input-output behavior expressed by the system function $F(\mathbf{x}, \mathbf{y})$ can be calculated. This function is shown in Figure 5.5d.

The basic approach can also be used to analyze an asynchronous sequential circuit. Such sequential circuits use logic gates as shown in Table 5.4, but loops cause a memory behavior that can be expressed by state variables s_i and state functions sf_i. The state variables indicate the state at the observed point in time and the state functions determine the values of the state reached at the following time originated by changes on the inputs. A cut in each loop can be used to model both the state variables s_i and state functions sf_i.

In the prepared system of equations that contains for each gate the associated equation taken from Table 5.4 occur variables x_i for the inputs, y_j for the outputs, s_k for the states, sf_k for the following states, and g_l for the remaining internal connections of the circuit. As solution of this system of equations we get a set of vectors $(\mathbf{x}, \mathbf{s}, \mathbf{sf}, \mathbf{g}, \mathbf{y})$. The associated characteristic function $F(\mathbf{x}, \mathbf{s}, \mathbf{sf}, \mathbf{g}, \mathbf{y})$ can again be simplified to the more compact global system function $F(\mathbf{x}, \mathbf{s}, \mathbf{sf}, \mathbf{y})$ of the asynchronous final state machine using an m-fold maximum with regard to $\mathbf{g}$:

$$F(\mathbf{x}, \mathbf{s}, \mathbf{sf}, \mathbf{y}) = \max_{\mathbf{g}}{}^m F(\mathbf{x}, \mathbf{s}, \mathbf{sf}, \mathbf{g}, \mathbf{y}) \ . \tag{5.66}$$

The alternative approach by means of the local phase lists can also be used and leads to the same result.

Example 5.10 Analysis of an Asynchronous Sequential Circuit. Figure 5.6a depicts an asynchronous sequential circuit. The behavior of this circuit has to be calculated. The Boolean variables x_1, x_2, s_1, s_2, sf_1, sf_2, g_1, g_2, g_3, and y_1, y_2 are already assigned in Figure 5.6a. Using these variables the equations of Figure 5.6b can be derived. Figure 5.6c shows the solution of this system of equations. The nine ternary vectors express the Boolean values on all wires for all combinations of the four input patterns and the four state patterns. Using (5.66) the simplified behavior of this final state machine can be calculated. It can be expressed by the system function $F(\mathbf{x}, \mathbf{s}, \mathbf{sf}, \mathbf{y})$ as shown in Figure 5.6d.

The same method can also be used for synchronous sequential circuits. Memory elements of such circuits are called flip-flops. They encapsulate the loop and provide the variables s_i and sf_i in the equation that describe their behavior. Hence, it is not necessary to cut in the model loop wires as basis for the system of equations. Table 5.5 shows the details of flip-flops which are often used.

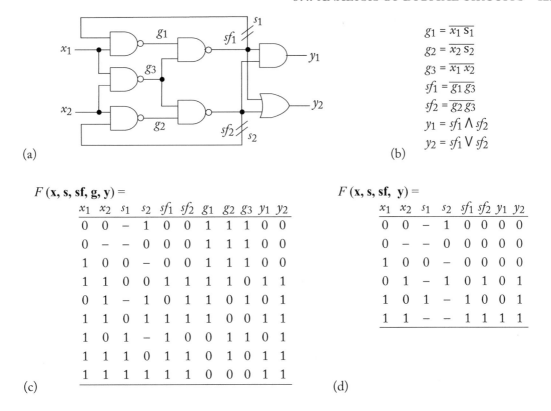

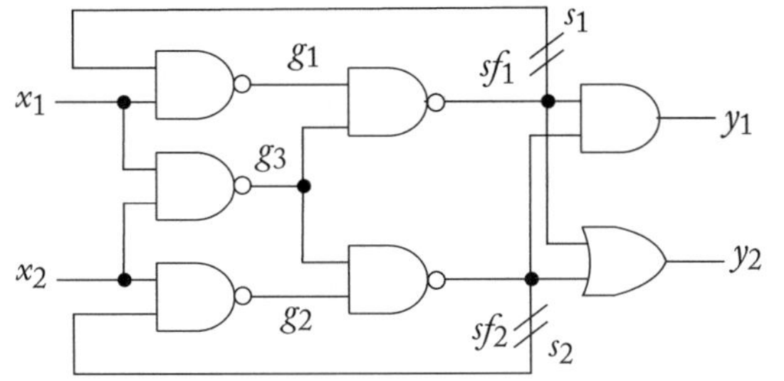

(a)

$$g_1 = \overline{x_1\, s_1}$$
$$g_2 = \overline{x_2\, s_2}$$
$$g_3 = \overline{x_1\, x_2}$$
$$sf_1 = \overline{g_1\, g_3}$$
$$sf_2 = \overline{g_2\, g_3}$$
$$y_1 = sf_1 \wedge sf_2$$
$$y_2 = sf_1 \vee sf_2$$

(b)

$F(\mathbf{x}, \mathbf{s}, \mathbf{sf}, \mathbf{g}, \mathbf{y}) =$

x_1	x_2	s_1	s_2	sf_1	sf_2	g_1	g_2	g_3	y_1	y_2
0	0	–	1	0	0	1	1	1	0	0
0	–	–	0	0	0	1	1	1	0	0
1	0	0	–	0	0	1	1	1	0	0
1	1	0	0	1	1	1	1	0	1	1
0	1	–	1	0	1	1	0	1	0	1
1	1	0	1	1	1	1	0	0	1	1
1	0	1	–	1	0	0	1	1	0	1
1	1	1	0	1	1	0	1	0	1	1
1	1	1	1	1	1	0	0	0	1	1

(c)

$F(\mathbf{x}, \mathbf{s}, \mathbf{sf}, \mathbf{y}) =$

x_1	x_2	s_1	s_2	sf_1	sf_2	y_1	y_2
0	0	–	1	0	0	0	0
0	–	–	0	0	0	0	0
1	0	0	–	0	0	0	0
0	1	–	1	0	1	0	1
1	0	1	–	1	0	0	1
1	1	–	–	1	1	1	1

(d)

Figure 5.6: Analysis of an asynchronous sequential circuit: (a) circuit structure; (b) system of equations; (c) global system function $F(\mathbf{x}, \mathbf{s}, \mathbf{sf}, \mathbf{g}, \mathbf{y})$; and (d) system function $F(\mathbf{x}, \mathbf{s}, \mathbf{sf}, \mathbf{g}, \mathbf{y})$ of the finite state machine.

The synchronous change of the state of all flip-flops of the circuit is triggered by a clock signal. The clock-inputs of all flip-flops are connected with the clock generator; however, it is not necessary to involve the clock variable in the analysis.

As basis for the calculation of the behavior of the synchronous circuit all connection wires are uniquely labeled: x_i for the inputs, y_j for the outputs, s_k respectively $\overline{s}_k$ for the two outputs of each flip-flop, and g_l for the other outputs of gates. Using the equations of Table 5.4 for the logic gates and from Table 5.5 for the flip-flops a system of equation has to be created and solved. The solution set of this system of equations contains the vectors $(\mathbf{x}, \mathbf{s}, \mathbf{sf}, \mathbf{g}, \mathbf{y})$. Alternatively, the same set of binary vectors is the solution of the intersection of the phase lists of the gates and flip-flops of the circuit. The associated characteristic function $F(\mathbf{x}, \mathbf{s}, \mathbf{sf}, \mathbf{g}, \mathbf{y})$ can again be simplified by (5.66) to the more compact global system function $F(\mathbf{x}, \mathbf{s}, \mathbf{sf}, \mathbf{y})$ of the synchronous final state machine.

Table 5.5: Flip-flops: name, symbol, equation, and phase list

Name	Symbol	Equation	Phase List
D-FF	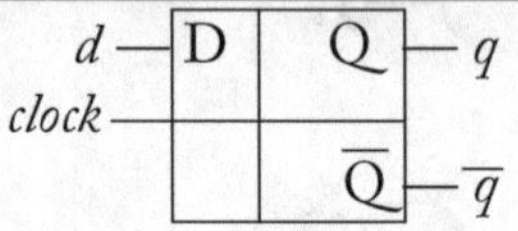	$qf = d$	$d\ q\ qf$ 0 – 0 1 – 1
T-FF	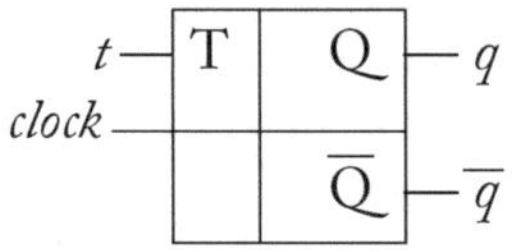	$qf = t \oplus q$	$t\ q\ qf$ 0 0 0 0 1 1 1 0 1 1 1 0
DE-FF	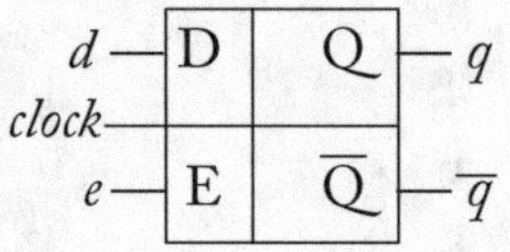	$qf = \bar{e}q \vee ed$	$d\ e\ q\ qf$ – 0 0 0 – 0 1 1 0 1 – 0 1 1 – 1
JK-FF	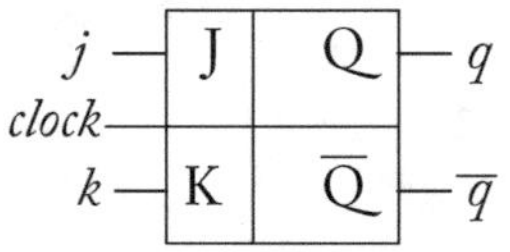	$qf = j\bar{q} \vee \bar{k}q$	$j\ k\ q\ qf$ 1 – 0 1 – 0 1 1 0 – 0 0 – 1 1 0

Example 5.11 Analysis of a Synchronous Sequential Circuit. Figure 5.7a depicts a synchronous sequential circuit. The behavior of this circuit has to be calculated. The Boolean variables x_1, x_2, s_1, s_2, g_1, and y_1, y_2 are already assigned in Figure 5.7a. Using these variables the equations of Figure 5.7b can be derived. Figure 5.7c shows the solution of this system of equations. The twelf ternary vectors express the Boolean values on all wires for all combinations of the four input patterns and the four state patterns. Using (5.66) the simplified behavior of this final state machine expressed by the system function $F(\mathbf{x}, \mathbf{s}, \mathbf{sf}, \mathbf{y})$ can be calculated. This function is shown in Figure 5.7d.

The universality of the suggested method becomes visible for reversible circuits that use different types of gates. A universal and often used library of reversible gates is NCT: NOT-gate, controlled NOT-gate (CNOT), and Toffoli gate. Table 5.6 shows the symbols, equations, and phase lists of these reversible gates.

Reversible gates do not change the values on the control lines but only on the target lines. Hence, only the sections between to gates on the same target line must be labeled by interme-

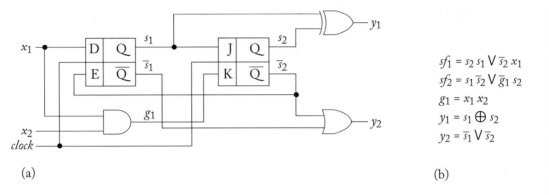

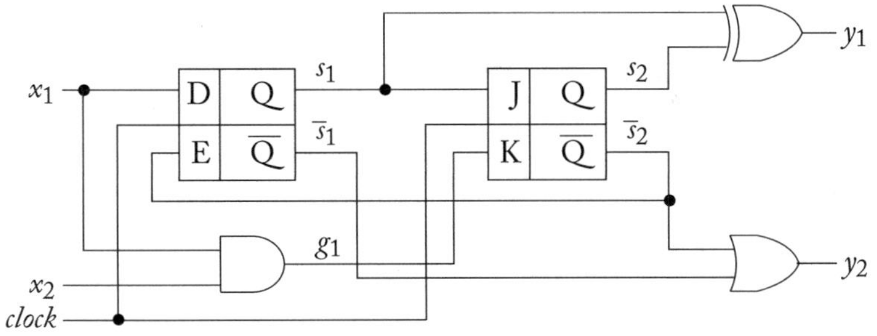

(a)

$$sf_1 = s_2\, s_1 \vee \overline{s_2}\, x_1$$
$$sf_2 = s_1\, \overline{s_2} \vee \overline{g_1}\, s_2$$
$$g_1 = x_1\, x_2$$
$$y_1 = s_1 \oplus s_2$$
$$y_2 = \overline{s_1} \vee \overline{s_2}$$

(b)

$F(\mathbf{x}, \mathbf{s}, \mathbf{sf}, g_1, \mathbf{y}) =$

x_1	x_2	s_1	s_2	sf_1	sf_2	g_1	y_1	y_2
1	0	1	0	1	1	0	1	1
1	1	1	0	1	1	1	1	1
1	0	0	0	1	0	0	0	1
1	1	0	0	1	0	1	0	1
0	1	1	1	1	1	0	0	0
–	0	1	1	1	1	0	0	0
1	1	1	1	1	0	1	0	0
0	–	1	0	0	1	0	1	1
0	–	0	0	0	0	0	0	1
0	1	0	1	0	1	0	1	1
–	0	0	1	0	1	0	1	1
1	1	0	1	0	0	1	1	1

(c)

$F(\mathbf{x}, \mathbf{s}, \mathbf{sf}, \mathbf{y}) =$

x_1	x_2	s_1	s_2	sf_1	sf_2	y_1	y_2
1	–	1	0	1	1	1	1
1	–	0	0	1	0	0	1
0	1	1	1	1	1	0	0
–	0	1	1	1	1	0	0
1	1	1	1	1	0	0	0
0	–	1	0	0	1	1	1
0	1	0	1	0	1	1	1
–	0	0	1	0	1	1	1
1	1	0	1	0	0	1	1

(d)

Figure 5.7: Analysis of a synchronous sequential circuit: (a) circuit structure; (b) system of equations; (c) global system function $F(\mathbf{x}, \mathbf{s}, \mathbf{sf}, \mathbf{g}, \mathbf{y})$; and (d) system function $F(\mathbf{x}, \mathbf{s}, \mathbf{sf}, \mathbf{g}, \mathbf{y})$ of the finite state machine.

diate variables g_k. The system of equations of a reversible circuit contains for each intermediate variable and each output variable one equation. Using variables x_i for the inputs and y_j for the outputs, the solution of this system of equations is a set of vectors $(\mathbf{x}, \mathbf{g}, \mathbf{y})$, and the associated characteristic function $F(\mathbf{x}, \mathbf{g}, \mathbf{y})$ can be reduced to the input-output behavior using (5.65).

Example 5.12 Analysis of a Reversible Circuit. Figure 5.8a depicts a reversible circuit. The behavior of this circuit has to be calculated. The Boolean variables x_1, x_2, x_3, g_1, g_2, and y_1, y_2, y_3 are already assigned in Figure 5.8a. Using these variables the equations of Figure 5.8b

Table 5.6: Reversible gates: name, symbol, equation, and phase list

Name	Symbol	Equation	Phase List
NOT	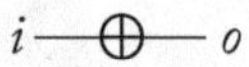	$o = \overline{i}$	$i\ o$ 0 1 1 0
CNOT	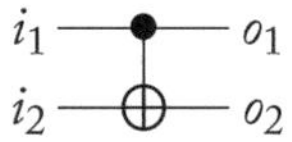	$o_1 = i_1$ $o_2 = i_1 \oplus i_2$	$i_1\ i_2\ o_1\ o_2$ 0 0 0 0 0 1 0 1 1 0 1 1 1 1 1 0
TOFFOLI	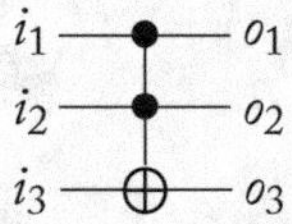	$o_1 = i_1$ $o_2 = i_2$ $o_3 = i_3 \oplus (i_1 \wedge i_2)$	$i_1\ i_2\ i_3\ o_1\ o_2\ o_3$ 0 0 0 0 0 0 0 1 0 0 1 0 1 0 0 1 0 0 1 1 0 1 1 1 0 0 1 0 0 1 0 1 1 0 1 1 1 0 1 1 1 1 1 1 1 1 0

can be derived. Figure 5.8c shows the solution of this system of equations. The eight binary vectors express the Boolean values on all sections of the three wires for all eight input patterns. Using (5.65) the simplified input-output behavior expressed by the system function $F(\mathbf{x}, \mathbf{y})$ can be calculated. This function is shown in Figure 5.8d. It can be seen that all eight combinations of three Boolean values occur both for the inputs (x_1, x_2, x_3) and the outputs (y_1, y_2, y_3).

As final variant of this universal approach we introduce the contact potential model. This model can be applied for networks of transistors. Transistors work like a switch that change the state (open, close) depending on an input signal. There are different types of metal-oxide-semiconductor-field-effect-transistors (MOS-FETs). Depending on the type of the channel (n-chanel or p-chanel) and working principle (enhancement or depletion) such transistors are either closed for the control signal 0 and open otherwise (opening contact) or open for the control signal 0 and closed otherwise (closing contact).

The contact potential model abstracts from these electronic details and uses for these two types of contacts Boolean variables for both the value of the control signal and the values at the pins of the contact. Table 5.7 shows the symbols, equations, and phase lists of these switches.

(a)

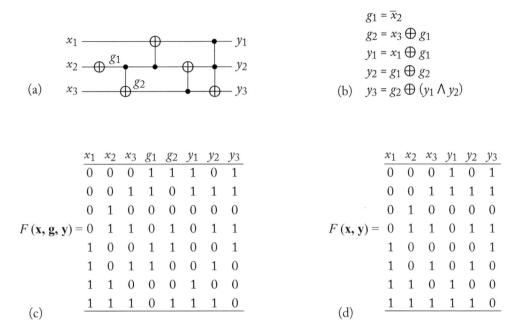

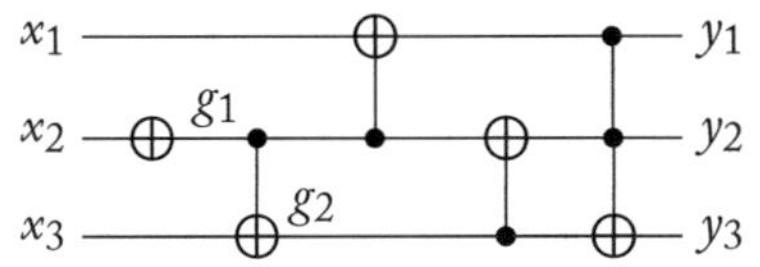

(b)

$$g_1 = \overline{x}_2$$
$$g_2 = x_3 \oplus g_1$$
$$y_1 = x_1 \oplus g_1$$
$$y_2 = g_1 \oplus g_2$$
$$y_3 = g_2 \oplus (y_1 \wedge y_2)$$

(c) $F(\mathbf{x}, \mathbf{g}, \mathbf{y}) =$

x_1	x_2	x_3	g_1	g_2	y_1	y_2	y_3
0	0	0	1	1	1	0	1
0	0	1	1	0	1	1	1
0	1	0	0	0	0	0	0
0	1	1	0	1	0	1	1
1	0	0	1	1	0	0	1
1	0	1	1	0	0	1	0
1	1	0	0	0	1	0	0
1	1	1	0	1	1	1	0

(d) $F(\mathbf{x}, \mathbf{y}) =$

x_1	x_2	x_3	y_1	y_2	y_3
0	0	0	1	0	1
0	0	1	1	1	1
0	1	0	0	0	0
0	1	1	0	1	1
1	0	0	0	0	1
1	0	1	0	1	0
1	1	0	1	0	0
1	1	1	1	1	0

Figure 5.8: Analysis of a reversible circuit: (a) circuit structure; (b) system of equations; (c) global system function $F(\mathbf{x}, \mathbf{g}, \mathbf{y})$; and (d) input-output system function $F(\mathbf{x}, \mathbf{y})$.

The switches of the contact potential model, as shown in Table 5.7 can be connected in an arbitrary manner. The switches are labeled by the input variables x_i. One or several potentials can be assigned to the outputs y_j. The assignment of the constant values 0 and 1 to certain potentials causes that the combination of switches behaves as logic circuit. Due to the freedom in the connection of the switches such circuits can be used as a bus-driver where for certain input values the output is neither connected with 0 nor with 1. However, for the same reason also short-circuit faults can occur for some input patterns.

Using well-parametrized equations of Table 5.7 for the switches, in addition with equations for the outputs and the constant potentials we get a system of equations. In this system of equations the variables x_i occur as inputs to control the switches, the variables y_j for the outputs and potentials p_k for all connection between the switches. As solution of this system of equations we get a set of vectors $(\mathbf{x}, \mathbf{p}, \mathbf{y})$. The associated characteristic function $F(\mathbf{x}, \mathbf{p}, \mathbf{y})$ can be simplified to the more compact global system function $F(\mathbf{x}, \mathbf{y})$ of the universal switching circuit using an m-fold maximum with regard to $\mathbf{p}$:

$$F(\mathbf{x}, \mathbf{y}) = \max_{\mathbf{p}}{}^{m} F(\mathbf{x}, \mathbf{p}, \mathbf{y}) \; . \tag{5.67}$$

Table 5.7: Switches of the contact potential model: name, symbol, equation, and phase list

Name	Symbol	Equation	Phase List		
			p_0	i	p_1
Opening contact	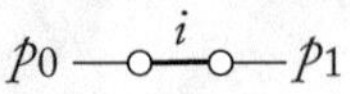	$\bar{i} \wedge (p_0 \odot p_1) \vee i = 1$	0	0	0
			1	0	1
			–	1	–
			p_0	i	p_1
Closing contact	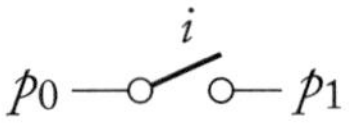	$i \wedge (p_0 \odot p_1) \vee \bar{i} = 1$	0	1	0
			1	1	1
			–	0	–

The alternative approach by means of the local phase lists can also be used and leads to the same result.

Example 5.13 Analysis of a Circuit of Switches Based on the Contact Potential Model. Figure 5.9a depicts a circuit of switches. The behavior of this circuit has to be calculated. The Boolean variables $x_1, \dots, x_4$ for the inputs, $p_0, \dots, p_4$ for the potentials, and y for the output as well a the constants 0 and 1 are already assigned in Figure 5.9a. Using these variables the equations of Figure 5.9b can be derived. Figure 5.9c shows the solution of this system of equations. The five ternary vectors express the Boolean values that specify all conflict-free states of the switches and the corresponding potentials. Using (5.67) the simplified input-output behavior expressed by the system function $F(\mathbf{x}, y)$ can be calculated. This function is shown in Figure 5.9d. A further analysis of this function is needed to distinguish between input patterns that cause a short-circuit, lead to open output (the output is isolated from both 0 and 1), or to determine the output value.

The function $f_{\text{conflict}}(\mathbf{x})$ is equal to 1 for such input patterns of a circuit of switches that cause a short-circuit. It can be calculated by

$$f_{\text{conflict}}(\mathbf{x}) = \overline{\max_{\mathbf{y}}{}^m F(\mathbf{x}, \mathbf{y})} \,, \tag{5.68}$$

because the input-output system function $F(\mathbf{x}, \mathbf{y})$ represents based on the contact potential model only the conflict-free phases. Applied to the example of Figure 5.9 we get:

$$\begin{aligned} f_{\text{conflict}}(\mathbf{x}) &= \overline{\max_{y} F(\mathbf{x}, y)} \\ &= \bar{x}_2 x_3 x_4 \vee x_1 \bar{x}_3 x_4 \vee x_1 \bar{x}_2 \bar{x}_4 \,. \end{aligned} \tag{5.69}$$

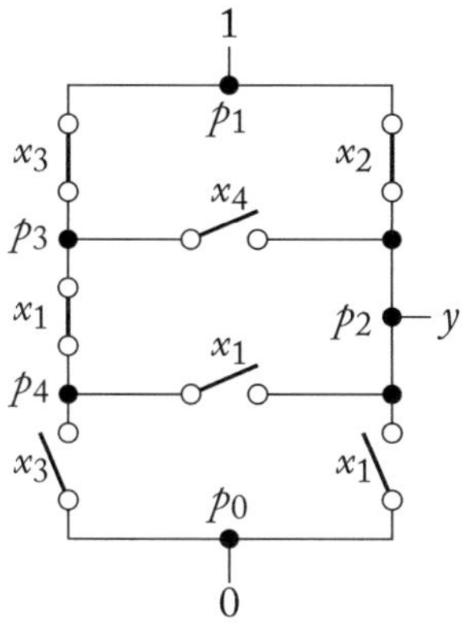

(a)

$$
\begin{aligned}
&\overline{x}_3 \wedge (p_1 \odot p_3) \vee x_3 = 1\\
&\overline{x}_2 \wedge (p_1 \odot p_2) \vee x_2 = 1\\
&x_4 \wedge (p_2 \odot p_3) \vee \overline{x}_4 = 1\\
&\overline{x}_1 \wedge (p_3 \odot p_4) \vee x_1 = 1\\
&x_1 \wedge (p_2 \odot p_4) \vee \overline{x}_1 = 1\\
&x_3 \wedge (p_0 \odot p_4) \vee \overline{x}_3 = 1\\
&x_1 \wedge (p_0 \odot p_2) \vee \overline{x}_1 = 1\\
&p_0 = 0\\
&p_1 = 1\\
&y = p_2
\end{aligned}
$$

(b)

$F(\mathbf{x}, \mathbf{p}, y) =$

x_1	x_2	x_3	x_4	p_0	p_1	p_2	p_3	p_4	y
–	1	1	–	0	1	0	0	0	0
0	–	1	0	0	1	1	0	0	1
1	1	–	0	0	1	0	1	0	0
0	1	0	0	0	1	0	1	1	0
0	–	0	–	0	1	1	1	1	1

(c)

$F(\mathbf{x}, y) =$

x_1	x_2	x_3	x_4	y
–	1	1	1	0
0	–	1	0	1
–	1	–	0	0
0	–	0	–	1

(d)

Figure 5.9: Analysis of a contact circuit: (a) circuit structure; (b) system of equations; (c) global system function $F(\mathbf{x}, \mathbf{p}, y)$; and (d) input-output system function $F(\mathbf{x}, y)$.

This function is equal to 1 for six input patterns. These are prohibited for the circuit of Figure 5.9a to avoid a short-circuit.

A circuit of switches can be constructed such that the output y_j is isolated from both 0 and 1 for certain input patterns. The function $f_{open}^{y_j}(\mathbf{x})$ is equal to 1 in these cases and can be calculated by

$$f_{open}^{y_j}(\mathbf{x}) = \min_{y_j} \left(\max_{\mathbf{y} \setminus y_j}{}^m F(\mathbf{x}, \mathbf{y}) \right) , \tag{5.70}$$

because the input-output system function contains these input patterns for both the output value $y_j = 0$ and $y_j = 1$. Applied to the example of Figure 5.9 we get:

$$
\begin{aligned}
f_{open}^{y}(\mathbf{x}) &= \min_{y} F(\mathbf{x}, y)\\
&= \overline{x}_1 x_2 \overline{x}_4 \; .
\end{aligned} \tag{5.71}
$$

This function is equal to 1 for two input patterns. Assuming that the output y is connected with a bus-wire, another device on the same bus-wire can determine the value of the bus for these two input patterns.

Input patterns of a circuit of switches that belong neither to $f_{\text{conflict}}(\mathbf{x})$ nor to $f_{\text{open}}(\mathbf{x})$ determine the output y_j and can be calculated by

$$f^{y_j}_{\text{determined}}(\mathbf{x}) = \overline{f^{y_j}_{\text{open}}(\mathbf{x})} \wedge \max_{\mathbf{y}}{}^m F(\mathbf{x}, \mathbf{y}) \ . \tag{5.72}$$

Applied to the example of Figure 5.9 we get:

$$\begin{aligned} f^{y}_{\text{determined}}(\mathbf{x}) &= \overline{f^{y}_{\text{open}}(\mathbf{x})} \wedge \max_{y} F(\mathbf{x}, y) \\ &= \overline{x}_1\overline{x}_2\overline{x}_4 \vee x_1x_2\overline{x}_4 \vee x_2x_3x_4 \vee \overline{x}_1\overline{x}_3x_4 \ . \end{aligned} \tag{5.73}$$

This function is equal to 1 for eight input patterns. The three functions (5.69), (5.71), and (5.73) are pairwise disjoint and cover the whole Boolean space $\mathbb{B}^4$ completely.

All logic gates and flip-flops introduced in this subsection uniquely determine the output value for given controlling values. Hence, the system equations

$$F(\mathbf{x}, \mathbf{y}) = 1 \ , \tag{5.74}$$

$$F(\mathbf{x}, \mathbf{s}, \mathbf{sf}, \mathbf{y}) = 1 \tag{5.75}$$

of the calculated system functions are uniquely solvable with regard to $\mathbf{y}$ and $\mathbf{sf}$. The change behavior of derived functions $y_j = f_j(\mathbf{x})$, $y_j = f_j(\mathbf{x}, \mathbf{s})$, and $sf_k = f_k(\mathbf{x}, \mathbf{s})$ can be analyzed be means of the derivative operations of the BDC.

Using the vectorial derivative operations all the input patterns $\mathbf{x}_0$ can be calculated for which the explored output changes its value (the vectorial derivative is equal to 1), the value 1 remains unchanged (the vectorial minimum is equal to 1), or the value 0 remains unchanged (the vectorial maximum is equal to 0).

If the set of variables $\mathbf{x}_0$ contains only the single variable x_i analog results can be calculated for the change of this variable using the associated single derivative operation. An important analysis question is whether the output really depends on all inputs. If the single derivative with regard to x_i is equal to 0 the explored function does not depend on this input variable; hence, a simpler circuit can be synthesized in such a case.

Using the m-fold derivative operations it can be analyzed whether the output value remains constant for arbitrary changes of the inputs $\mathbf{x}_0$ for dedicated input patterns of the other inputs $\mathbf{x}_1$. A constant value of 1 appears on the output for arbitrary values of the inputs $\mathbf{x}_0$ if the remaining inputs $\mathbf{x}_1$ satisfy that the m-fold minimum of the explored function with regard to $\mathbf{x}_0$ is equal to 1. The reverse case (a constant value of 0 appears at the output) is determined for such patterns of $\mathbf{x}_1$ for which the m-fold maximum of the explored function with regard to $\mathbf{x}_0$ is equal to 0. If it has to be determined whether the output remains is either constant 0 or constant 1 for arbitrary values of the inputs $\mathbf{x}_0$ the associated input pattern $\mathbf{x}_1$ satisfy that the Δ-operation of the explored function with regard to $\mathbf{x}_0$ is equal to 1.

The analysis questions mentioned above could be answered using exactly one derivative operation. The combination of derivative operation in one equation extends the possibilities of the analysis. A practically significant example is the detection of hazards. A hazard describes the possibility that a glitch occurs. A glitch is an actual occurrence of a spurious signal in a circuit. Assuming that a sequential circuit counts the changes from zero to one of one signal, a glitch would be lead to a wrong result.

The cause of a hazard can be the simultaneous change of the input values (x_i, x_j) of a combinational circuit. Depending on the value at the begin and the end of this change we distinguish between:

- (010) - static 0-hazard, and
- (101) - static 1-hazard.

The sequence in the above parentheses determine the time series of function values that can occur on the output of a circuit when input values (x_i, x_j) change simultaneously. The inverse value in the middle is the spurious signal in the transition that starts and ends with the same signal value.

The static 0-hazard requires the both the value at start and the end of the simultaneous change of the input values (x_i, x_j) is equal to 0; this can be expressed by the complement of the vectorial maximum. The spurious signal during this transition requires that the subspace specified by x_i and x_j contains at least one value 1; this can be expressed by a 2-fold maximum. Hence, the condition for the static 0-hazard is:

$$\overline{\max_{(x_i,x_j)} f(x_i, x_j, \mathbf{x}_1)} \wedge \max_{(x_i,x_j)}{}^2 f(x_i, x_j, \mathbf{x}_1) = 1 \,. \tag{5.76}$$

The condition for the static 1-hazard combines the vectorial minimum (that determines equal values 1 at the begin and end of the simultaneous change of (x_i, x_j)) and the complement of the 2-fold minimum (that determines at least one value 0 in the subspace specified by x_i and x_j). Hence, the condition for the static 1-hazard is:

$$\min_{(x_i,x_j)} f(x_i, x_j, \mathbf{x}_1) \wedge \overline{\min_{(x_i,x_j)}{}^2 f(x_i, x_j, \mathbf{x}_1)} = 1 \,. \tag{5.77}$$

Both types of static hazards can commonly be detected. The common condition requires that the function does not change their value in the case of a simultaneous change of the input values (x_i, x_j) (the vectorial derivative must be equal to 0) and that the subspace specified by x_i and x_j contains different values (the delta-operation must be equal to 1). Hence, the condition for both static hazards is:

$$\overline{\frac{\partial f(x_i, x_j, \mathbf{x}_1)}{\partial (x_i, x_j)}} \wedge \Delta_{(x_i,x_j)} f(x_i, x_j, \mathbf{x}_1) = 1 \,. \tag{5.78}$$

There are many other properties of the functions realized by a circuit. Examples are the extension of the above analysis of static function hazards to dynamic or to structural hazards. The analysis of properties is also needed in other fields of applications. For instance, it can be necessary to verify:

- in the field of *cryptography* whether a given function is a bent function, which has a maximal distance to all linear functions;
- in the field of *testing* whether a faulty signal in the middle of a circuit is observable at least at one of the outputs;
- in the field of *synthesis* whether a circuit structure for the specified system function exists; or
- in the field of *synthesis of combinational circuits* whether a decomposition of a special type is possible.

We answer some of these questions in the next subsections.

5.5 SYNTHESIS OF DIGITAL CIRCUITS

The synthesis of digital circuits is the reversal of their analysis. A circuit structure is wanted for a given behavioral description. The allowed behavior of a combinational circuit can be specified by a set of binary vectors $(\mathbf{x}, \mathbf{y})$ that contains all output values $\mathbf{y}$ for the specified input values $\mathbf{x}$. This set defines the characteristic function $F(\mathbf{x}, \mathbf{y})$ of the system to be designed, and the solution of the system equation

$$F(\mathbf{x}, \mathbf{y}) = 1 \tag{5.79}$$

is identical with the given set of binary vectors.

As shown in Section 5.4, the system function $F(\mathbf{x}, \mathbf{y})$ of a combinational circuit is uniquely defined. However, due to the freedom in the specification of the allowed input-output combinations, the system functions $F(\mathbf{x}, \mathbf{y})$ can be specified such that not all input-output combinations of an arbitrary combinational circuit belong to the allowed set. The reason for this is that a combinational circuit has for *each* input pattern an associated output pattern. This raises the question: Is the given system functions $F(\mathbf{x}, \mathbf{y})$ *realizable* with an combinational circuit? This question is equivalent to the question whether the associated system Equation (5.79) is solvable with regard to all output variables $\mathbf{y}$. Hence, we can adopt the results of Section 5.4 as follows.

Theorem 5.14 Realizability of Combinational Circuits. *The behavior described by a system Equation* (5.79) *can be realized by a combinational circuit if and only if*

$$\max_{\mathbf{y}}{}^m F(\mathbf{x}, \mathbf{y}) = 1\ . \tag{5.80}$$

If Theorem 5.14 is not satisfied then the solution set of the equation

$$\max_{\mathbf{y}}{}^m F(\mathbf{x}, \mathbf{y}) = 0 \tag{5.81}$$

for the given function $F(\mathbf{x},\mathbf{y})$ contains all input pattens which must be extended by at least one acceptable output pattern and added to the set of the allowed binary vectors $(\mathbf{x},\mathbf{y})$ to establish a sufficient system function $F(\mathbf{x},\mathbf{y})$ for the combinational circuit to be designed.

If Theorem 5.14 is satisfied it can be verified by (5.34) whether the function $f_j(\mathbf{x})$ of the output y_i is uniquely determined and this functions can be calculated in this case by means of (5.35). In the case that the given system function $F(\mathbf{x},\mathbf{y})$ is realizable, but not unique, the function $f_j(\mathbf{x})$ of the output y_i can be chosen out of a lattice of Boolean functions defined by the mark functions $f_{qj}(\mathbf{x})$ (5.36) and $f_{rj}(\mathbf{x})$ (5.37). As explained in Section 5.2, the system function $F(\mathbf{x},\mathbf{y})$ must be restricted for the chosen function $f_j(\mathbf{x})$ of this lattice by (5.38) in order to avoid inadmissible combinations with functions of lattices for other outputs.

The system function $F(\mathbf{x},\mathbf{y})$ depends on all input variables $\mathbf{x}$ and output variables $\mathbf{y}$ needed to describe the behavior of the circuit to synthesize. However, not each output must depend on all inputs. This raises the problem to simplify all functions $f_j(\mathbf{x})$ such that all variables $x_i \in \mathbf{x}$ are removed from the set of variables $\mathbf{x}$ on which the verified function $f_j(\mathbf{x})$ does not depend. Using the split of the variables

$$\mathbf{x} = (x_i, \mathbf{x}_1) , \tag{5.82}$$

the function $f_j(\mathbf{x})$ is independent of x_i if

$$\frac{\partial f_j(x_i, \mathbf{x}_1)}{\partial x_i} = 0 . \tag{5.83}$$

If the condition (5.83) for independence of f_j from x_i is satisfied, the simplified function $f_j(\mathbf{x}_1)$ can be calculated by

$$f_j(\mathbf{x}_1) = \max_{x_i} f_j(x_i, \mathbf{x}_1) . \tag{5.84}$$

This evaluation and conditional simplification can be iteratively applied for all variables $x_i \in \mathbf{x}$.

The possibility that at least one function of a lattice is not depending on a selected input variable x_i increases the more function the lattice contains. It is not necessary to evaluate all functions of a lattice $f_k(\mathbf{x}) \in L\langle f_q(\mathbf{x}), f_r(\mathbf{x})\rangle$ separately; this lattice contains at least one function $f_k(\mathbf{x})$ which is not depending on x_i if

$$f_q^{\partial x_i}(\mathbf{x}_1) = \max_{x_i} f_q(x_i, \mathbf{x}_1) \wedge \max_{x_i} f_r(x_i, \mathbf{x}_1) = 0 , \tag{5.85}$$

that means, the ON-set function of the derivative of this lattice with regard to x_i must be equal to 0, see (3.59). Using the split of variables (5.82), we get in the case that (5.85) is satisfied the mark functions of the simplified lattice $L\langle f_q(\mathbf{x}_1), f_r(\mathbf{x}_1)\rangle$ by

$$f_q(\mathbf{x}_1) = \max_{x_i} f_q(x_i, \mathbf{x}_1) , \tag{5.86}$$

$$f_r(\mathbf{x}_1) = \max_{x_i} f_r(x_i, \mathbf{x}_1) . \tag{5.87}$$

A lattice of functions is more general than a single function. A single function can be expressed by a lattice that consists of this function only. The mapping rules for a given function $f(\mathbf{x})$ into a lattice of functions $L\langle f_q(\mathbf{x}), f_r(\mathbf{x})\rangle$ are:

$$f_q(\mathbf{x}) = f(\mathbf{x}) \,, \tag{5.88}$$
$$f_r(\mathbf{x}) = \overline{f(\mathbf{x})} \,. \tag{5.89}$$

If all independent variables x_i are removed from a lattice by means of (5.85), (5.86), and (5.87), then all functions of the simplified lattice depend only on the remaining variables. We assume that n variables remain and denote these variables for simplicity again with $\mathbf{x}$. Even though each function of such a lattice depends on all variables $\mathbf{x}$, the lattice can contain functions that are independent of the simultaneous change of the subset $\mathbf{x}_0$ of these variables. This is the case when the ON-set function of the vectorial derivative of this lattice with regard to $\mathbf{x}_0$ is equal to 0:

$$f_q^{\partial \mathbf{x}_0}(\mathbf{x}) = \max_{\mathbf{x}_0} f_q(\mathbf{x}) \wedge \max_{\mathbf{x}_0} f_r(\mathbf{x}) = 0 \,. \tag{5.90}$$

If the condition (5.90) is satisfied, the lattice can be restricted by an extended independence function that is characterized by the unique independence matrix in which the detected independent directions of change $\mathbf{x}_0$ have to be merged using Algorithm 3.8. This simplification can be applied for all $2^n - n - 1$ directions in which more than one variable changes simultaneously. If a new independent direction of change is found the rank of the independence matrix increases, and this means that the lattice is restricted to simpler functions for the remaining synthesis without any circuit effort.

The synthesis of sequential circuits requires slight extensions of the method presented above for the combinational circuits. The allowed behavior of a sequential circuit can be specified by a set of binary vectors $(\mathbf{x}, \mathbf{s}, \mathbf{sf}, \mathbf{y})$ that contains all following state values $\mathbf{sf}$ and output values $\mathbf{y}$ for the specified input values $\mathbf{x}$ and the values of the state $\mathbf{s}$. This set defines the characteristic function $F(\mathbf{x}, \mathbf{s}, \mathbf{sf}, \mathbf{y})$ of the system to design and the solution of the system equation

$$F(\mathbf{x}, \mathbf{s}, \mathbf{sf}, \mathbf{y}) = 1 \tag{5.91}$$

is identical with the given set of binary vectors. The freedom in the specification of the allowed input-state-state-output function combinations can lead to system functions $F(\mathbf{x}, \mathbf{s}, \mathbf{sf}, \mathbf{y})$ for which no sequential circuits exist. The reason for this is again that a sequential circuit generates for each combination of the input and state values one combination of the state functions and outputs. Hence, we have to answer the question: Is the given system functions $F(\mathbf{x}, \mathbf{s}, \mathbf{sf}, \mathbf{y})$ *realizable* with a sequential circuit? This question is equivalent to the question whether the associated system Equation (5.91) is solvable with regard to all variables **sf** and **y**. Hence, we can adopt the results of Section 5.4 as follows:

Theorem 5.15 Realizability of Sequential Circuits. *The behavior described by a system Equation (5.91) can be realized by a sequential circuit if and only if*

$$\max_{(\mathbf{sf},\mathbf{y})}{}^{m} F(\mathbf{x},\mathbf{s},\mathbf{sf},\mathbf{y}) = 1 \,. \tag{5.92}$$

Both the functions for the next state $sf_j = h_j(\mathbf{x},\mathbf{s})$ and the functions for the output $y_j = f_j(\mathbf{x},\mathbf{s})$ depend on the input variables **x** and the state variables **s**. Besides of the special meaning of the variables the synthesis of asynchronous sequential circuits can reuse the synthesis approach introduced for combinational circuits. The use of flip-flops in synchronous sequential circuits faces us with the problem to find the control functions of the inputs of the flip-flops.

This problem can easily be solved for the D-flip-flop: due to the according flip-flop equation of the first row in Table 5.5 we can substitute the next state function $sf_j = h_j(\mathbf{x},\mathbf{s})$ for qf and get:

$$d_j = h_j(\mathbf{x},\mathbf{s}) \,. \tag{5.93}$$

Because of the linearity of the flip-flop equation of the T-FF (see second row of Table 5.5) we can separate t_j and get:

$$t_j = s_j \oplus h_j(\mathbf{x},\mathbf{s}) \,. \tag{5.94}$$

In the other cases, the functions for the next state $sf_j = h_j(\mathbf{x},\mathbf{s})$ must be substituted into the adopted flip-flop equation which has to be solved with regard to the variables of the flip-flop inputs. Hence, we have to solve a sequence of two equations with regard to certain variables. This two-step approach can be reduced to a single step when the dependence of flip-flop inputs is mapped into the system function.

Assuming that $F_j^{FF}(s_j, sf_j, i_j^1, i_j^2)$ is the characteristic function generated from the adopted solution set of the used flip-flop of Table 5.5, the mapping of the behavior from sf_j to the flip-flop inputs i_j^1, i_j^2 is achieved by:

$$F(\mathbf{x},\mathbf{s},i_j^1,i_j^2) = \max_{(\mathbf{sf},\mathbf{y})}{}^{m} \left(F(\mathbf{x},\mathbf{s},\mathbf{sf},\mathbf{y}) \wedge F_j^{FF}(s_j, sf_j, i_j^1, i_j^2)\right) \,. \tag{5.95}$$

Now, the adopted system equation for the flip-flop j

$$F(\mathbf{x},\mathbf{s},i_j^1,i_j^2) = 1 \tag{5.96}$$

can be solved with regard to the variables of the flip-flop inputs i_j^1 and i_j^2, and in the consecutive synthesis the functions $i_j^1 = i_j^1(\mathbf{x},\mathbf{s})$ and $i_j^2 = i_j^2(\mathbf{x},\mathbf{s})$ are selected and immediately used to restrict the conflict-free remaining behavior:

$$F(\mathbf{x},\mathbf{s},i_j^1,i_j^2) = \left(i_j^1 \odot i_j^1(\mathbf{x},\mathbf{s})\right) \wedge F(\mathbf{x},\mathbf{s},i_j^1,i_j^2) \,, \tag{5.97}$$

$$F(\mathbf{x},\mathbf{s},i_j^1,i_j^2) = \left(i_j^2 \odot i_j^2(\mathbf{x},\mathbf{s})\right) \wedge F(\mathbf{x},\mathbf{s},i_j^1,i_j^2) \,. \tag{5.98}$$

The synthesis of the flip-flop input of the other flip-flops can be done in the same manner using the restricted system function:

$$F(\mathbf{x}, \mathbf{s}, \mathbf{sf}, \mathbf{y}) = \max_{(i_j^1, i_j^2)}{}^2 \left(F(\mathbf{x}, \mathbf{s}, i_j^1, i_j^2) \wedge F(\mathbf{x}, \mathbf{s}, \mathbf{sf}, \mathbf{y}) \right) . \tag{5.99}$$

Algorithm 5.9 shows a general procedure for the synthesis of a synchronous sequential circuit of a finite state machine determined by the system function $F(\mathbf{x}, \mathbf{s}, \mathbf{sf}, \mathbf{y})$ and the characteristic functions $F_j^{FF}(s_j, sf_j, \mathbf{i}_j)$ of the used flip-flops. The results are

- either all control functions of the flip-flops as well as all output functions or
- the message that the system function is not realizable.

The benefits of this universal approach are:

- both deterministic and non-deterministic system functions can be used;
- the realizability is verified based on the given system function;
- arbitrary flip-flops can be used;
- adopted system functions can directly be solved with regard to the input variables of the flip-flop as well as the output variables;
- the functions to realize in the circuit can be chosen out of the calculated lattices of Boolean functions without any restrictions; and
- the system function of the realized finite state machine is known when the synthesis has been finished.

Example 5.16 Synthesis of a Synchronous Sequential Circuit.
The non-deterministic finite state machine of Figure 5.10a combines the behavior of a modulo 2 counter for $x = 0$ and a modulo 3 counter for $x = 1$. The state $s = (00)$ is not needed for these behaviors. All edges from this state reach one of the needed states. The selection of one of these edges is utilized to optimize the searched circuit structure for which two DE-flip-flops must be used. Figure 5.10b shows the TVL in ODA-form of the associated system function $F(x, s_1, s_2, sf_1, sf_2, y)$.

The test for realizability in line 1 of Algorithm 5.9 is satisfied for the system function of Figure 5.10b. Using the adopted TVL of DE-FF of the third row of Table 5.5 we get in the first sweep of the while-loop in lines 6–12:

$$\begin{aligned} i_{q1}^1(x, s_1, s_2) &= d_{q1}(x, s_1, s_2) = \overline{s}_1 s_2 , \\ i_{r1}^1(x, s_1, s_2) &= d_{r1}(x, s_1, s_2) = s_1 s_2 . \end{aligned}$$

Algorithm 5.9 SFSM($F(\mathbf{x}, \mathbf{s}, \mathbf{sf}, \mathbf{y})$, $F_j^{FF}(s_j, sf_j, \mathbf{i}_j)$): Synthesis of a Finite State Machine

Input : $F(\mathbf{x}, \mathbf{s}, \mathbf{sf}, \mathbf{y})$: system function (behavior)
Input : $F_j^{FF}(s_j, sf_j, \mathbf{i}_j)$: characteristic functions of the used flip-flops
Output : $i_j^k(\mathbf{x}, \mathbf{s})$: control function of the input k of flip-flop specified by $F_j^{FF}(\mathbf{i}_j, \mathbf{x}, \mathbf{s})$
Output : $f_j(\mathbf{x}, \mathbf{s})$: function of the output y_j

if $\max_{(\mathbf{sf},\mathbf{y})}{}^m F(\mathbf{x}, \mathbf{s}, \mathbf{sf}, \mathbf{y}) = 1$ (5.92) **then**
  $j \leftarrow 1$
  while $j \le |\mathbf{sf}|$ **do** {for all flip-flops}
    $F(\mathbf{x}, \mathbf{s}, \mathbf{i}_j) \leftarrow \max_{(\mathbf{sf},\mathbf{y})}{}^m \left(F(\mathbf{x}, \mathbf{s}, \mathbf{sf}, \mathbf{y}) \wedge F_j^{FF}(s_j, sf_j, \mathbf{i}_j)\right)$ (5.95)
    $k \leftarrow 1$
    while $k \le |\mathbf{i}_j|$ **do** {for all inputs of the flip-flop j}
      $i_{qj}^k(\mathbf{x}, \mathbf{s}) \leftarrow \overline{\max_{(\mathbf{i}_j)}{}^m \left(\overline{i}_j^k \wedge F(\mathbf{x}, \mathbf{s}, \mathbf{i}_j)\right)}$ (5.36)
      $i_{rj}^k(\mathbf{x}, \mathbf{s}) \leftarrow \max_{(\mathbf{i}_j)}{}^m \left(i_j^k \wedge F(\mathbf{x}, \mathbf{s}, \mathbf{i}_j)\right)$ (5.37)
      synthesis of $i_j^k(\mathbf{x}, \mathbf{s})$ selected from $L\langle i_{qj}^k(\mathbf{x}, \mathbf{s}), i_{rj}^k(\mathbf{x}, \mathbf{s})\rangle$
      $F(\mathbf{x}, \mathbf{s}, \mathbf{i}_j) \leftarrow (i_j^k \odot i_j^k(\mathbf{x}, \mathbf{s})) \wedge F(\mathbf{x}, \mathbf{s}, \mathbf{i}_j)$ (5.97)
      $k \leftarrow k + 1$
    end while
    $F(\mathbf{x}, \mathbf{s}, \mathbf{sf}, \mathbf{y}) \leftarrow \max_{(\mathbf{i}_j)}{}^m \left(F(\mathbf{x}, \mathbf{s}, \mathbf{i}_j) \wedge F(\mathbf{x}, \mathbf{s}, \mathbf{sf}, \mathbf{y})\right)$ (5.99)
    $j \leftarrow j + 1$
  end while
  $F(\mathbf{x}, \mathbf{s}, \mathbf{y}) \leftarrow \max_{(\mathbf{sf})}{}^m (F(\mathbf{x}, \mathbf{s}, \mathbf{sf}, \mathbf{y}))$
  $j \leftarrow 1$
  while $j \le |\mathbf{y}|$ **do** {for all outputs y_j}
    $f_{qj}(\mathbf{x}, \mathbf{s}) \leftarrow \overline{\max_{(\mathbf{y})}{}^m \left(\overline{y}_j \wedge F(\mathbf{x}, \mathbf{s}, \mathbf{y})\right)}$ (5.36)
    $f_{rj}(\mathbf{x}, \mathbf{s}) \leftarrow \max_{(\mathbf{y})}{}^m \left(y_j \wedge F(\mathbf{x}, \mathbf{s}, \mathbf{y})\right)$ (5.37)
    synthesis of $f_j(\mathbf{x}, \mathbf{s})$ selected from $L\langle f_{qj}(\mathbf{x}, \mathbf{s}), f_{rj}(\mathbf{x}, \mathbf{s})\rangle$
    $F(\mathbf{x}, \mathbf{s}, \mathbf{y}) \leftarrow (y_j \odot f_j(\mathbf{x}, \mathbf{s})) \wedge F(\mathbf{x}, \mathbf{s}, \mathbf{y})$
    $j \leftarrow j + 1$
  end while
  realized finite state machine: $F(\mathbf{x}, \mathbf{s}, \mathbf{sf}, \mathbf{y}) \leftarrow F(\mathbf{x}, \mathbf{s}, \mathbf{sf}, \mathbf{y}) \wedge F(\mathbf{x}, \mathbf{s}, \mathbf{y})$
else
  the system function is not realizable
end if

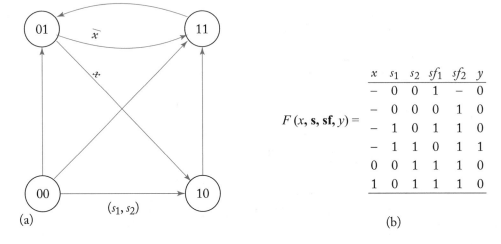

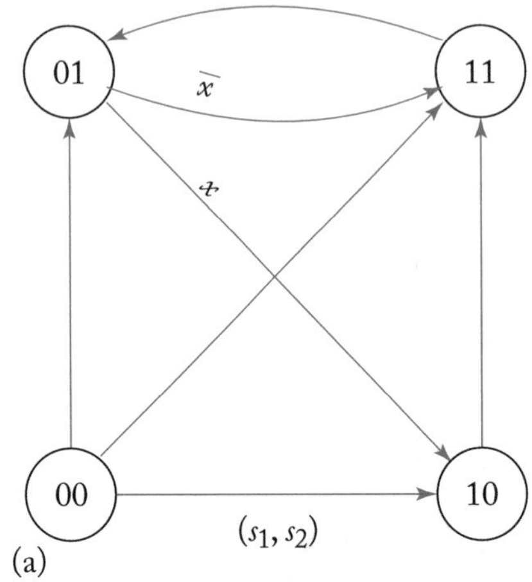

$F(x, \mathbf{s}, \mathbf{sf}, y) =$

x	s_1	s_2	sf_1	sf_2	y
–	0	0	1	–	0
–	0	0	0	1	0
–	1	0	1	1	0
–	1	1	0	1	1
0	0	1	1	1	0
1	0	1	1	1	0

(b)

Figure 5.10: Non-deterministic finite state machine: (a) graph and (b) system function.

(a) $d_1(x, \mathbf{s})$

s_1 s_2	$x=0$	$x=1$
0 0	Φ	Φ
0 1	1	1
1 1	0	0
1 0	Φ	Φ

$d_1 = \overline{s}_1$

(b) $e_1(x, \mathbf{s})$

s_1 s_2	$x=0$	$x=1$
0 0	Φ	Φ
0 1	1	1
1 1	1	1
1 0	0	0

$e_1 = s_2$

(c) $d_2(x, \mathbf{s})$

s_1 s_2	$x=0$	$x=1$
0 0	1	1
0 1	Φ	Φ
1 1	Φ	0
1 0	1	1

$d_2 = \overline{s}_2$

(d) $e_2(x, \mathbf{s})$

s_1 s_2	$x=0$	$x=1$
0 0	1	1
0 1	0	1
1 1	0	0
1 0	1	1

$e_2 = \overline{s}_2 \vee x\,\overline{s}_1$

(e) $y(x, \mathbf{s})$

s_1 s_2	$x=0$	$x=1$
0 0	0	0
0 1	0	0
1 1	1	1
1 0	0	0

$y = s_1\,s_2$

Figure 5.11: Lattices and the chosen functions of the flip-flop inputs and the output: (a) d_1; (b) e_1; (c) d_2; (d) e_2; and (e) y.

Figure 5.11a shows the Karnaugh-map of this lattice and the chosen function which is used in line 10 of Algorithm 5.9 to restrict the system function of the first DE-FF. In the second sweep of the while-loop in lines 6–12 we get:

$$i_{q1}^2(x, s_1, s_2) = e_{q1}(x, s_1, s_2) = s_2 ,$$
$$i_{r1}^2(x, s_1, s_2) = e_{r1}(x, s_1, s_2) = s_1\overline{s}_2 .$$

Figure 5.11b shows the Karnaugh-map of this lattice and the chosen function which is used in line 10 of Algorithm 5.9 to restrict the system function of the first DE-FF furthermore.

The control functions of the second DE-FF are calculated in the same way in the second sweep of the while-loop in lines 3–15. The found mark functions of the lattices are:

$$\begin{aligned}
i^1_{q2}(x, s_1, s_2) &= d_{q2}(x, s_1, s_2) = \overline{s}_2 \;, \\
i^1_{r2}(x, s_1, s_2) &= d_{r2}(x, s_1, s_2) = x\overline{s}_1 s_2 \;, \\
i^2_{q2}(x, s_1, s_2) &= e_{q2}(x, s_1, s_2) = \overline{s}_2 \vee x\overline{s}_1 \;, \\
i^2_{r2}(x, s_1, s_2) &= e_{r2}(x, s_1, s_2) = s_2(\overline{x} \vee s_1) \;,
\end{aligned}$$

where the chosen function of Figure 5.11c is used to determine the lattice of the control function $e_2(x, s_1, s_2)$ which is completely specified as can be seen in Figure 5.11d.

After the restriction of the system function in lines 13 and 16 of Algorithm 5.9 the output function is determined in the single sweep of the while-loop in lines 18–24. Due to the mark functions

$$f_{q1}(x, s_1, s_2) = s_1 s_2 \;, \tag{5.100}$$

$$f_{r1}(x, s_1, s_2) = \overline{s}_1 \vee \overline{s}_2 \;, \tag{5.101}$$

the output function $y = f_1(x, s_1, s_2) = s_1 s_2$ in this example is uniquely defined.

The system function of the realized circuit is finally calculated in line 25 and shown in Figure 5.12a. The associated graph in Figure 5.12b shows that out of the three permissible transitions beginning in node (00), the transition (00) $\rightarrow$ (01) is chosen in Algorithm 5.9 to synthesize the simple circuit structure of Figure 5.12c.

5.6 TEST OF DIGITAL CIRCUITS

The necessity to test digital circuits was an important practical problem that lead to initial studies of derivatives and later on to the comprehensive exploration of the Boolean Differential Calculus. A fault in a digital circuit can change a fault-free value into a faulty one. The stuck-at fault model is simple and frequently used. It maps physical errors to two types of stuck-at faults. We introduce the logic variable T into the model to express which type of stuck-at fault is considered. Table 5.8 summarizes the relationship between the fault in the circuit, the type of the fault in the stuck-at fault model, and the model variable T.

Table 5.8: Stuck-at fault model

Faulty Behavior in the Circuit	Stuck-at Fault Model	T
Constant value 0	Stuck-at-0 fault (*s-a-*0)	0
Constant value 1	Stuck-at-1 fault (*s-a-*1)	1

To test a combinational circuit, test patterns are necessary. The test by means of all 2^n input patterns of a circuit with n inputs is mostly not possible due to the exponential number

$F(x, \mathbf{s}, \mathbf{sf}, y) =$

x	s_1	s_2	sf_1	sf_2	y
–	0	0	0	1	0
–	1	0	1	1	0
–	1	1	0	1	1
0	0	1	1	1	0
1	0	1	1	1	0

(a)

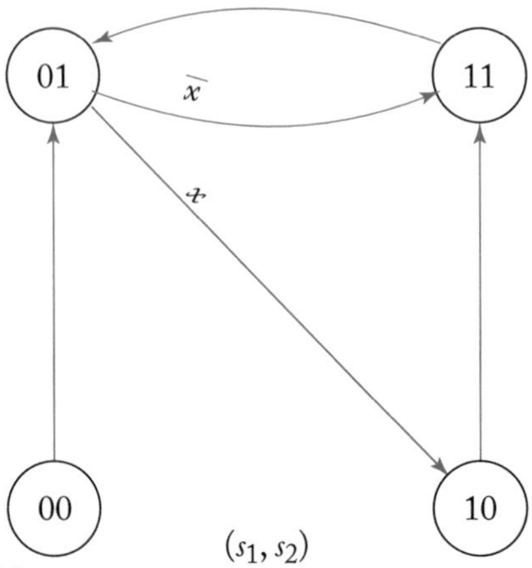

(b)

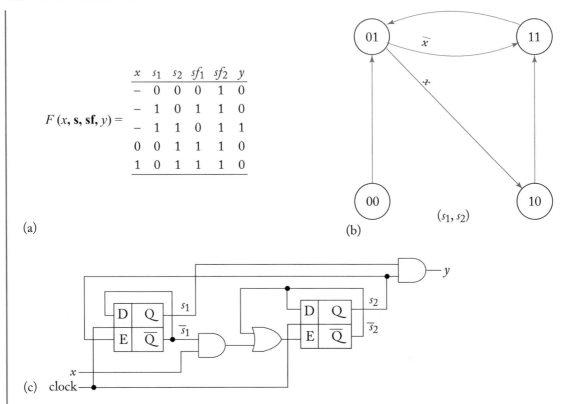

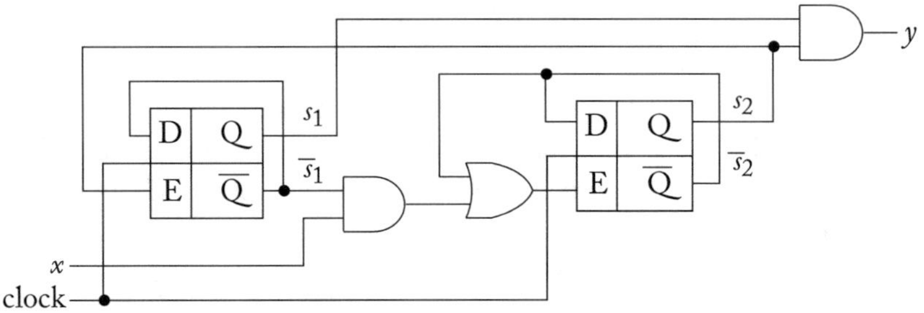

(c)

Figure 5.12: Synthesized synchronous sequential circuit: (a) deterministic system function $F(x, \mathbf{s}, \mathbf{sf}, y)$; (b) associated realized graph; and (c) circuit structure of the finite state machine.

of patterns. Hence, a small number of test patterns is needed that detect all possible faults on all pins of the gates of the circuit. Typically, access is restricted to the inputs and outputs of the circuit. The following two approaches can be used to find test patterns that check an internal wire.

• **Sensible Path**

The idea of the generation of test patterns based on the model *Sensible Path* is that the change of the value of one input variable causes changes on all pins of the gates located along a path from this input to one output of the circuit; if the change of the output value is observed it can be concluded that no stuck-at fault exists on any gate-pin located on this path. The condition that a physical path through the circuit is a *sensible* one is that the signals along the path are not determined by such inputs of the gates where the gate but not the input belongs to the path.

A NOT-gate cannot inhibit the sensibility of a path because only its input value and no other control value influences its output. Hence, only AND-, OR-, and EXOR-gates along the selected sensible path must be taken into consideration for the calculation of all test patterns of this path. Figure 5.13 shows the model of a *Sensible Path* through a combinational circuit. The chosen path from the input x_i to the output $f(\mathbf{x})$ is indicated by bold lines and the control functions $g_j(\mathbf{x})$ are distinguished regarding the gate they control.

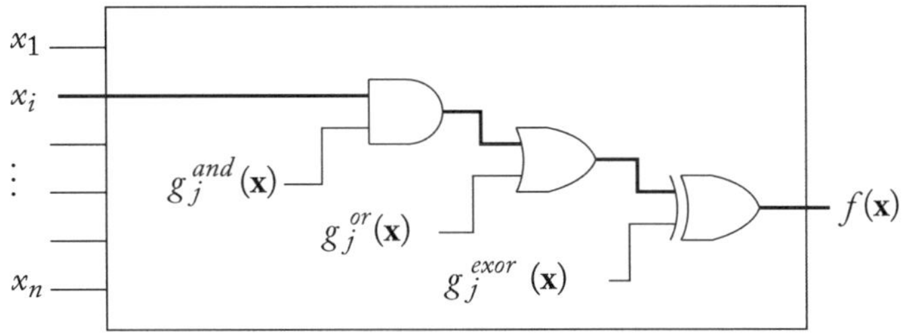

Figure 5.13: Sensible path.

The change of the output $f(\mathbf{x})$ is caused by the change of the input x_i if:

- all functions $g_j^{and}(\mathbf{x})$ are equal to 1 and do not change their values if x_i changes,
- all functions $g_j^{or}(\mathbf{x})$ are equal to 0 and do not change their values if x_i changes,
- all functions $g_j^{exor}(\mathbf{x})$ do not change their values if x_i changes, and
- the function $f(\mathbf{x})$ changes its value if x_i changes:

$$\begin{aligned}
\bigwedge_j \left(g_j^{and}(\mathbf{x}) \wedge \overline{\frac{\partial g_j^{and}(\mathbf{x})}{\partial x_i}} \right) \wedge \\
\bigwedge_j \left(\overline{g_j^{or}(\mathbf{x})} \wedge \overline{\frac{\partial g_j^{or}(\mathbf{x})}{\partial x_i}} \right) \wedge \\
\bigwedge_j \left(\overline{\frac{\partial g_j^{exor}(\mathbf{x})}{\partial x_i}} \right) \wedge \\
\frac{\partial f(\mathbf{x})}{\partial x_i} = 1 \,.
\end{aligned} \tag{5.102}$$

The values of the functions $g_j^{exor}(\mathbf{x})$ have no influence to the sensibility of the selected path because for each unchanged value of $g_j^{exor}(\mathbf{x})$ the change of the input of an EXOR-gate located on the path causes the change of the output of this gate in the fault-free case.

The benefit of the model *Sensible Path* is that one pair of patterns $(x_i = 0, \mathbf{x_0} = \mathbf{c_0})$ and $(x_i = 1, \mathbf{x_0} = \mathbf{c_0})$ is sufficient to detect all stuck-at faults on all gate-pins which are located on the path. The drawback of this model is that the solution of (5.102) can be empty even though for each pin on the path both at least one stuck-at-0 and stuck-at-1 test pattern exist. The reason for that is that the pair of patterns is specified by unchanged values $\mathbf{x}_0 = \mathbf{c}_0$ and changes only the value of x_i. This drawback does not occur in the models for test patterns generation explored next.

• **Sensible Point—Inner Connection**

Using this model *all* test patterns for an arbitrary sensible point located on an inner connection can be calculated. The variable s is used to determine the selected sensible point. Figure 5.14 shows the circuit structure of this model. The sensible point is indicated by the thick connection line between the two sub-circuits. The variable s identifies the sensible point.

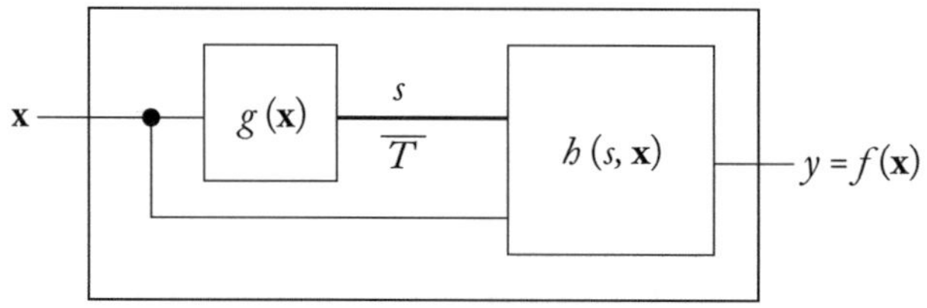

Figure 5.14: Sensible point—inner connection.

Preparing this model, the variable s is substituted for the input pin of the gate belonging to the sub-circuit $h(s, \mathbf{x})$ for which all test patterns must be calculated. Due to the direct connection between the output of the controlling gate of this point and the associated input pin of the controlled gate the found test patterns detect all possible stuck-at-T faults of both the output of the sub-circuit $g(\mathbf{x})$ and the input s of the sub-circuit $h(s, \mathbf{x})$.

Three conditions must be satisfied for a test pattern to detect a *stuck-at-T* fault.

1. It is necessary to control the circuit such that the function value of $g(\mathbf{x})$ is the complement of the stuck-at-T fault to be checked at the wire s. Only in this case a *stuck-at-T* fault can cause a change of the circuit output. The solutions of (5.103) enable this effect of a *stuck-at-T* fault. Therefore, we call Equation (5.103) *fault controllability condition*:

$$g(\mathbf{x}) \oplus T = 1 \,. \tag{5.103}$$

2. Due to our assumption that only the output of the circuit is directly observable, the effect of an internal fault at the wire s must influence the function value at the circuit output. Hence, the second necessary condition for test patterns is the *fault observability*. The value at the wire s is observable by the value at the circuit output for all solution vectors $\mathbf{x}$ of (5.104):

$$\frac{\partial h(s, \mathbf{x})}{\partial s} = 1 \,. \tag{5.104}$$

3. It is necessary to compare the observed output value y with the expected function value of $f(\mathbf{x})$ to evaluate whether there exists a *stuck-at-T* fault at the wire s or not. The expected

value y is related to the input pattern $\mathbf{x}$ in the solutions of (5.105). These solutions allows the *fault evaluation*:

$$f(\mathbf{x}) \odot y = 1 \; . \tag{5.105}$$

The three necessary conditions (5.103), (5.104), and (5.105) are collectively sufficient and can be combined to build the necessary and sufficient condition (5.106). The solutions of Equation (5.106) comprise the set of all test patterns $(\mathbf{x}, y, T)$ of the sensible point s. Each of these patterns detects a *stuck-at-T* fault if the circuit does not show the value y controlled by the input $\mathbf{x}$:

$$(g(\mathbf{x}) \oplus T) \wedge \frac{\partial h(s, \mathbf{x})}{\partial s} \wedge (f(\mathbf{x}) \odot y) = 1 \; . \tag{5.106}$$

• Sensible Point—Inner Branch

The model of the *Sensible Point* explored above is restricted to inner connections and a single output of the combinational circuit. Here we generalize this model to combinational circuits with

- an arbitrary number of outputs and
- an *inner branch* at the sensible point.

Inner branches complicate the calculation of test patterns because faulty values caused by a single stuck-at fault can be combined to a correct value at a gate where they control different inputs of the same gate. Figure 5.15 shows the circuit structure used for this generalized test model.

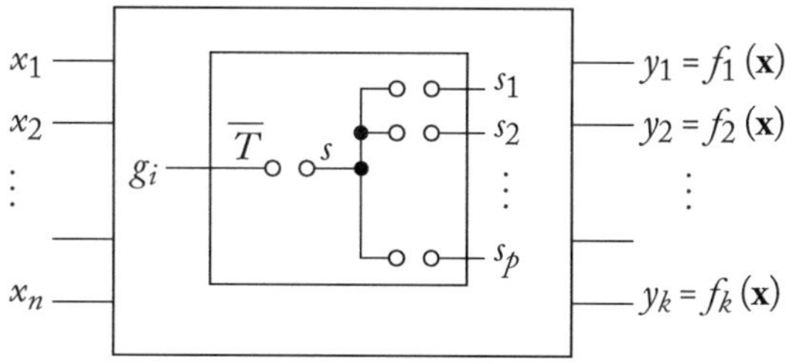

Figure 5.15: Sensible point—inner branch.

All test patterns for the output of the controlling gate g_i as well as for each sink s_j of the selected inner branch as sensible point can be calculated if the connections to this branch are separated from each other and labeled as shown in Figure 5.15. The calculation uses given system functions $F_i^G(\mathbf{x}, \mathbf{y}, \mathbf{g}, \mathbf{s})$ where variables $\mathbf{x}$ indicate the controllable inputs of the circuit, $\mathbf{y}$ the observable outputs of the circuit, $\mathbf{g}$ the inner not directly observable outputs of the gates, and $\mathbf{s}$ the sinks of the branch as shown in Figure 5.15.

The *type* of the fault (stuck-at-0 or stuck-at-1) that can be detected by the calculated test patterns is determined by:

$$F^T(g_i, T) = g_i \oplus T \; . \tag{5.107}$$

The system function $F^C(\mathbf{x}, \mathbf{y}, \mathbf{s}, T)$ describes the *complete* behavior (both the faulty and the correct one) of the circuit with regard to the modeled sensible point (inner branch):

$$F^C(\mathbf{x}, \mathbf{y}, \mathbf{s}, T) = \max_{\mathbf{g}}{}^m \left(F^T(g_i, T) \wedge \bigwedge_{i=1}^{l} F_i^G(\mathbf{x}, \mathbf{y}, \mathbf{g}, \mathbf{s}) \right) . \tag{5.108}$$

The correct behavior on the explored *branch* is determined by:

$$F^B(\mathbf{s}, T) = T \wedge \bigwedge_{j=1}^{p} \overline{s}_j \vee \overline{T} \wedge \bigwedge_{j=1}^{p} s_j . \tag{5.109}$$

Using two previously defined functions, the system function of the *required* fault-free behavior can be specified:

$$F^R(\mathbf{x}, \mathbf{y}) = \max_{(\mathbf{s},T)}{}^m \left(F^C(\mathbf{x}, \mathbf{y}, \mathbf{s}, T) \wedge F^B(\mathbf{s}, T) \right) . \tag{5.110}$$

It is possible to distinguish between faults at the output controlling gate and each input of the controlled gate of an inner branch. We introduce the variable *place* for the specification of the place for which the test patterns are calculated:

$$place = \begin{cases} T & \text{the place of the fault to evaluate is the source of the inner branch} \\ s_j & \text{the place of the fault to evaluate is the sink } s_j \text{ of the inner branch} \end{cases} . \tag{5.111}$$

The following formula shows the strength of the Boolean Differential Calculus to express complicated practical problems in a compact and well understandable manner. Using the above definitions all test patterns for the selected inner branch are the result of the following equation:

$$\frac{\partial^p \left(F^C(\mathbf{x}, \mathbf{y}, \mathbf{s}, T) \wedge \max_{place} F^B(\mathbf{s}, T) \right)}{\partial s_1 \ldots \partial s_p} \wedge F^R(\mathbf{x}, \mathbf{y}) = 1 . \tag{5.112}$$

The single maximum in (5.112) enables the fault impact of exactly one place of the branch. Whether this fault changes at least one of the outputs of the circuit is detected by the p-fold derivative. The function $F^R(\mathbf{x}, \mathbf{y})$ in (5.112) contributes the correct output values for the found test patterns.

It should be mentioned that the bi-decomposition that will be explored in the next section allows us to compute circuits that are completely testable with regard to *stuck-at-T* faults at all inputs, outputs and gate-connections Posthoff and Steinbach [2004]. These multi-level circuits have a very small depth Mishchenko et al. [2001]. Furthermore, these test patterns can be generated in parallel to the synthesis of the circuit Steinbach and Stöckert [1994], so that only a small overhead of about 10% solves the complex problem of test pattern generation.

5.7 SYNTHESIS BY BI-DECOMPOSITIONS

Basically, there are two approaches to synthesize the structure of a combinational circuit. These are either covering methods or decomposition methods.

Covering methods use preferably prime conjunctions to cover all function values 1. This leads theoretically to a two-level circuit. However, both the AND-gates of the first level and the OR-gate of the second level must be realized by trees of these gates in the case of typical functions of many inputs due to the restricted number of inputs of the gates. This split of single gates into sub-trees leads to the loss of a minimal circuit structure.

The bi-decomposition belongs to the second group of synthesis methods. The bi-decomposition utilizes the behavior of logic gates (OR, AND, or EXOR) with two inputs that the function on the output of the gate combines the functions of the two associated inputs. If the functions on both inputs are simpler than the given function an iterative application of such decompositions leads directly to a minimal multilevel circuit.

The strong bi-decomposition uses as condition for the simplification that the decomposition functions on the inputs of the gate depend on less variables than the function on the output. The strong bi-decomposition reaches this requirement by dividing the set of all variables $\mathbf{x}$ into three disjoint subsets $\mathbf{x}_a$, $\mathbf{x}_b$, and $\mathbf{x}_c$ where $\mathbf{x}_a$ is only used in the decomposition function $g(\mathbf{x}_a, \mathbf{x}_c)$, $\mathbf{x}_b$ is only used in the decomposition function $h(\mathbf{x}_b, \mathbf{x}_c)$, and the remaining variables $\mathbf{x}_c$ are commonly used by both decomposition functions. The larger the number of variables in the sets $\mathbf{x}_a$ and $\mathbf{x}_b$ the simpler are the decomposition functions $g(\mathbf{x}_a, \mathbf{x}_c)$ and $h(\mathbf{x}_b, \mathbf{x}_c)$. We use the term *disjoint* bi-decomposition for the welcome case that $\mathbf{x}_c = \emptyset$ because all variables are assigned either to the set $\mathbf{x}_a$ or $\mathbf{x}_b$.

Definition 5.17 Strong Bi-Decompositions.
For three disjoint subsets of variables $\mathbf{x}_a$, $\mathbf{x}_b$, and $\mathbf{x}_c$, a strong OR-bi-decomposition

$$f(\mathbf{x}_a, \mathbf{x}_b, \mathbf{x}_c) = g(\mathbf{x}_a, \mathbf{x}_c) \vee h(\mathbf{x}_b, \mathbf{x}_c) , \tag{5.113}$$

a strong AND-bi-decomposition

$$f(\mathbf{x}_a, \mathbf{x}_b, \mathbf{x}_c) = g(\mathbf{x}_a, \mathbf{x}_c) \wedge h(\mathbf{x}_b, \mathbf{x}_c) , \tag{5.114}$$

or a strong EXOR-bi-decomposition

$$f(\mathbf{x}_a, \mathbf{x}_b, \mathbf{x}_c) = g(\mathbf{x}_a, \mathbf{x}_c) \oplus h(\mathbf{x}_b, \mathbf{x}_c) \tag{5.115}$$

decomposes a Boolean function $f(\mathbf{x}_a, \mathbf{x}_b, \mathbf{x}_c)$ into the decomposition functions $g(\mathbf{x}_a, \mathbf{x}_c)$ and $h(\mathbf{x}_b, \mathbf{x}_c)$.

Figure 5.16 shows the circuit structures of these three types of strong bi-decompositions.

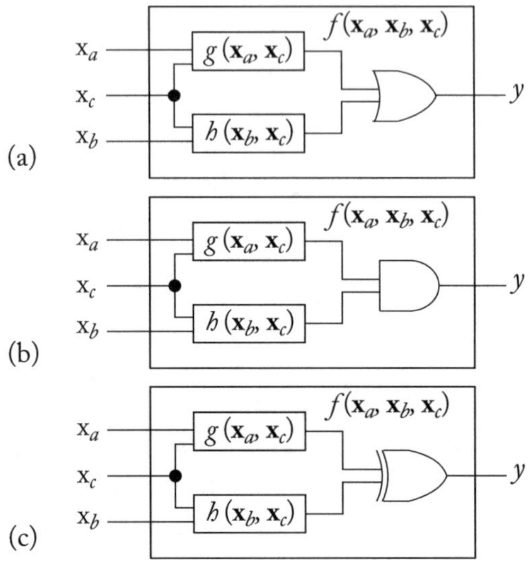

Figure 5.16: Strong bi-decompositions of the function $y = f(\mathbf{x}_a, \mathbf{x}_b, \mathbf{x}_c)$ into the decomposition functions $g(\mathbf{x}_a, \mathbf{x}_c)$ and $h(\mathbf{x}_b, \mathbf{x}_c)$ using: (a) an OR-gate; (b) an AND-gate; and (c) an EXOR gate.

x_{b1}	x_{b2}	$h(\mathbf{x}_b)$
0	0	0
0	1	1
1	1	0
1	0	1

x_{b1}	x_{b2}	$f(\mathbf{x}_a, \mathbf{x}_b)$				
0	0	0	1	0	0	
0	1	1	1	1	1	
1	1	0	1	0	0	
1	0	1	1	1	1	
		0	1	1	0	x_{a2}
		0	0	1	1	x_{a1}

$g(\mathbf{x}_a)$				
0	1	0	0	
0	1	1	0	x_{a2}
0	0	1	1	x_{a1}

Figure 5.17: Disjoint strong OR-bi-decomposition of $f(\mathbf{x}_a, \mathbf{x}_b)$ into $g(\mathbf{x}_a)$ and $h(\mathbf{x}_b)$.

The existence of a strong bi-decomposition of a function $f(\mathbf{x})$ with regard to dedicated sets of variables $\mathbf{x}_a \subseteq \mathbf{x}$ and $\mathbf{x}_b \subseteq \mathbf{x}$, for each of the three gates OR, AND, or EXOR is a property of the given function $f(\mathbf{x})$. We explore these properties by means of simple examples.

Example 5.18 Strong OR-Bi-Decomposition.
Figure 5.17 shows the function $f(\mathbf{x}_a, \mathbf{x}_b)$ and the associated decomposition functions $g(\mathbf{x}_a)$ and $h(\mathbf{x}_b)$ of a strong OR-bi-decomposition. The set $\mathbf{x}_c$ is empty; due to the missing common variables such bi-decompositions are called *disjoint*.

The allocation of the variables $\mathbf{x}_a$ below the Karnaugh-map of $f(\mathbf{x}_a, \mathbf{x}_b)$ in Figure 5.17 and the variables $\mathbf{x}_b$ to the left help to see that the disjunction of the shown decomposition functions $g(\mathbf{x}_a)$ and $h(\mathbf{x}_b)$ results in the given function $f(\mathbf{x}_a, \mathbf{x}_b)$. The value 1 of $g(\mathbf{x}_a)$ appears in each row of the Karnaugh-map of $f(\mathbf{x}_a, \mathbf{x}_b)$ because $1 \vee x = 1$. For the same reason the value 1 of $h(\mathbf{x}_b)$ appears in each column of the Karnaugh-map of $f(\mathbf{x}_a, \mathbf{x}_b)$.

A strong OR-bi-decomposition with regard to the dedicated sets of variables $\mathbf{x}_a$ and $\mathbf{x}_b$ does not exists for all functions $f(\mathbf{x}_a, \mathbf{x}_b)$. The reason for that is the conclusion of the following observations. If $f(\mathbf{x}_a = \mathbf{c}_a, \mathbf{x}_b = \mathbf{c}_b) = 0$ then $g(\mathbf{x}_a = \mathbf{c}_a) = 0$ and $h(\mathbf{x}_b = \mathbf{c}_b) = 0$ because only $0 \vee 0 = 0$. Assuming we change one value of the function $f(\mathbf{x}_a, \mathbf{x}_b)$ in Figure 5.17 from 0–1: $f(0,0,0,0) = 1$. For this changed function we have:

- $g(0,0) = 0$ due to $f(0,0,1,1) = 0$,
- $h(0,0) = 0$ due to $f(1,1,0,0) = 0$, so that
- $f(0,0,0,0) = 1$ cannot be the result of a disjunction of $g(0,0) = 0$ and $h(0,0) = 0$.

An analog contradiction causes the change of one value of the function $f(\mathbf{x}_a, \mathbf{x}_b)$ in Figure 5.17 from 1–0: $f(0,0,0,1) = 0$. For this changed function we have:

- $g(1,1) = 0$ due to $f(1,1,0,0) = 0$,
- $h(0,1) = 0$ due to $f(0,0,0,1) = 0$, so that
- $f(1,1,0,1) = 1$ cannot be the result of a disjunction of $g(1,1) = 0$ and $h(0,1) = 0$.

Hence, the condition of a disjoint strong OR-bi-decomposition is that no function value 1 of f occurs in both the projection of the function values 0 in the direction of $\mathbf{x}_a$ and $\mathbf{x}_b$:

$$f(\mathbf{x}_a, \mathbf{x}_b) \wedge \max_{\mathbf{x}_a}{}^m \overline{f(\mathbf{x}_a, \mathbf{x}_b)} \wedge \max_{\mathbf{x}_b}{}^m \overline{f(\mathbf{x}_a, \mathbf{x}_b)} = 0 \,. \tag{5.116}$$

Due to the duality of the Boolean Algebras $(\mathbb{B}^n, \wedge, \vee, \overline{}, \mathbf{0}, \mathbf{1})$, and $(\mathbb{B}^n, \vee, \wedge, \overline{}, \mathbf{1}, \mathbf{0})$ the roles of the values 0 and 1 are exchanged as can be seen in Example 5.19.

Example 5.19 Strong AND-Bi-Decomposition.
Figure 5.18 shows the function $f(\mathbf{x}_a, \mathbf{x}_b)$ and the associated decomposition functions $g(\mathbf{x}_a)$ and

$h(\mathbf{x}_b)$ of a strong AND-bi-decomposition. The set $\mathbf{x}_c$ is again empty, so that, due to the missing common variables $\mathbf{x}_c$, this strong AND-bi-decomposition is *disjoint*.

x_{b1}	x_{b2}	$h(\mathbf{x}_b)$
0	0	1
0	1	1
1	1	0
1	0	1

x_{b1}	x_{b2}	$f(\mathbf{x}_a, \mathbf{x}_b)$			
0	0	1	0	1	0
0	1	1	0	1	0
1	1	0	0	0	0
1	0	1	0	1	0
	x_{a2}	0	1	1	0
	x_{a1}	0	0	1	1

	$g(\mathbf{x}_a)$			
	1	0	1	0
x_{a2}	0	1	1	0
x_{a1}	0	0	1	1

Figure 5.18: Disjoint strong AND-bi-decomposition of $f(\mathbf{x}_a, \mathbf{x}_b)$ into $g(\mathbf{x}_a)$ and $h(\mathbf{x}_b)$.

Using the same allocation of the variables $\mathbf{x}_a$ and $\mathbf{x}_b$ as in Figure 5.17 we see in Figure 5.18 that a value 0 of $g(\mathbf{x}_a)$ appears in each row of the Karnaugh-map of $f(\mathbf{x}_a, \mathbf{x}_b)$ because $0 \wedge x = 0$ and the value 0 of $h(\mathbf{x}_b)$ appears in each column of the Karnaugh-map of $f(\mathbf{x}_a, \mathbf{x}_b)$ for the same reason.

A strong AND-bi-decomposition with regard to the dedicated sets of variables $\mathbf{x}_a$ and $\mathbf{x}_b$ requires also that the function $f(\mathbf{x}_a, \mathbf{x}_b)$ satisfies a special condition for the following reason.

If $f(\mathbf{x}_a = \mathbf{c}_a, \mathbf{x}_b = \mathbf{c}_b) = 1$ than $g(\mathbf{x}_a = \mathbf{c}_a) = 1$ and $h(\mathbf{x}_b = \mathbf{c}_b) = 1$ because only $1 \wedge 1 = 1$. Assuming we change one value of the function $f(\mathbf{x}_a, \mathbf{x}_b)$ in Figure 5.18 from 1–0: $f(0,0,0,0) = 0$. For this changed function we have:

- $g(0,0) = 1$ due to $f(0,0,0,1) = 1$,
- $h(0,0) = 1$ due to $f(1,1,0,0) = 1$, so that
- $f(0,0,0,0) = 0$ cannot be the result of a conjunction of $g(0,0) = 1$ and $h(0,0) = 1$.

An analog contradiction causes the change of one value of the function $f(\mathbf{x}_a, \mathbf{x}_b)$ in Figure 5.17 from 0–1: $f(0,0,1,1) = 1$. For this changed function we have:

- $g(1,1) = 1$ due to $f(1,1,0,0) = 1$,
- $h(1,1) = 1$ due to $f(0,0,1,1) = 1$, so that

- $f(1,1,1,1) = 0$ cannot be the result of a conjunction of $g(1,1) = 1$ and $h(1,1) = 1$.

Hence, the condition of a disjoint strong AND-bi-decomposition is that no function value 0 of f occurs in both projections of the function values 1 in the direction of $\mathbf{x}_a$ and $\mathbf{x}_b$:

$$\overline{f(\mathbf{x}_a, \mathbf{x}_b)} \wedge \max_{\mathbf{x}_a}{}^m f(\mathbf{x}_a, \mathbf{x}_b) \wedge \max_{\mathbf{x}_b}{}^m f(\mathbf{x}_a, \mathbf{x}_b) = 0 \,. \tag{5.117}$$

The EXOR-operation belongs to the Boolean Ring and not to a Boolean Algebra like the OR- and the AND-operation. This leads to significantly different conditions of the EXOR-bi-decomposition. Example 5.20 supports the comprehension of the EXOR-bi-decomposition by the construction of all functions of two variables for which an EXOR-bi-decomposition exists.

Example 5.20 All EXOR-Bi-Decompositions of two Variables.
Figure 5.19 depicts all functions $f(x_{a1}, x_{b1}) = g(x_{a1}) \oplus h(x_{b1})$. Based on this complete exploration the condition of an EXOR-bi-decomposition can be found.

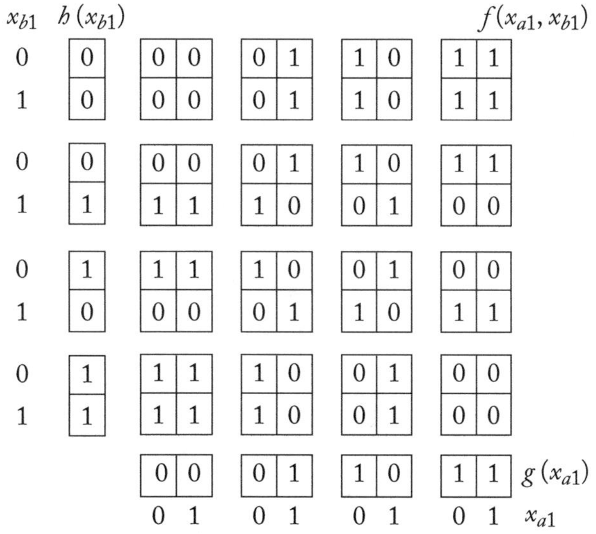

Figure 5.19: All functions $f(x_{a1}, x_{b1})$ of two variables for which an EXOR-bi-decomposition exists.

It can be seen in Figure 5.19 that all eight functions of two variables with an even number of function values 1 occur as a result of $g(x_{a1}) \oplus h(x_{b1})$ for all functions $g(x_{a1})$ and $h(x_{b1})$. This property can be expressed by the 2-fold derivative

$$\frac{\partial^2 f(x_{a1}, x_{b1})}{\partial x_{a1}\,\partial x_{b1}} = 0 \,. \tag{5.118}$$

The conditions for the disjoint strong bi-decompositions introduced above can easily be generalized for all three types of non-disjoint strong bi-decompositions. The know conditions must be satisfied for each subspace $\mathbf{x}_c = \mathbf{c}_0$ separately. Theorem 5.21 summarizes these generalizations. In the case of the strong EXOR-bi-decomposition an extension of the single variable x_{b1} to a set of variables $\mathbf{x}_b$ is possible. It is easy to prove that the condition of the EXOR-bi-decomposition of a function is that in each subspace $\mathbf{x}_c = \mathbf{c}_0$ the change of x_{a1} must either change the function value for all assignments to $\mathbf{x}_b$ or for none of them; with other words, the change of f depending on x_{a1} must be constant with regard to $\mathbf{x}_b$ within each subspace $\mathbf{x}_c = \mathbf{c}_0$.

Theorem 5.21 Conditions for Strong Bi-Decompositions of a Function.
A Boolean function $f(\mathbf{x}_a, \mathbf{x}_b, \mathbf{x}_c)$ is strongly bi-decomposable with regard to the dedicated sets of variables $\mathbf{x}_a$ and $\mathbf{x}_b$ for an OR-gate if and only if

$$f(\mathbf{x}_a, \mathbf{x}_b, \mathbf{x}_c) \wedge \max_{\mathbf{x}_a}{}^m \overline{f(\mathbf{x}_a, \mathbf{x}_b, \mathbf{x}_c)} \wedge \max_{\mathbf{x}_b}{}^m \overline{f(\mathbf{x}_a, \mathbf{x}_b, \mathbf{x}_c)} = 0\,, \tag{5.119}$$

for an AND-gate if and only if

$$\overline{f(\mathbf{x}_a, \mathbf{x}_b, \mathbf{x}_c)} \wedge \max_{\mathbf{x}_a}{}^m f(\mathbf{x}_a, \mathbf{x}_b, \mathbf{x}_c) \wedge \max_{\mathbf{x}_b}{}^m f(\mathbf{x}_a, \mathbf{x}_b, \mathbf{x}_c) = 0\,, \tag{5.120}$$

or for an EXOR-gate with regard to the single variable $\mathbf{x}_a = x_{a1}$ and the set of variables $\mathbf{x}_b$ if and only if

$$\Delta_{\mathbf{x}_b} \frac{\partial f(x_{a1}, \mathbf{x}_b, \mathbf{x}_c)}{\partial x_{a1}} = 0\,. \tag{5.121}$$

Knowing that at least one of the bi-decompositions exists the question arises how the decomposition functions can be calculated. Different pairs of decomposition functions can be used to generate the given function $f(\mathbf{x}_a, \mathbf{x}_b, \mathbf{x}_c)$. Here we present one of them for each type of the bi-decompositions.

The decomposition functions of a strong OR-bi-decomposition can be calculated by

$$g(\mathbf{x}_a, \mathbf{x}_c) = \min_{\mathbf{x}_b}{}^m f(\mathbf{x}_a, \mathbf{x}_b, \mathbf{x}_c)\,, \tag{5.122}$$

$$h(\mathbf{x}_b, \mathbf{x}_c) = \min_{\mathbf{x}_a}{}^m f(\mathbf{x}_a, \mathbf{x}_b, \mathbf{x}_c)\,. \tag{5.123}$$

The decomposition function $g(\mathbf{x}_a, \mathbf{x}_c)$ of a strong OR-bi-decomposition is equal to 1 if all assignments of $\mathbf{x}_b$ within a subspace $(\mathbf{x}_a, \mathbf{x}_c) = \mathbf{c}_0$ are equal to 1. The associated decomposition function $h(\mathbf{x}_b, \mathbf{x}_c)$ is based on an analog condition for $\mathbf{x}_a$ within a subspace $(\mathbf{x}_b, \mathbf{x}_c) = \mathbf{c}_0$.

The correctness of this solution can easily be verified. The substitution of (5.122) and (5.123) leads for an OR-bi-decomposition to

$$f(x_{a1}, \mathbf{x}_b, \mathbf{x}_c) = \min_{\mathbf{x}_b}{}^m f(\mathbf{x}_a, \mathbf{x}_b, \mathbf{x}_c) \vee \min_{\mathbf{x}_a}{}^m f(\mathbf{x}_a, \mathbf{x}_b, \mathbf{x}_c) \,. \tag{5.124}$$

Using (5.124) for the first element in the conjunction of the condition of the OR-bi-decomposition (5.119) and applying the rule (2.135) to the other two elements results in

$$\begin{aligned}
\left(\min_{\mathbf{x}_b}{}^m f(\mathbf{x}_a, \mathbf{x}_b, \mathbf{x}_c) \vee \min_{\mathbf{x}_a}{}^m f(\mathbf{x}_a, \mathbf{x}_b, \mathbf{x}_c)\right) \wedge \\
\overline{\min_{\mathbf{x}_a}{}^m f(\mathbf{x}_a, \mathbf{x}_b, \mathbf{x}_c)} \wedge \overline{\min_{\mathbf{x}_b}{}^m f(\mathbf{x}_a, \mathbf{x}_b, \mathbf{x}_c)} &= 0 \,, \\
\left(\min_{\mathbf{x}_b}{}^m f(\mathbf{x}_a, \mathbf{x}_b, \mathbf{x}_c) \wedge \overline{\min_{\mathbf{x}_a}{}^m f(\mathbf{x}_a, \mathbf{x}_b, \mathbf{x}_c)} \wedge \overline{\min_{\mathbf{x}_b}{}^m f(\mathbf{x}_a, \mathbf{x}_b, \mathbf{x}_c)}\right) \vee \\
\left(\min_{\mathbf{x}_a}{}^m f(\mathbf{x}_a, \mathbf{x}_b, \mathbf{x}_c) \wedge \overline{\min_{\mathbf{x}_a}{}^m f(\mathbf{x}_a, \mathbf{x}_b, \mathbf{x}_c)} \wedge \overline{\min_{\mathbf{x}_b}{}^m f(\mathbf{x}_a, \mathbf{x}_b, \mathbf{x}_c)}\right) &= 0 \,, \\
0 &= 0 \,,
\end{aligned} \tag{5.125}$$

because the conjunction of an arbitrary function and its complement is equal to 0.

The decomposition functions of a strong AND-bi-decomposition can be calculated by

$$g(\mathbf{x}_a, \mathbf{x}_c) = \max_{\mathbf{x}_b}{}^m f(\mathbf{x}_a, \mathbf{x}_b, \mathbf{x}_c) \,, \tag{5.126}$$

$$h(\mathbf{x}_b, \mathbf{x}_c) = \max_{\mathbf{x}_a}{}^m f(\mathbf{x}_a, \mathbf{x}_b, \mathbf{x}_c) \,. \tag{5.127}$$

The decomposition function $g(\mathbf{x}_a, \mathbf{x}_c)$ of a strong AND-bi-decomposition is equal to 1 if at least one assignment of $\mathbf{x}_b$ within a subspace $(\mathbf{x}_a, \mathbf{x}_c) = \mathbf{c}_0$ is equal to 1. Again, the associated decomposition function $h(\mathbf{x}_b, \mathbf{x}_c)$ is based on an analog condition for $\mathbf{x}_a$ within a subspace $(\mathbf{x}_b, \mathbf{x}_c) = \mathbf{c}_0$. The proof of correctness of these decomposition function can be done in the same manner.

The decomposition functions of the EXOR-bi-decomposition cannot independently be chosen as in the case of the OR- of the AND-bi-decomposition. There are also several decomposition functions for the EXOR-bi-decomposition, but one chosen decomposition function determines the other one. One of these possibilities to calculate the decomposition functions of a strong EXOR-bi-decomposition is

$$g(x_{a1}, \mathbf{x}_c) = \max_{\mathbf{x}_b}{}^m \left(f(x_{a1}, \mathbf{x}_b, \mathbf{x}_c) \wedge \bigwedge_{i=1}^{m} x_{bi}\right) \,, \tag{5.128}$$

$$h(\mathbf{x}_b, \mathbf{x}_c) = \max_{x_{a1}} \left(f(x_{a1}, \mathbf{x}_b, \mathbf{x}_c) \oplus g(x_{a1}, \mathbf{x}_c)\right) \,. \tag{5.129}$$

Unfortunately, there are functions for which no bi-decomposition exists. This drawback can be restricted when the function to decompose can be chosen out of a lattice of many functions. It is not necessary to check each function of the lattice for the different types of the bi-decompositions separately. Theorem 5.22 provides the conditions to verify whether the lattice contains at least one function for which a bi-decomposition of the selected type with regard to the dedicated sets of of variables $\mathbf{x}_a$ and $\mathbf{x}_b$ exists. These formulas generalize the formulas for single functions but take into account that contradictions concerning the bi-decomposition can only be caused by the mark functions $f_q(\mathbf{x}_a, \mathbf{x}_b, \mathbf{x}_c)$ and $f_r(\mathbf{x}_a, \mathbf{x}_b, \mathbf{x}_c)$.

Theorem 5.22 Conditions for Strong Bi-Decompositions of a Lattice.
A lattice of Boolean functions $L\langle f_q(\mathbf{x}_a, \mathbf{x}_c), f_r(\mathbf{x}_a, \mathbf{x}_c)\rangle$ contains at least one function $f(\mathbf{x}_a, \mathbf{x}_b, \mathbf{x}_c)$ that is strongly bi-decomposable with regard to the dedicated sets of variables $\mathbf{x}_a$ and $\mathbf{x}_b$ for an OR-gate if and only if

$$f_q(\mathbf{x}_a, \mathbf{x}_b, \mathbf{x}_c) \wedge \max_{\mathbf{x}_a}{}^m f_r(\mathbf{x}_a, \mathbf{x}_b, \mathbf{x}_c) \wedge \max_{\mathbf{x}_b}{}^m f_r(\mathbf{x}_a, \mathbf{x}_b, \mathbf{x}_c) = 0 , \tag{5.130}$$

for an AND-gate if and only if

$$f_r(\mathbf{x}_a, \mathbf{x}_b, \mathbf{x}_c) \wedge \max_{\mathbf{x}_a}{}^m f_q(\mathbf{x}_a, \mathbf{x}_b, \mathbf{x}_c) \wedge \max_{\mathbf{x}_b}{}^m f_q(\mathbf{x}_a, \mathbf{x}_b, \mathbf{x}_c) = 0 , \tag{5.131}$$

or for an EXOR-gate with $\mathbf{x}_a = x_{a1}$ and $\mathbf{x}_b$ if and only if

$$\max_{\mathbf{x}_b}{}^m f_q^{\partial x_{a1}}(\mathbf{x}_b, \mathbf{x}_c) \wedge f_r^{\partial x_{a1}}(\mathbf{x}_b, \mathbf{x}_c) = 0 , \tag{5.132}$$

where $f_q^{\partial x_{a1}}(\mathbf{x}_b, \mathbf{x}_c)$ and $f_r^{\partial x_{a1}}(\mathbf{x}_b, \mathbf{x}_c)$ are the mark functions of the single derivative of the given lattice with regard to x_{a1}; see (3.59) and (3.60).

The lattices of the decomposition functions can be calculated for a strong OR-bi-decomposition by

$$g_q(\mathbf{x}_a, \mathbf{x}_c) = \max_{\mathbf{x}_b}{}^m \left(f_q(\mathbf{x}_a, \mathbf{x}_b, \mathbf{x}_c) \wedge \max_{\mathbf{x}_a}{}^m f_r(\mathbf{x}_a, \mathbf{x}_b, \mathbf{x}_c)\right) , \tag{5.133}$$

$$g_r(\mathbf{x}_a, \mathbf{x}_c) = \max_{\mathbf{x}_b}{}^m f_r(\mathbf{x}_a, \mathbf{x}_b, \mathbf{x}_c) , \tag{5.134}$$

$$h_q(\mathbf{x}_b, \mathbf{x}_c) = \max_{\mathbf{x}_a}{}^m \left(f_q(\mathbf{x}_a, \mathbf{x}_b, \mathbf{x}_c) \wedge \overline{g(\mathbf{x}_a, \mathbf{x}_c)}\right) , \tag{5.135}$$

$$h_r(\mathbf{x}_b, \mathbf{x}_c) = \max_{\mathbf{x}_a}{}^m f_r(\mathbf{x}_a, \mathbf{x}_b, \mathbf{x}_c) , \tag{5.136}$$

where $g(\mathbf{x}_a, \mathbf{x}_c)$ in (5.135) is the chosen decomposition function of the lattice with the mark functions (5.133) and (5.134).

The lattices of the decomposition functions can be calculated for a strong AND-bi-decomposition by

$$g_q(\mathbf{x}_a, \mathbf{x}_c) = \max_{\mathbf{x}_b}{}^m f_q(\mathbf{x}_a, \mathbf{x}_b, \mathbf{x}_c) , \tag{5.137}$$

$$g_r(\mathbf{x}_a, \mathbf{x}_c) = \max_{\mathbf{x}_b}{}^m \left(f_r(\mathbf{x}_a, \mathbf{x}_b, \mathbf{x}_c) \wedge \max_{\mathbf{x}_a}{}^m f_q(\mathbf{x}_a, \mathbf{x}_b, \mathbf{x}_c) \right) , \tag{5.138}$$

$$h_q(\mathbf{x}_b, \mathbf{x}_c) = \max_{\mathbf{x}_a}{}^m f_q(\mathbf{x}_a, \mathbf{x}_b, \mathbf{x}_c) , \tag{5.139}$$

$$h_r(\mathbf{x}_b, \mathbf{x}_c) = \max_{\mathbf{x}_a}{}^m \left(f_r(\mathbf{x}_a, \mathbf{x}_b, \mathbf{x}_c) \wedge g(\mathbf{x}_a, \mathbf{x}_c) \right) , \tag{5.140}$$

where $g(\mathbf{x}_a, \mathbf{x}_c)$ in (5.140) is the chosen decomposition function of the lattice with the mark functions (5.137) and (5.138).

A possible decomposition function $g(x_{a1}, \mathbf{x}_c)$ of a strong EXOR-bi-decomposition is uniquely specified by

$$g(x_{a1}, \mathbf{x}_c) = x_{a1} \wedge \max_{\mathbf{x}_b}{}^m f_q^{\partial x_{a1}}(\mathbf{x}_b, \mathbf{x}_c), \tag{5.141}$$

and the associated decomposition function $h(\mathbf{x}_b, \mathbf{x}_c)$ can be chosen from the lattice with the mark functions

$$h_q(\mathbf{x}_b, \mathbf{x}_c) = \max_{x_{a1}}((\overline{g(x_{a1}, \mathbf{x}_c)} \wedge f_q(x_{a1}, \mathbf{x}_b, \mathbf{x}_c)) \vee (g(x_{a1}, \mathbf{x}_c) \wedge f_r(x_{a1}, \mathbf{x}_b, \mathbf{x}_c))) , \tag{5.142}$$

$$h_r(\mathbf{x}_b, \mathbf{x}_c) = \max_{x_{a1}}((\overline{g(x_{a1}, \mathbf{x}_c)} \wedge f_r(x_{a1}, \mathbf{x}_b, \mathbf{x}_c)) \vee (g(x_{a1}, \mathbf{x}_c) \wedge f_q(x_{a1}, \mathbf{x}_b, \mathbf{x}_c))) . \tag{5.143}$$

More details about these strong bi-decompositions are given in Posthoff and Steinbach [2004] and Steinbach and Posthoff [2010a]. A very welcome property of these calculations of decomposition functions is shown in Steinbach and Stöckert [1994]; the synthesized multilevel circuits are completely testable with regard to stuck-at faults at each pin of all gates, and the test patterns can be generated on the fly needing only an additional computation time of about 10% [Steinbach and Stöckert, 1994].

Even if the the function to decompose can be chosen out of a large number of functions of a lattice it remains the problem that for not any of these functions a strong bi-decomposition may exist. This gap has been closed by the weak bi-decomposition which was suggested for the first time by Le [1989]. Figure 5.20 shows the circuit structures of the weak bi-decompositions.

The formal difference to the strong bi-decomposition is that the set of variables $\mathbf{x}_b$ is empty for the weak bi-decomposition. A basic requirement for each bi-decomposition is that both decomposition functions must be simpler than the given function to decompose. It is possible that this requirement is not achieved by the weak EXOR-bi-decomposition due to the following equivalent transformation of linear equations:

$$f(\mathbf{x}_a, \mathbf{x}_c) = g(\mathbf{x}_a, \mathbf{x}_c) \oplus h(\mathbf{x}_c) , \tag{5.144}$$

$$g(\mathbf{x}_a, \mathbf{x}_c) = f(\mathbf{x}_a, \mathbf{x}_c) \oplus h(\mathbf{x}_c) . \tag{5.145}$$

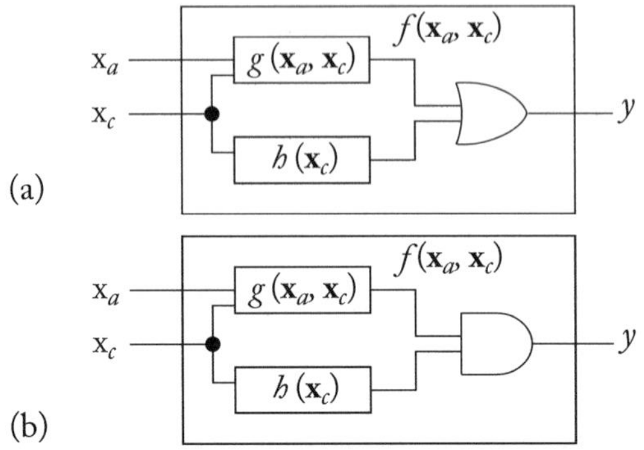

Figure 5.20: Weak bi-decompositions of the function $y = f(\mathbf{x}_a, \mathbf{x}_c)$ into the decomposition functions $g(\mathbf{x}_a, \mathbf{x}_c)$ and $h(\mathbf{x}_c)$ using: (a) an OR-gate and (b) an AND-gate.

The decomposition function $g(\mathbf{x}_a, \mathbf{x}_c)$ of the weak EXOR-bi-decomposition can be calculated for an arbitrary function $h(\mathbf{x}_c)$ without any condition to the given function $f(\mathbf{x}_a, \mathbf{x}_c)$. The comparison of (5.144) and (5.145) reveals that the functions $f(\mathbf{x}_a, \mathbf{x}_c)$ and $g(\mathbf{x}_a, \mathbf{x}_c)$ appear in exchanged roles so that the complexity of $g(\mathbf{x}_a, \mathbf{x}_c)$ can be smaller but also larger than the complexity of $f(\mathbf{x}_a, \mathbf{x}_c)$. Therefore we omit the the weak EXOR-bi-decomposition.

The decomposition functions $g(\mathbf{x}_a, \mathbf{x}_c)$ of weak bi-decompositions using an OR-gate or an AND-gate also depend on the same variables as the given function $f(\mathbf{x}_a, \mathbf{x}_c)$. However, these functions can be chosen from a lattice of functions if a certain condition is satisfied. In case of a given lattice of $L\langle f_q(\mathbf{x}_a, \mathbf{x}_c), f_r(\mathbf{x}_a, \mathbf{x}_c)\rangle$ the weak OR- or AND-bi-decomposition originates a larger lattice of $L\langle g_q(\mathbf{x}_a, \mathbf{x}_c), g_r(\mathbf{x}_a, \mathbf{x}_c)\rangle$ which increases the potential for simplifications of the associated sub-circuit.

The conditions of weak OR- or AND-bi-decompositions can be specified for both a single function $f(\mathbf{x}_a, \mathbf{x}_c)$ or a lattice $\langle f_q(\mathbf{x}_a, \mathbf{x}_c), f_r(\mathbf{x}_a, \mathbf{x}_c)\rangle$. A single function can be mapped to the special case that a lattice contains only one function. Hence, Theorem 5.23 provides the more general conditions to verify whether a lattice contains at least one function for which an OR- or AND-bi-decomposition with regard to the dedicated set of of variables $\mathbf{x}_a$ exists.

Theorem 5.23 Conditions for Weak Bi-Decompositions of a Lattice.
A lattice of Boolean functions $L\langle f_q(\mathbf{x}_a, \mathbf{x}_c), f_r(\mathbf{x}_a, \mathbf{x}_c)\rangle$ contains at least one functions $f(\mathbf{x}_a, \mathbf{x}_c)$ that is weakly bi-decomposable with regard to the dedicated set of variables $\mathbf{x}_a$ for an OR-gate if and only if

$$f_q(\mathbf{x}_a, \mathbf{x}_c) \wedge \overline{\max_{\mathbf{x}_a}{}^m f_r(\mathbf{x}_a, \mathbf{x}_c)} \neq 0 , \tag{5.146}$$

or for an AND-gate if and only if

$$f_r(\mathbf{x}_a, \mathbf{x}_c) \wedge \overline{\max_{\mathbf{x}_a}{}^m f_q(\mathbf{x}_a, \mathbf{x}_c)} \neq 0 . \tag{5.147}$$

The lattices of the decomposition functions can be calculated for a weak OR-bi-decomposition by

$$g_q(\mathbf{x}_a, \mathbf{x}_c) = f_q(\mathbf{x}_a, \mathbf{x}_c) \wedge \max_{\mathbf{x}_a}{}^m f_r(\mathbf{x}_a, \mathbf{x}_c) \,, \tag{5.148}$$

$$g_r(\mathbf{x}_a, \mathbf{x}_c) = f_r(\mathbf{x}_a, \mathbf{x}_c) \,, \tag{5.149}$$

$$h_q(\mathbf{x}_c) = \max_{\mathbf{x}_a}{}^m \left(f_q(\mathbf{x}_a, \mathbf{x}_c) \wedge \overline{g(\mathbf{x}_a, \mathbf{x}_c)} \right) \,, \tag{5.150}$$

$$h_r(\mathbf{x}_c) = \max_{\mathbf{x}_a}{}^m f_r(\mathbf{x}_a, \mathbf{x}_c) \,, \tag{5.151}$$

where $g(\mathbf{x}_a, \mathbf{x}_c)$ in (5.150) is the chosen decomposition function of the lattice with the mark functions (5.148) and (5.149). The m-fold maximum in (5.148) decreases $g_q(\mathbf{x}_a, \mathbf{x}_c)$ in comparison to $f_q(\mathbf{x}_a, \mathbf{x}_c)$ so that the lattice of the decomposition function $g(\mathbf{x}_a, \mathbf{x}_c)$ contains more functions than the lattice of $f(\mathbf{x}_a, \mathbf{x}_c)$.

Example 5.24 Weak OR-Bi-Decomposition.
Figure 5.21 shows an example of a weak OR-bi-decomposition.

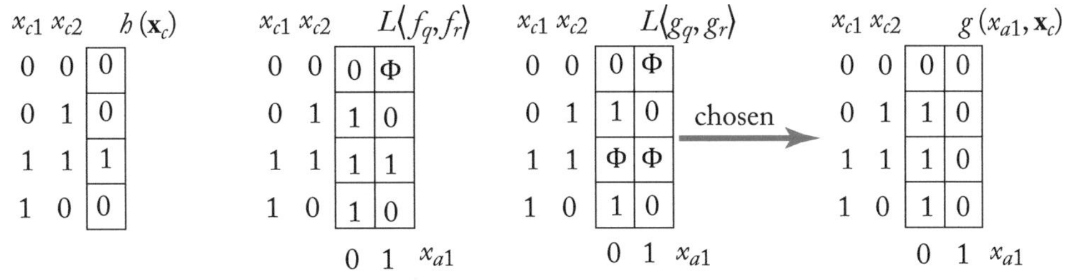

Figure 5.21: Weak OR-bi-decomposition of the lattice $L\langle f_q(x_{a1}, \mathbf{x}_c), f_r(x_{a1}, \mathbf{x}_c)\rangle$.

The left-hand side of (5.146) is equal to $x_{c1}x_{c2}$ for the lattice of f shown in the second Karnaugh-map of Figure 5.21; hence, the condition of the weak OR-bi-decomposition is satisfied. Using (5.148) we get the mark function

$$g_q(x_{a1}, \mathbf{x}_c) = \overline{x}_{a1}(x_{c1} \oplus x_{c2})$$

which is smaller than $f_q(x_{a1}, \mathbf{x}_c)$. Due to the additional two don't-cares in the third Karnaugh-map of Figure 5.21 the decomposition function

$$g(x_{a1}, \mathbf{x}_c) = \overline{x}_{a1}(x_{c1} \vee x_{c2})$$

can be chosen (see the right-most Karnaugh-map of Figure 5.21). Using this decomposition function $g(x_{a1}, \mathbf{x}_c)$ the mark function (5.151) is the complement of (5.150) so that the decomposition function $h(\mathbf{x}_c)$ is uniquely specified:

$$h(\mathbf{x}_c) = x_{c1} \wedge x_{c2} \ .$$

The lattices of the decomposition functions can be calculated for a weak AND-bi-decomposition by

$$g_q(\mathbf{x}_a, \mathbf{x}_c) = f_q(\mathbf{x}_a, \mathbf{x}_c) \ , \tag{5.152}$$

$$g_r(\mathbf{x}_a, \mathbf{x}_c) = f_r(\mathbf{x}_a, \mathbf{x}_b, \mathbf{x}_c) \wedge \max_{\mathbf{x}_a}{}^m f_q(\mathbf{x}_a, \mathbf{x}_c) \ , \tag{5.153}$$

$$h_q(\mathbf{x}_c) = \max_{\mathbf{x}_a}{}^m f_q(\mathbf{x}_a, \mathbf{x}_c) \ , \tag{5.154}$$

$$h_r(\mathbf{x}_c) = \max_{\mathbf{x}_a}{}^m (f_r(\mathbf{x}_a, \mathbf{x}_c) \wedge g(\mathbf{x}_a, \mathbf{x}_c)) \ , \tag{5.155}$$

where $g(\mathbf{x}_a, \mathbf{x}_c)$ in (5.155) is the chosen decomposition function of the lattice with the mark functions (5.152) and (5.153).

Example 5.25 Weak AND-Bi-Decomposition.
Figure 5.22 shows the details of a weak AND-bi-decomposition.

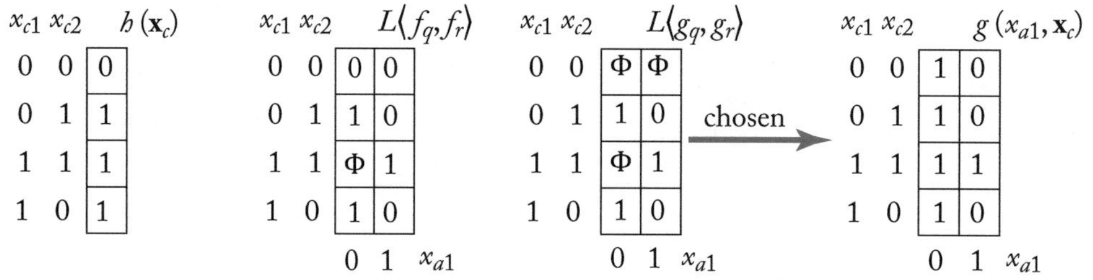

Figure 5.22: Weak AND-bi-decomposition of the lattice $L\langle f_q(x_{a1}, \mathbf{x}_c), f_r(x_{a1}, \mathbf{x}_c)\rangle$.

The condition of the weak AND-bi-decomposition is satisfied because the left-hand side of (5.147) is equal to $\overline{x}_{c1}\overline{x}_{c2}$ for the lattice of f shown in the second Karnaugh-map of Figure 5.22. Using (5.153)

$$g_r(x_{a1}, \mathbf{x}_c) = x_{a1}(x_{c1} \oplus x_{c2})$$

we get the extended lattice of g, shown in the third Karnaugh-map of Figure 5.22. Here, the decomposition function

$$g(x_{a1}, \mathbf{x}_c) = \overline{x}_{a1} \vee x_{c1}x_{c2}$$

can be chosen as shown in the right-most Karnaugh-map of Figure 5.22. Using this decomposition function $g(x_{a1}, \mathbf{x}_c)$ we get again mark functions (5.152) and (5.155) of a uniquely specified decomposition function $h(\mathbf{x}_c)$:

$$h(\mathbf{x}_c) = x_{c1} \vee x_{c2} .$$

Commonly with the strong EXOR-bi-decompositions the two weak bi-decompositions ensure the completeness of this method for the synthesis of a combinational circuit for each Boolean function.

Theorem 5.26 Completeness of the Bi-Decomposition.
A lattice of Boolean functions

$$L\left\langle f_q(x_{a1}, \mathbf{x}_c), f_r(x_{a1}, \mathbf{x}_c)\right\rangle = L\left\langle f_r(x_{a1}, \mathbf{x}_b, \mathbf{x}_{c0}), f_r(x_{a1}, \mathbf{x}_b, \mathbf{x}_{c0})\right\rangle$$

that contains only functions $f(x_{a1}, \mathbf{x}_c)$ which are neither weak OR-bi-decomposable nor weak AND-bi-decomposable with regard to x_{a1} includes at least one function $f(x_{a1}, \mathbf{x}_b, \mathbf{x}_{c0})$ that is strong EXOR-bi-decomposable with regard to x_{a1} and $\mathbf{x}_b$.

Proof. Due to the monotony (2.118) of the m-fold maximum, both types of weak bi-decompositions with regard to the set of variables $\mathbf{x}_a$ imply the bi-decompositions of the same type with regard to the single variables $x_{a1} \in \mathbf{x}_a$. Hence, only weak bi-decompositions with regard to the single variables x_{a1} must be taken into account.

The lattice $L\left\langle f_q(x_{a1}, \mathbf{x}_c), f_r(x_{a1}, \mathbf{x}_c)\right\rangle$ does *not* contain a function $f(x_{a1}, \mathbf{x}_c)$ that is weak OR-bi-decomposable if

$$f_q(x_{a1}, \mathbf{x}_c) \wedge \overline{\max_{x_{a1}} f_r(x_{a1}, \mathbf{x}_c)} = 0 . \tag{5.156}$$

Equivalent equations to (5.156) are:

$$f_q(x_{a1}, \mathbf{x}_c) \wedge \min_{x_{a1}} \overline{f_r(x_{a1}, \mathbf{x}_c)} = 0 ,$$
$$f_q(x_{a1}, \mathbf{x}_c) \wedge \min_{x_{a1}}(f_q(x_{a1}, \mathbf{x}_c) \vee f_\varphi(x_{a1}, \mathbf{x}_c)) = 0 . \tag{5.157}$$

The function $f_\varphi(x_{a1}, \mathbf{x}_c))$ can be removed from the left-hand side of (5.157) because this equation implies

$$f_q(x_{a1}, \mathbf{x}_c) \wedge \min_{x_{a1}} f_q(x_{a1}, \mathbf{x}_c) = 0 , \tag{5.158}$$

which can be simplified to

$$\min_{x_{a1}} f_q(x_{a1}, \mathbf{x}_c) = 0 . \tag{5.159}$$

Using analog transformations we get from

$$f_r(x_{a1}, \mathbf{x}_c) \wedge \overline{\max_{x_{a1}} f_q(x_{a1}, \mathbf{x}_c)} = 0 \tag{5.160}$$

the sufficient condition

$$\min_{x_{a1}} f_r(x_{a1}, \mathbf{x}_c) = 0 \tag{5.161}$$

that the lattice $L\langle f_q(x_{a1}, \mathbf{x}_c), f_r(x_{a1}, \mathbf{x}_c)\rangle$ does *not* contain a function $f(x_{a1}, \mathbf{x}_c)$ that is weak AND-bi-decomposable.

The substitution of (5.159) and (5.161) into (3.60) results in

$$f_r^{\partial x_{a1}}(\mathbf{x}_c) = \min_{x_i} f_q(x_{a1}, \mathbf{x}_c) \vee \min_{x_i} f_r(x_{a1}, \mathbf{x}_c) = 0 \tag{5.162}$$

and expresses the assumptions of Theorem 5.26 that the lattice contains only functions $f(x_{a1}, \mathbf{x}_c)$ which are neither weak OR-bi-decomposable nor weak AND-bi-decomposable with regard to x_{a1}. Splitting $\mathbf{x}_c = (\mathbf{x}_b, \mathbf{x}_{c0})$ in (5.162) and replacing $\mathbf{x}_c$ by $\mathbf{x}_{c0}$ in (5.132) leads to the adopted condition of the strong EXOR-bi-decomposition

$$\max_{\mathbf{x}_b}{}^m f_q^{\partial x_{a1}}(\mathbf{x}_b, \mathbf{x}_{c0}) \wedge f_r^{\partial x_{a1}}(\mathbf{x}_b, \mathbf{x}_{c0}) = 0\,. \tag{5.163}$$

The final substitution of (5.162) into (5.163) shows that with the assumptions of Theorem 5.26 the condition of the strong EXOR-bi-decomposition with regard to x_{a1} and $\mathbf{x}_b$ is satisfied:

$$\max_{\mathbf{x}_b}{}^m f_q^{\partial x_{a1}}(\mathbf{x}_b, \mathbf{x}_{c0}) \wedge 0 = 0\,. \tag{5.164}$$

□

Example 5.27 demonstrates that the complete synthesis of a function selected from a lattice of four functions for which no strong bi-decomposition exists is reached by the common use of both the strong and the weak bi-decompositions.

Example 5.27 Complete Synthesis Using Weak and Strong Bi-Decomposition.
A multilevel combinational circuit must be synthesized for an arbitrary function of the given lattice $L\langle f_q(\mathbf{x}), f_r(\mathbf{x})\rangle$ of five variables that is specified by the following mark functions:

$$\begin{aligned} f_q(\mathbf{x}) = {} & \overline{x}_1\overline{x}_2x_3\overline{x}_5 \vee \overline{x}_1x_2\overline{x}_3\overline{x}_5 \vee x_1x_2x_3\overline{x}_5 \vee \overline{x}_1x_2x_3x_5 \vee x_1x_2\overline{x}_3\overline{x}_4x_5 \vee \\ & x_1\overline{x}_2x_3\overline{x}_4x_5 \vee \overline{x}_1x_3x_4x_5 \vee \overline{x}_1x_2x_4x_5 \vee x_1x_2x_4\overline{x}_5\,, \end{aligned} \tag{5.165}$$

$$\begin{aligned} f_r(\mathbf{x}) = {} & \overline{x}_2\overline{x}_3\overline{x}_4 \vee \overline{x}_1x_2x_3\overline{x}_5 \vee x_1x_2\overline{x}_3\overline{x}_4\overline{x}_5 \vee x_1\overline{x}_2x_3\overline{x}_5 \vee \overline{x}_1\overline{x}_2\overline{x}_4x_5 \vee \\ & \overline{x}_1x_2\overline{x}_3\overline{x}_4x_5 \vee x_1x_2x_3\overline{x}_4x_5 \vee \overline{x}_1\overline{x}_2\overline{x}_3x_4x_5 \vee x_1x_4x_5\,. \end{aligned} \tag{5.166}$$

We assume that AND-, OR-, or EXOR-gates with two optionally negated inputs can be used as elements for this circuit.

Figure 5.23a shows the Karnaugh-map of this lattice. The synthesized circuit is depicted in Figure 5.23b where the indices of the decomposition functions g and h are assigned in the order of the applied bi-decompositions.

(a) $L\langle f_q(\mathbf{x}), f_r(\mathbf{x})\rangle$

x_4 x_5									
0 0	0	1	0	1	0	1	0	0	
0 1	0	0	1	0	1	0	1	0	
1 1	0	1	1	1	0	0	0	0	
1 0	Φ	1	0	1	1	1	0	Φ	
	0	1	1	0	0	1	1	0	x_3
	0	0	1	1	1	1	0	0	x_2
	0	0	0	0	1	1	1	1	x_1

chosen →

(c) $f(\mathbf{x})$

x_4 x_5									
0 0	0	1	0	1	0	1	0	0	
0 1	0	0	1	0	1	0	1	0	
1 1	0	1	1	1	0	0	0	0	
1 0	0	1	0	1	1	1	0	0	
	0	1	1	0	0	1	1	0	x_3
	0	0	1	1	1	1	0	0	x_2
	0	0	0	0	1	1	1	1	x_1

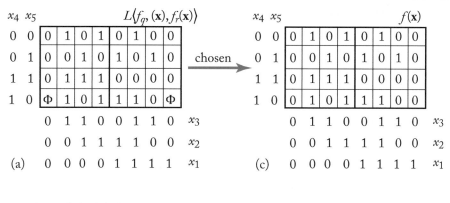

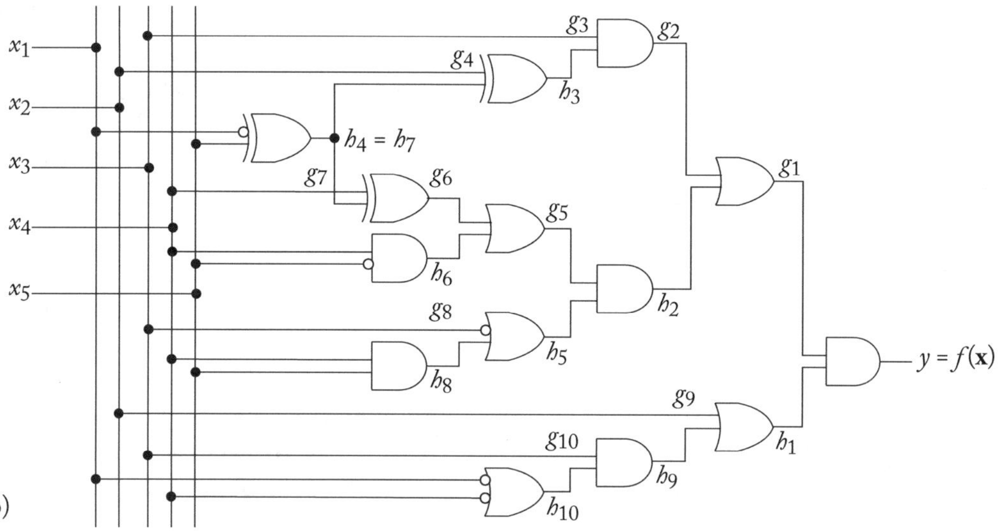

(b)

Figure 5.23: Complete synthesis by weak and strong bi-decompositions: (a) Karnaugh-map of the given lattice; (b) structure of the circuit; and (c) Karnaugh-map of the realized function.

The check for all three types of strong bi-decompositions and all ten pairs of variables by means of the conditions of Theorem 5.22 confirms: there is no strong bi-decomposition for the four functions of the given lattice. Due to Theorem 5.26 we apply the possible weak AND-bi-decomposition with regard to $x_{a1} = x_5$. It can be seen in Figure 5.23b that the decomposition function g_1 depends on all five variables but the decomposition function h_1 of this weak AND-bi-decomposition is independent of x_5.

This weak AND-bi-decomposition extends the lattice of the decomposition function g_1 such that a strong OR-bi-decomposition with regard to x_2 and x_4 can be performed. Figure 5.23b shows that the decomposition function g_2 does not depend on x_4 and h_2 not on

x_2. A disjoint strong AND-bi-decomposition separates $g_3 = x_3$ from the variables x_1, x_4, and x_5 which determine the decomposition function h_3. A disjoint strong EXOR-bi-decomposition into $g_4 = x_2$ and $h_4 = \overline{x}_1 \oplus x_5$ terminates this branch of the circuit.

The strong AND-bi-decomposition of h_2 with regard to x_1 and x_3 determines the completely specified function g_5 for which no strong bi-decomposition exists. The weak OR-bi-decomposition of g_5 with regard to $x_{a1} = x_1$ leads to g_6 for which a disjoint strong EXOR-bi-decomposition with regard to x_4 and (x_1, x_5) exists such that the function $h_4 = \overline{x}_1 \oplus x_5$ can be reused as function h_7.

A disjoint strong OR-bi-decomposition of the completely specified function h_5 with regard to x_3 and (x_4, x_5) terminates this branch of the strong AND-bi-decomposition of h_2 as well as the strong OR-bi-decomposition of g_1.

It remains the synthesis of the lattice of h_1 for which a disjoint strong OR-bi-decomposition with regard to x_2 and (x_1, x_3, x_4) exists. For the associated decomposition function h_9 exists a disjoint strong AND-bi-decomposition with regard to x_3 and $(x_1 x_4)$ so that h_{10} terminates the synthesis of the combinational circuit.

The synthesized circuit of Figure 5.23b realizes the function $y = f(\mathbf{x})$ for which the Karnaugh-map is shown in Figure 5.23c. Obviously, we have for this circuit: $f(\mathbf{x}) = f_q(\mathbf{x})$.

In order to evaluate the circuit synthesized in Example 5.27 using strong and weak bi-decompositions we mapped a minimal disjunctive form of the same lattice (see Figure 5.24a) to a circuit where also two-input gates with optionally negated inputs are used. There are six minimal disjunctive forms of the function where both don't-cares are assigned to the value 1. Figure 5.24b show the circuit of that minimal form where as much as possible two-input gates could be reused. It should be mentioned that the circuit structure of the minimal disjunctive form of $f_q(\mathbf{x})$ would need one gate more.

Example 5.27 demonstrates the benefits of the synthesis using strong and weak bi-decompositions. We take circuit of Figure 5.24 that maps a minimal disjunctive form of one function of the lattice to the smallest number of gates for comparison.

- The expression of the minimal disjunctive form consists of nine conjunctions, one with three variables, six with four variables and the remaining two contain all five variables. The mapping of this disjunctive form leads to a circuit of two-input AND- and OR-gates shown in Figure 5.24 that consists of 29 gates where as much as possible two input gates are reused. Most of the paths of this circuit belong to the longest one which contains six gates each.

- Figure 5.23b shows that only 14 two-input gates represent a function of the same lattice synthesized by means of strong and weak bi-decompositions where the longest path of 6 levels appears only twice.

(a) $L\langle f_q, (\mathbf{x}), f_r(\mathbf{x})\rangle$

x_4 x_5								
0 0	0	1	0	1	0	1	0	0
0 1	0	0	1	0	1	0	1	0
1 1	0	1	1	1	0	0	0	0
1 0	Φ	1	0	1	1	1	0	Φ
x_3	0	1	1	0	0	1	1	0
x_2	0	0	1	1	1	1	0	0
x_1	0	0	0	0	1	1	1	1

chosen →

(c) $f(\mathbf{x})$

x_4 x_5								
0 0	0	1	0	1	0	1	0	0
0 1	0	0	1	0	1	0	1	0
1 1	0	1	1	1	0	0	0	0
1 0	0	1	0	1	1	1	0	1
x_3	0	1	1	0	0	1	1	0
x_2	0	0	1	1	1	1	0	0
x_1	0	0	0	0	1	1	1	1

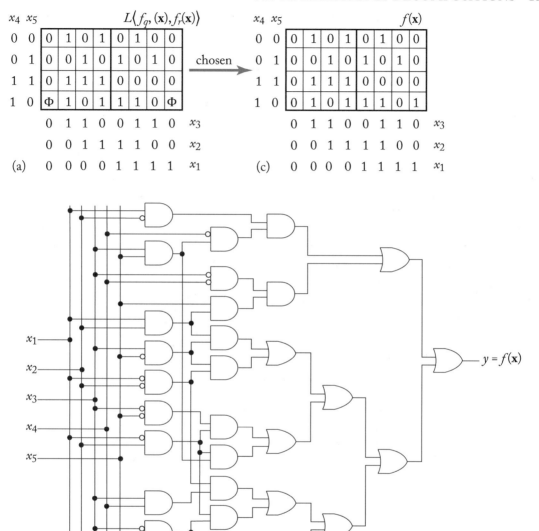

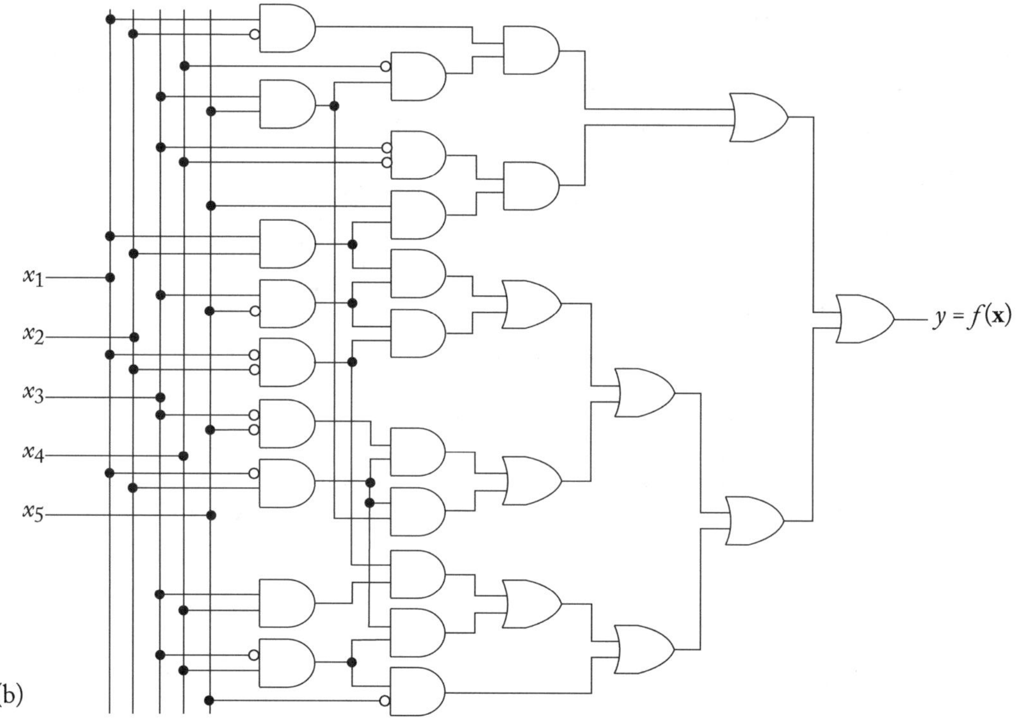

(b)

Figure 5.24: Complete synthesis based on a minimal disjunctive form: (a) Karnaugh-map of the given lattice; (b) structure of the circuit; and (c) Karnaugh-map of the realized function.

The reached reduction to less than one half of two-input gates of the minimal disjunctive form confirms the benefit of strong and weak bi-decompositions that the utilization of the properties of the synthesized function significantly reduces the needed area and power consumption. The length of the longest path determines the delay of a circuit. Due to the needed two weak bi-decompositions the longest paths have an equal length of six in both circuits synthesized by the different approaches.

Weak bi-decompositions assure on the one hand the completeness of the synthesis by bi-decompositions but can cause on the other hand *unbalanced* circuits. It can be seen in Figure 5.23b that the decomposition function g_1 of the weak AND-bi-decomposition consists of ten gates on five levels in comparison to the decomposition function h_1 of this weak AND-bi-decomposition with only three gates on three levels. Both the decomposition function g_1 of the weak AND-bi-decomposition and the decomposition function g_5 of the weak OR-bi-decomposition belong to the longest path of six levels. Hence, a complete synthesis that uses strong bi-decompositions and avoids as many as possible weak bi-decompositions is desirable.

The vectorial bi-decomposition contributes to this aim. A vectorial bi-decomposition can be realized using an AND-, an OR-, or an EXOR-gate with two inputs. These additional types of bi-decompositions were presented for the first time by Steinbach [2015]. A generalization of vectorial bi-decompositions for lattices of Boolean functions became possible based on the derivative operations for generalized lattices introduced in Chapter 3. Conditions and rules to calculate the generalized lattices of the decomposition function for vectorial bi-decompositions were published for the first time by Steinbach and Posthoff [2016].

The vectorial bi-decompositions consequently extend the strong bi-decompositions. Figure 5.25 shows the structure of the three types of vectorial bi-decompositions and summarizes the required conditions for simplifications.

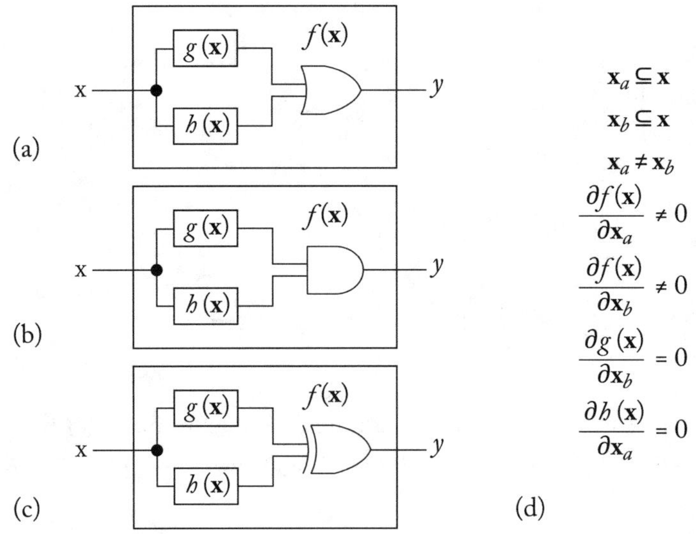

Figure 5.25: Vectorial bi-decompositions of the function $y = f(\mathbf{x})$ into the decomposition functions $g(\mathbf{x})$ and $h(\mathbf{x})$ using: (a) an OR-gate; (b) an AND-gate; (c) an EXOR-gate; and (d) conditions for simplifications.

The main condition for each synthesis approach by decomposition is that the decomposition functions are simpler than the given function to decompose. The strong bi-decompositions satisfy this condition because each decomposition function depends at least on one variable less than the given function. In a bi-decomposition of $f(x_i, x_j, \mathbf{x}_c)$ into $g(x_i, \mathbf{x}_0)$ and $h(x_j, \mathbf{x}_c)$ we can express this property using single derivatives:

$$\frac{\partial f(x_i, x_j, \mathbf{x}_c)}{\partial x_i} \neq 0\,, \qquad \frac{\partial f(x_i, x_j, \mathbf{x}_c)}{\partial x_j} \neq 0\,, \tag{5.167}$$

$$\frac{\partial g(x_i, \mathbf{x}_c)}{\partial x_j} = 0\,, \qquad \frac{\partial h(x_j, \mathbf{x}_c)}{\partial x_i} = 0\,. \tag{5.168}$$

All these derivatives with regard to a single variable are only a special case of the vectorial derivatives. Replacing the single variable x_i by the set of variables $\mathbf{x}_a$ and x_j by $\mathbf{x}_b$, we get the generalized specification of the vectorial bi-decompositions:

$$\frac{\partial f(\mathbf{x})}{\partial \mathbf{x}_a} \neq 0\,, \qquad \frac{\partial f(\mathbf{x})}{\partial \mathbf{x}_b} \neq 0\,, \tag{5.169}$$

$$\frac{\partial g(\mathbf{x})}{\partial \mathbf{x}_b} = 0\,, \qquad \frac{\partial h(\mathbf{x})}{\partial \mathbf{x}_a} = 0\,. \tag{5.170}$$

It is known that in general all variables $(\mathbf{x})$ appear in the result of a vectorial derivative operation with regard to either $\mathbf{x}_a$ or $\mathbf{x}_b$. However, in the case that all formulas of (5.169) and (5.170) are satisfied, we have

$$\mathbf{rank}(g(\mathbf{x})) > \mathbf{rank}(f(\mathbf{x})) \tag{5.171}$$

$$\mathbf{rank}(h(\mathbf{x})) > \mathbf{rank}(f(\mathbf{x})) \tag{5.172}$$

so that both the function g and h are simpler than the function f. Taking into account that there are $2^n - 1$ vectorial derivatives but only n single derivatives for a Boolean space $\mathbb{B}^n$, the vectorial bi-decompositions significantly extend the approach of bi-decompositions.

The vectorial bi-decomposition of a single function $f(\mathbf{x})$ is the special case for $f_q(\mathbf{x}) = f(\mathbf{x})$ and $f_r(\mathbf{x}) = \overline{f(\mathbf{x})}$. Hence, the following exploration of vectorial bi-decompositions for lattices of functions comprises vectorial bi-decompositions for a single function as well. A formal specification of all three types of vectorial bi-decompositions is given in the following definition.

Definition 5.28 Vectorial Bi-Decomposition for Generalized Lattices. A lattice

$$L\langle f_q(\mathbf{x}), f_r(\mathbf{x}), f^{id}(\mathbf{x})\rangle \tag{5.173}$$

contains at least one function $f_i(\mathbf{x})$ that is OR-, AND-, or EXOR-bi-decomposable with regard to the subsets of variables $\mathbf{x}_a \subseteq \mathbf{x}$ and $\mathbf{x}_b \subseteq \mathbf{x}$, $\mathbf{x}_a \neq \mathbf{x}_b$, if

1. $f_i(\mathbf{x})$ can be expressed by

$$f_i(\mathbf{x}) = g(\mathbf{x}) \bullet h(\mathbf{x}) \tag{5.174}$$

where $\bullet \in \{\vee, \wedge, \oplus\}$;

2. the decomposition functions $g(\mathbf{x})$ and $h(\mathbf{x})$ satisfy

$$\frac{\partial g(\mathbf{x})}{\partial \mathbf{x}_b} = 0 , \tag{5.175}$$

$$\frac{\partial h(\mathbf{x})}{\partial \mathbf{x}_a} = 0 ; \tag{5.176}$$

3. and the functions of the given lattice depend on the simultaneous change of both $\mathbf{x}_a$ and $\mathbf{x}_b$:

$$\mathrm{MIDC}(\mathrm{IDM}(f), \mathbf{x}_a) \neq \mathbf{0} , \tag{5.177}$$

$$\mathrm{MIDC}(\mathrm{IDM}(f), \mathbf{x}_b) \neq \mathbf{0} . \tag{5.178}$$

Both the condition to check whether a generalized lattice contains at least one function that is vectorial OR-bi-decomposable and formulas to calculate the generalized lattices of the decomposition function g and h are given in the following theorem.

Theorem 5.29 Vectorial OR-Bi-Decomposition for a Generalized Lattice.
A lattice specified by (5.173) *that satisfies* (5.177) *and* (5.178) *contains at least one function* $f(\mathbf{x})$ *that is vectorial OR-bi-decomposable with regard to the dedicated sets* $\mathbf{x}_a \subseteq \mathbf{x}$ *and* $\mathbf{x}_b \subseteq \mathbf{x}$, $\mathbf{x}_a \neq \mathbf{x}_b$, *if and only if*

$$f_q(\mathbf{x}) \wedge \max_{\mathbf{x}_a} f_r(\mathbf{x}) \wedge \max_{\mathbf{x}_b} f_r(\mathbf{x}) = 0 . \tag{5.179}$$

Possible decomposition functions $g_i(\mathbf{x})$ *can be chosen from the lattice* $L\left\langle g_q(\mathbf{x}), g_r(\mathbf{x}), g^{id}(\mathbf{x})\right\rangle$ *with*

$$g_q(\mathbf{x}) = \max_{\mathbf{x}_b}(f_q(\mathbf{x}) \wedge \max_{\mathbf{x}_a} f_r(\mathbf{x})) , \tag{5.180}$$

$$g_r(\mathbf{x}) = \max_{\mathbf{x}_b} f_r(\mathbf{x}) , \tag{5.181}$$

$$g^{id}(\mathbf{x}) = f^{id}(\mathbf{x}) \vee \frac{\partial f(\mathbf{x})}{\partial \mathbf{x}_b} , \tag{5.182}$$

so that

$$IDM(g) = UM(IDM(f), \mathbf{x}_b) . \tag{5.183}$$

Assuming that the function $g(\mathbf{x})$ *is selected from this lattice to realize the circuit, possible decomposition functions* $h_j(\mathbf{x})$ *can be chosen from the lattice* $L\left\langle h_q(\mathbf{x}), h_r(\mathbf{x}), h^{id}(\mathbf{x})\right\rangle$ *with*

$$h_q(\mathbf{x}) = \max_{\mathbf{x}_a}(f_q(\mathbf{x}) \wedge \overline{g(\mathbf{x})}) , \tag{5.184}$$

$$h_r(\mathbf{x}) = \max_{\mathbf{x}_a} f_r(\mathbf{x}) , \tag{5.185}$$

$$h^{id}(\mathbf{x}) = f^{id}(\mathbf{x}) \vee \frac{\partial f(\mathbf{x})}{\partial \mathbf{x}_a} , \tag{5.186}$$

so that

$$IDM(h) = UM(IDM(f), \mathbf{x}_a) \,. \tag{5.187}$$

The directions of change of a vectorial bi-decomposition are specified by the subsets of variables $\mathbf{x}_a$ and $\mathbf{x}_b$. These two directions of change determine quadruples of nodes in the Boolean space. Such quadruples of nodes are also evaluated for strong bi-decompositions with regard to single variables x_i and x_j. The formulas of Theorem 5.29 transform the known condition of the strong OR-bi-decomposition with regard to x_i and x_j to the vectorial OR-bi-decomposition with regard to $\mathbf{x}_a$ and $\mathbf{x}_b$. The complexities of the decomposition lattices L_g and L_h are reduced due to the general requirements (5.175) and (5.176) of each vectorial bi-decomposition.

The duality between the Boolean Algebras $(\mathbb{B}^n, \wedge, \vee, ^{-}, \mathbf{0}, \mathbf{1})$ and $(\mathbb{B}^n, \vee, \wedge, ^{-}, \mathbf{1}, \mathbf{0})$ is the basis for the similarity between the vectorial OR- and AND-bi-decomposition. Hence, Theorem 5.30 generalizes the strong AND-bi-decomposition with regard to the variables x_i and x_j to the vectorial AND-bi-decomposition with regard to the subsets of variables $\mathbf{x}_a$ and $\mathbf{x}_b$ in the same manner as before in Theorem 5.29 for the OR-bi-decomposition.

Theorem 5.30 Vectorial AND-Bi-Decomposition for a Generalized Lattice.
A lattice specified by (5.173) that satisfies (5.177) and (5.178) contains at least one function $f(\mathbf{x})$ that is vectorial AND-bi-decomposable with regard to the dedicated sets $\mathbf{x}_a \subseteq \mathbf{x}$ and $\mathbf{x}_b \subseteq \mathbf{x}$, $\mathbf{x}_a \neq \mathbf{x}_b$, if and only if

$$f_r(\mathbf{x}) \wedge \max_{\mathbf{x}_a} f_q(\mathbf{x}) \wedge \max_{\mathbf{x}_b} f_q(\mathbf{x}) = 0 \,. \tag{5.188}$$

Possible decomposition functions $g_i(\mathbf{x})$ can be chosen from the lattice $L\langle g_q(\mathbf{x}), g_r(\mathbf{x}), g^{id}(\mathbf{x})\rangle$ with

$$g_q(\mathbf{x}) = \max_{\mathbf{x}_b} f_q(\mathbf{x}) \,, \tag{5.189}$$

$$g_r(\mathbf{x}) = \max_{\mathbf{x}_b}(f_r(\mathbf{x}) \wedge \max_{\mathbf{x}_a} f_q(\mathbf{x})) \,, \tag{5.190}$$

$$g^{id}(\mathbf{x}) = f^{id}(\mathbf{x}) \vee \frac{\partial f(\mathbf{x})}{\partial \mathbf{x}_b} \,, \tag{5.191}$$

so that

$$IDM(g) = UM(IDM(f), \mathbf{x}_b) \,. \tag{5.192}$$

Assuming that the function $g(\mathbf{x})$ is selected from this lattice to realize the circuit, possible decomposition functions $h_j(\mathbf{x})$ can be chosen from the lattice $L\langle h_q(\mathbf{x}), h_r(\mathbf{x}), h^{id}(\mathbf{x})\rangle$ with

$$h_q(\mathbf{x}) = \max_{\mathbf{x}_a} f_q(\mathbf{x}) \,, \tag{5.193}$$

$$h_r(\mathbf{x}) = \max_{\mathbf{x}_a}(f_r(\mathbf{x}) \wedge g(\mathbf{x})) \,, \tag{5.194}$$

$$h^{id}(\mathbf{x}) = f^{id}(\mathbf{x}) \vee \frac{\partial f(\mathbf{x})}{\partial \mathbf{x}_a} \,, \tag{5.195}$$

so that

$$IDM(h) = UM(IDM(f), \mathbf{x}_a) . \tag{5.196}$$

Finally, we generalize the strong EXOR-bi-decomposition with regard to the variables x_i and x_j to the vectorial EXOR-bi-decomposition with regard of the subsets of variables $\mathbf{x}_a$ and $\mathbf{x}_b$.

Theorem 5.31 Vectorial EXOR-Bi-Decomposition for a Generalized Lattice.
A lattice specified by (5.173) that satisfies (5.177) and (5.178) contains at least one function $f(\mathbf{x})$ that is vectorial EXOR-bi-decomposable with regard to the dedicated sets $\mathbf{x}_a \subseteq \mathbf{x}$ and $\mathbf{x}_b \subseteq \mathbf{x}$, $\mathbf{x}_a \neq \mathbf{x}_b$, if and only if

$$\max_{\mathbf{x}_b} \left(f_q^{\partial \mathbf{x}_a}(\mathbf{x}) \right) \wedge f_r^{\partial \mathbf{x}_a}(\mathbf{x}) = 0 . \tag{5.197}$$

Possible decomposition functions $g_i(\mathbf{x})$ can be chosen from the lattice $L\left\langle g_q(\mathbf{x}), g_r(\mathbf{x}), g^{id}(\mathbf{x})\right\rangle$ with $x_{bi} \in \mathbf{x}_b$

$$g_q(\mathbf{x}) = x_{bi} \wedge \max_{\mathbf{x}_b} \left(f_q^{\partial \mathbf{x}_a}(\mathbf{x}) \right) , \tag{5.198}$$

$$g_r(\mathbf{x}) = \max_{\mathbf{x}_b} \left(f_r^{\partial \mathbf{x}_a}(\mathbf{x}) \right) , \tag{5.199}$$

$$g^{id}(\mathbf{x}) = f^{id}(\mathbf{x}) \vee \frac{\partial f(\mathbf{x})}{\partial \mathbf{x}_b} , \tag{5.200}$$

so that

$$IDM(g) = UM(IDM(f), \mathbf{x}_b) . \tag{5.201}$$

Assuming that the function $g(\mathbf{x})$ is selected from this lattice to realize the circuit, possible decomposition functions $h_j(\mathbf{x})$ can be chosen from the lattice $L\left\langle h_q(\mathbf{x}), h_r(\mathbf{x}), h^{id}(\mathbf{x})\right\rangle$ with

$$h_q(\mathbf{x}) = \max_{\mathbf{x}_a} \left((f_q(\mathbf{x}) \wedge \overline{g(\mathbf{x})}) \vee (f_r(\mathbf{x}) \wedge g(\mathbf{x})) \right) , \tag{5.202}$$

$$h_r(\mathbf{x}) = \max_{\mathbf{x}_a} \left((f_r(\mathbf{x}) \wedge \overline{g(\mathbf{x})}) \vee (f_q(\mathbf{x}) \wedge g(\mathbf{x})) \right) , \tag{5.203}$$

$$h^{id}(\mathbf{x}) = f^{id}(\mathbf{x}) \vee \frac{\partial f(\mathbf{x})}{\partial \mathbf{x}_a} , \tag{5.204}$$

so that

$$IDM(h) = UM(IDM(f), \mathbf{x}_a) . \tag{5.205}$$

The known condition of the strong EXOR-bi-decomposition with regard to the single variables x_i and x_j

$$\frac{\partial^2 f(\mathbf{x})}{\partial x_j \partial x_i} = \frac{\partial}{\partial x_j} \left(\frac{\partial f(\mathbf{x})}{\partial x_i} \right) = 0 \tag{5.206}$$

can be generalized to the condition (5.207)

$$\frac{\partial}{\partial \mathbf{x}_b}\left(\frac{\partial f(\mathbf{x})}{\partial \mathbf{x}_a}\right) = 0 \tag{5.207}$$

of a vectorial EXOR-bi-decomposition with regard to the dedicated sets of variables $\mathbf{x}_a$ and $\mathbf{x}_b$. Condition (5.197) is the generalization of (5.207) for the more general case of a lattice of functions.

The remaining formulas of Theorem 5.31 adapt the known formulas for the strong EXOR-bi-decomposition of a lattice with regard to single variables x_i and x_j to the new vectorial EXOR-bi-decomposition of a generalized lattice with regard to $\mathbf{x}_a$ and $\mathbf{x}_b$.

Example 5.32 Complete Synthesis Using Vectorial and Strong Bi-Decomposition. A multilevel combinational circuit must be synthesized for an arbitrary function of the given lattice $L\langle f_q(\mathbf{x}), f_r(\mathbf{x})\rangle$ of five variables that is specified by the mark functions f_q (5.165) and f_r (5.166) introduced in Example 5.27. For a direct comparison we assume again that AND-, OR-, or EXOR-gates with two optionally negated inputs can be used as elements for this circuit.

Figure 5.26a repeats the Karnaugh-map of this lattice. The synthesized circuit is depicted in Figure 5.26b where the indices of the decomposition functions g and h are again assigned in the order of applied bi-decomposition.

We know from Example 5.27 that no strong bi-decomposition for any function of the given lattice exists. However, there is a vectorial OR-bi-decomposition with regard to $\mathbf{x}_a = (x_2, x_3)$ and $\mathbf{x}_b = (x_1, x_5)$. It can be seen in Figure 5.26b that both the decomposition function g_1 and h_1 depend on all five variables, but both g_1 and h_1 are simpler than all functions of the given lattice because g_1 does not depend on the simultaneous change of x_1 and x_5 and h_1 does not depend on the simultaneous change of x_2 and x_3.

It is known from Steinbach [2016] that the decomposition function of a vectorial bi-decomposition with regard to two variables in one dedicated set satisfies the condition of a strong EXOR-bi-decomposition with regard to these variables. The lattice of the decomposition function g_1 even contains a function for which a disjoint strong AND-bi-decomposition with regard to $\mathbf{x}_a = (x_1, x_5)$ and $\mathbf{x}_b = (x_2, x_3, x_4)$ exists. One more disjoint strong AND-bi-decomposition with regard to $\mathbf{x}_a = x_2$ and $\mathbf{x}_b = (x_3, x_4)$ of the decomposition function h_2 terminates the synthesis of g_1.

A disjoint strong AND-bi-decomposition with regard to $\mathbf{x}_a = (x_2, x_3)$ and $\mathbf{x}_b = (x_1, x_4, x_5)$ followed by a disjoint strong EXOR-bi-decomposition with regard to $\mathbf{x}_a = x_1$ and $\mathbf{x}_b = (x_4, x_5)$ determines the synthesis of the circuit structure of h_1.

The synthesized circuit of Figure 5.26b realizes the function $y = f(\mathbf{x})$ for which the Karnaugh-map is shown in Figure 5.26c. Obviously, we have again: $f(\mathbf{x}) = f_q(\mathbf{x})$. That means the circuits of Figure 5.23b and Figure 5.26b realize the same function.

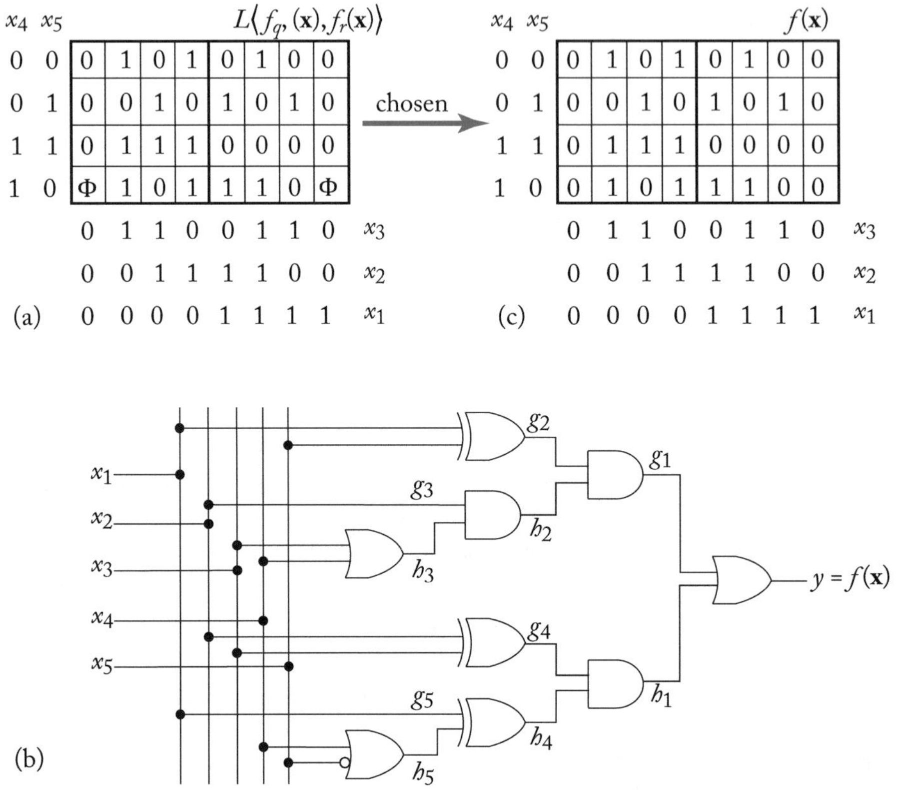

Figure 5.26: Complete synthesis by vectorial and strong bi-decompositions: (a) Karnaugh-map of the given lattice; (b) structure of the circuit; and (c) Karnaugh-map of the realized function.

Example 5.32 demonstrates the benefits of the synthesis using a vectorial bi-decomposition together with all types of strong bi-decompositions. The used vectorial OR-bi-decomposition avoids in this example weak bi-decompositions at all. We compare the circuit of Figure 5.23b synthesized by strong and weak bi-decompositions with the circuit of Figure 5.26b that additionally utilize vectorial bi-decompositions.

- Figure 5.23b shows that 14 two-input gates with a longest path of 6 levels represent the function $f(\mathbf{x}) = f_q(\mathbf{x})$ (5.165) synthesized by means of strong and weak bi-decompositions.
- The additional use of a vectorial OR-bi-decomposition leads to the circuit of Figure 5.26b. Even though both the decomposition functions g_1 and h_1 determined by the vectorial OR-bi-decomposition depend on all 5 variables, the whole circuit only needs 9 gates on 4 levels.

Taking the number of gates as a raw measure of the power consumption and the number of levels as a raw measure of the delay, the benefit of the synthesis by bi-decompositions including the vectorial one becomes clearly recognizable for these important properties of combinational circuits.

SUMMARY

We demonstrated in this chapter that definitions, axioms, and theorems of Boolean calculations together with the Boolean Differential Calculus are strong tools to solve many tasks around digital circuits and logic systems efficiently.

- The Boolean Differential Calculus simplifies the detection of properties of Boolean functions.
- Each system of Boolean equations can be transformed into a single equivalent Boolean equation. The solution of a Boolean equation with regard to variables provides by means of the Boolean Differential Calculus the reverse direction of this task.
- Edges describe the change between the vertices of the graph. Differentials of Boolean variables are an appropriate aid to describe such edges.
- The general analysis task of a digital circuit is the calculation of the behavior for a given structure of the circuit. Boolean equations solve this task for several types of digital circuits and provide a white-box description. A more compact black-box description of the behavior can be computed using a derivative operation of the Boolean Differential Calculus.
- The operations of the Boolean Differential Calculus facilitate the answer to many special analysis questions, as, e.g., hazard detection, influence of selected variables on certain outputs of the circuit, degree of linearity, conflicts in bus structures, and many more.
- The system equation is a general model to describe the behavior of a combinational or sequential circuit. The synthesis of combinational circuits requires the solution of the system equation with regard to the output variables. The synthesis of sequential circuits additionally requires the solution of flip-flop equations with regard to the control variables. All these tasks can efficiently by solved by means of the Boolean Differential Calculus even if lattices of Boolean functions are utilized for optimization reasons.
- The test of digital circuits requires the evaluation whether the change between a fault-free and a faulty circuit can be detected. Again, derivative operations of Boolean Differential Calculus helps to solve this problem.
- The bi-decomposition leads to circuit structures with a short delay due to the optimal utilization of the properties of the function for which a circuit structure is needed. The

knowledge of the Boolean Differential Calculus was the prerequisite to find this powerful synthesis method and to prove their completeness.

- Recently, the vectorial bi-decomposition was suggested for further improvements. Lattices of Boolean functions increase the search space to detect an optimal circuit structure. The complete evaluation of such lattices by derivative operations for lattices of Boolean functions improves this powerful method even more.

These examples of applications should support the brainstorming to solve further so far open problems by means of the Boolean Differential Calculus.

EXERCISES

5.1 A combinational circuit has the three inputs x_1, x_2, and x_3 and the two outputs y_1 and y_2. Its allowed behavior has been determined by the system equation

$$\begin{aligned} F(\mathbf{x}, \mathbf{y}) &= \overline{x}_2\overline{x}_3 y_1 \vee \overline{x}_1\overline{x}_2 x_3 \overline{y}_1 y_2 \vee \overline{x}_1 x_2 x_3 y_1 y_2 \vee x_2\overline{x}_3\overline{y}_1 y_2 \vee x_1 x_2 x_3 \overline{y}_1\overline{y}_2 \vee x_1\overline{x}_2 y_1 \\ &= 1 \,. \end{aligned} \tag{5.208}$$

(a) Is the circuit realizable? Is there a solution with regard to y_1 and y_2?

(b) Are the functions $y_1 = f_1(\mathbf{x})$ and $y_2 = f_2(\mathbf{x})$ uniquely specified?

(c) Calculate these functions or the mark functions of the lattice from which they can be chosen. Select a simple function in the case of a lattice.

5.2 Calculate all test patterns for the source and both sinks of the inner branch in the combinational circuit shown in Figure 5.27.

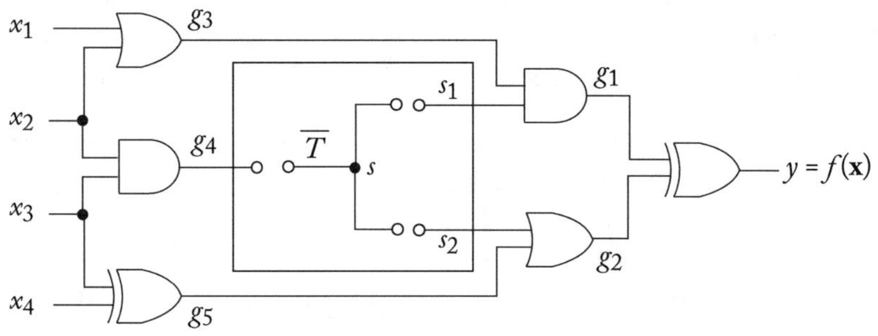

Figure 5.27: Combinational circuit with a highlighted inner branch.

5.3 The behavior of a combinational circuit is specified by the mark functions

$$f_q(\mathbf{x}) = \overline{x}_1\overline{x}_2\overline{x}_3\overline{x}_4 \vee \overline{x}_1 x_2\overline{x}_3 x_4 \vee x_1 x_2 x_4 \vee x_1\overline{x}_2 x_3\overline{x}_4 \tag{5.209}$$

$$f_r(\mathbf{x}) = \overline{x}_1\overline{x}_2 x_4 \vee \overline{x}_1 x_2 x_3 \vee x_2\overline{x}_3\overline{x}_4 \vee x_1\overline{x}_2 x_4 \,. \tag{5.210}$$

(a) For which pairs of variables contains this lattice at least one function of an OR-, AND-, or EXOR-bi-decomposition.

(b) For which of the found bi-decompositions can one or more variables from the so far common set $\mathbf{x}_c$ moved into the dedicated sets $\mathbf{x}_a$ or $\mathbf{x}_b$ so that a more compact bi-decomposition can be realized? Is there a disjoint bi-decomposition?

(c) Calculate the decomposition functions for the most compact bi-decomposition and draw the found circuit structure.

5.4 The special case of a disjoint bi-decomposition where the function g is equal to the variable x_i or its complement is also denoted as *separation of a variable*. Figure 5.28 shows the circuit structure where the variable x_i is separated from the function $f(x_i, \mathbf{x}_1)$ using an OR-gate.

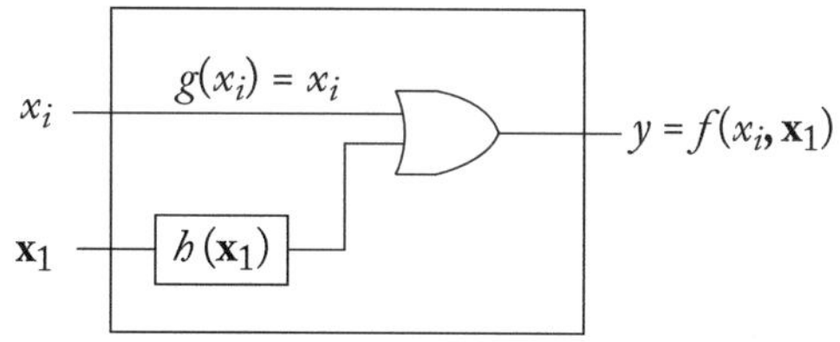

Figure 5.28: Separation of the variable x_i from the function $f(x_i, \mathbf{x}_1)$ using an OR-gate.

(a) Determine a condition using a derivative operation that must be satisfied so that the separation of x_i from the function $f(x_i, \mathbf{x}_1)$ can be realized using an OR-gate, as shown in Figure 5.28. Verify the correctness of the determined condition.

(b) How can the OR-separation function $h(\mathbf{x}_1)$ be calculated when this separation is possible? Verify the correctness of the formula that specifies the separation function $h(\mathbf{x}_1)$.

Such conditions for the separation of x_i or $\overline{x}_i$ and the associated formulas to calculate $h(\mathbf{x}_1)$ can be determined for OR-, AND-, and EXOR-gates in additional exercises. The generalization for lattices even more extends the skills to use the Boolean Differential Calculus to solve practical problems.

CHAPTER 6

Solutions of the Exercises

6.1 SOLUTIONS OF CHAPTER 1

1.1 Identical values in the columns 5, 7, and 8 of the following table confirm the associativity of the conjunction.

x_1	x_2	x_3	$(x_2 \vee x_3)$	$x_1 \vee (x_2 \vee x_3)$	$(x_1 \vee x_2)$	$(x_1 \vee x_2) \vee x_3$	$x_1 \vee x_2 \vee x_3$
0	0	0	0	**0**	0	**0**	**0**
0	0	1	1	**1**	0	**1**	**1**
0	1	0	1	**1**	1	**1**	**1**
0	1	1	1	**1**	1	**1**	**1**
1	0	0	0	**1**	1	**1**	**1**
1	0	1	1	**1**	1	**1**	**1**
1	1	0	1	**1**	1	**1**	**1**
1	1	1	1	**1**	1	**1**	**1**

1.2 The following table contains both intermediate results and the wanted function.

x_1	x_2	x_3	x_4	$(x_1 \vee x_2)$	$(x_1 \oplus x_3)$	$(x_2 \oplus x_4)$	$((x_1 \oplus x_3) \vee (x_2 \oplus x_4))$	f
0	0	0	0	1	0	1	1	1
0	0	0	1	1	0	0	0	0
0	0	1	0	1	1	1	1	1
0	0	1	1	1	1	0	1	1
0	1	0	0	0	0	0	0	0
0	1	0	1	0	0	1	1	0
0	1	1	0	0	1	0	1	0
0	1	1	1	0	1	1	1	0
1	0	0	0	1	1	1	1	1
1	0	0	1	1	1	0	1	1
1	0	1	0	1	0	1	1	1
1	0	1	1	1	0	0	0	0
1	1	0	0	1	1	0	1	1
1	1	0	1	1	1	1	1	1
1	1	1	0	1	0	0	0	0
1	1	1	1	1	0	1	1	1

1.3 (a) The lattice consists of $2^3 = 8$ functions due to the three don't-cares.

(b) The disjunctive normal forms of the eight functions of the lattice are:

$$\begin{aligned}
f_1(x_1, x_2, x_3) &= \overline{x}_1\overline{x}_2\overline{x}_3 \vee \overline{x}_1\overline{x}_2x_3 \vee x_1x_2x_3 \,, \\
f_2(x_1, x_2, x_3) &= \overline{x}_1\overline{x}_2\overline{x}_3 \vee \overline{x}_1\overline{x}_2x_3 \vee x_1x_2x_3 \vee \overline{x}_1x_2x_3 \,, \\
f_3(x_1, x_2, x_3) &= \overline{x}_1\overline{x}_2\overline{x}_3 \vee \overline{x}_1\overline{x}_2x_3 \vee x_1x_2x_3 \vee x_1\overline{x}_2x_3 \,, \\
f_4(x_1, x_2, x_3) &= \overline{x}_1\overline{x}_2\overline{x}_3 \vee \overline{x}_1\overline{x}_2x_3 \vee x_1x_2x_3 \vee x_1x_2\overline{x}_3 \,, \\
f_5(x_1, x_2, x_3) &= \overline{x}_1\overline{x}_2\overline{x}_3 \vee \overline{x}_1\overline{x}_2x_3 \vee x_1x_2x_3 \vee \overline{x}_1x_2x_3 \vee x_1\overline{x}_2x_3 \,, \\
f_6(x_1, x_2, x_3) &= \overline{x}_1\overline{x}_2\overline{x}_3 \vee \overline{x}_1\overline{x}_2x_3 \vee x_1x_2x_3 \vee \overline{x}_1x_2x_3 \vee x_1x_2\overline{x}_3 \,, \\
f_7(x_1, x_2, x_3) &= \overline{x}_1\overline{x}_2\overline{x}_3 \vee \overline{x}_1\overline{x}_2x_3 \vee x_1x_2x_3 \vee x_1\overline{x}_2x_3 \vee x_1x_2\overline{x}_3 \,, \\
f_8(x_1, x_2, x_3) &= \overline{x}_1\overline{x}_2\overline{x}_3 \vee \overline{x}_1\overline{x}_2x_3 \vee x_1x_2x_3 \vee \overline{x}_1x_2x_3 \vee x_1\overline{x}_2x_3 \vee x_1x_2\overline{x}_3 \,.
\end{aligned}$$

(c) The simplified disjunctive forms of the eight functions of the lattice are:

$$\begin{aligned}
f_1(x_1, x_2, x_3) &= \overline{x}_1\overline{x}_2 \vee x_1x_2x_3 \,, \\
f_2(x_1, x_2, x_3) &= \overline{x}_1\overline{x}_2 \vee x_2x_3 \,, \\
f_3(x_1, x_2, x_3) &= \overline{x}_1\overline{x}_2 \vee x_1x_3 \,, \\
f_4(x_1, x_2, x_3) &= \overline{x}_1\overline{x}_2 \vee x_1x_2 \,,
\end{aligned}$$

$$f_5(x_1,x_2,x_3) = \overline{x}_1\overline{x}_2 \vee x_3 \ ,$$
$$f_6(x_1,x_2,x_3) = \overline{x}_1\overline{x}_2 \vee x_2x_3 \vee x_1x_2 \ ,$$
$$f_7(x_1,x_2,x_3) = \overline{x}_1\overline{x}_2 \vee x_1x_3 \vee x_1x_2 \ ,$$
$$f_8(x_1,x_2,x_3) = \overline{x}_1\overline{x}_2 \vee x_3 \vee x_1x_2 \ .$$

(d) The solution of (b) shows: $\min(f_1,\ldots,f_8) = f_1$ and $\max(f_1,\ldots,f_8) = f_8$. The wanted pairs of functions are:

$$\begin{array}{lll} f_1(x_1,x_2,x_3),\ f_8(x_1,x_2,x_3): & f_1 \wedge f_8 = f_1 \ , & f_1 \vee f_8 = f_8 \ , \\ f_2(x_1,x_2,x_3),\ f_7(x_1,x_2,x_3): & f_2 \wedge f_7 = f_1 \ , & f_2 \vee f_7 = f_8 \ , \\ f_3(x_1,x_2,x_3),\ f_6(x_1,x_2,x_3): & f_3 \wedge f_6 = f_1 \ , & f_3 \vee f_6 = f_8 \ , \\ f_4(x_1,x_2,x_3),\ f_5(x_1,x_2,x_3): & f_4 \wedge f_5 = f_1 \ , & f_4 \vee f_5 = f_8 \ . \end{array}$$

(e) The solution of (c) shows that $f_5(x_1,x_2,x_3) = \overline{x}_1\overline{x}_2 \vee x_3$ is the simplest disjunctive form of these eight functions.

1.4 The left-hand side of Equation (1.73) is equal to

$$\begin{array}{ll} 0: & S_0^l(x_1,x_2,x_3) = \{(000),(001),(011),(100),(101)\} \ , \\ 1: & S_1^l(x_1,x_2,x_3) = \{(010),(110),(111)\} \ . \end{array}$$

The right-hand side of Equation (1.73) is equal to

$$\begin{array}{ll} 0: & S_0^r(x_1,x_3,x_4) = \{(000),(011)\} \ , \\ 1: & S_1^r(x_1,x_3,x_4) = \{(001),(010),(100),(101),(110),(111)\} \ . \end{array}$$

Hence, the BVL of the solution set S and the simplified orthogonal TVL are

$$S = \begin{array}{cccc} x_1 & x_2 & x_3 & x_4 \\ \hline 0 & 0 & 0 & 0 \\ 0 & 0 & 1 & 1 \\ 0 & 1 & 1 & 1 \\ 0 & 1 & 0 & 1 \\ 1 & 1 & 0 & 0 \\ 1 & 1 & 0 & 1 \\ 1 & 1 & 1 & 0 \\ 1 & 1 & 1 & 1 \\ \hline \end{array} = \begin{array}{cccc} x_1 & x_2 & x_3 & x_4 \\ \hline 0 & 0 & 0 & 0 \\ 0 & - & 1 & 1 \\ 0 & 1 & 0 & 1 \\ 1 & 1 & - & - \\ \hline \end{array} .$$

1.5 (a) The second and the fourth ternary vector of the TVL

x_1	x_2	x_3	x_4
1	–	0	0
–	1	1	–
0	–	0	–
–	–	1	1

are not orthogonal to each other, because not any component contains a 01 combination. The common intersection vector of these two ternary vectors is (-111).

(b) Removing the intersection vector from one of these vectors (the fourth vector is used) leads from the D-form to the wanted orthogonal TVL in ODA-form.

$$D(f_1) = \begin{array}{cccc} x_1 & x_2 & x_3 & x_4 \\ \hline 1 & - & 0 & 0 \\ - & 1 & 1 & - \\ 0 & - & 0 & - \\ - & - & 1 & 1 \\ \hline \end{array} \quad \Longrightarrow \quad ODA(f_1) = \begin{array}{cccc} x_1 & x_2 & x_3 & x_4 \\ \hline 1 & - & 0 & 0 \\ - & 1 & 1 & - \\ 0 & - & 0 & - \\ - & 0 & 1 & 1 \\ \hline \end{array} .$$

(c) Removing the intersection vector from both of these vectors leads from the A-form to the wanted orthogonal TVL in ODA-form.

$$A(f_2) = \begin{array}{cccc} x_1 & x_2 & x_3 & x_4 \\ \hline 1 & - & 0 & 0 \\ - & 1 & 1 & - \\ 0 & - & 0 & - \\ - & - & 1 & 1 \\ \hline \end{array} \quad \Longrightarrow \quad ODA(f_2) = \begin{array}{cccc} x_1 & x_2 & x_3 & x_4 \\ \hline 1 & - & 0 & 0 \\ - & 1 & 1 & 0 \\ 0 & - & 0 & - \\ - & 0 & 1 & 1 \\ \hline \end{array} .$$

6.2 SOLUTIONS OF CHAPTER 2

2.1 The transformation of the given expression into an antivalence form simplifies the calculation of the vectorial derivative:

$$\begin{aligned}
f(x_1,x_2,x_3,x_4) &= (x_1\overline{x}_2 \oplus \overline{x}_3x_4) \vee \overline{x}_1x_4 \\
&= (x_1\overline{x}_2 \oplus \overline{x}_3x_4) \oplus \overline{x}_1x_4 \oplus (x_1\overline{x}_2 \oplus \overline{x}_3x_4)\overline{x}_1x_4 \\
&= x_1\overline{x}_2 \oplus \overline{x}_3x_4 \oplus \overline{x}_1x_4 \oplus \overline{x}_1\overline{x}_3x_4 \\
&= x_1\overline{x}_2 \oplus \overline{x}_3x_4 \oplus \overline{x}_1x_3x_4 \,.
\end{aligned}$$

Using Definition (2.1) we get the vectorial derivative:

$$\begin{aligned}
\frac{\partial f(\mathbf{x})}{\partial(x_1,x_3)} &= f(x_1,x_2,x_3,x_4) \oplus f(\overline{x}_1,x_2,\overline{x}_3,x_4) \\
&= x_1\overline{x}_2 \oplus \overline{x}_3x_4 \oplus \overline{x}_1x_3x_4 \oplus \overline{x}_1\overline{x}_2 \oplus x_3x_4 \oplus x_1\overline{x}_3x_4 \\
&= \overline{x}_2(x_1 \oplus \overline{x}_1) \oplus x_4(\overline{x}_3 \oplus x_3) \oplus x_4(\overline{x}_1x_3 \oplus x_1\overline{x}_3) \\
&= \overline{x}_2 \oplus x_4 \oplus x_4x_1 \oplus x_4x_3 \\
&= \overline{x}_2 \oplus \overline{x}_1x_4 \oplus x_3x_4 \,.
\end{aligned}$$

2.2 (a) A function of 4 variables has $\binom{4}{2} = 6$ different pairs of variables.

(b) Three vectorial derivatives of (2.9) are sufficient to verify whether (2.156) is symmetric with regard to all pairs of variables; the symmetry of the other three directions of change follows implicitly.

(c)

$$\begin{aligned}
\left((x_1 \oplus x_2) \wedge \frac{\partial f(x_1,x_2,x_3,x_4)}{\partial(x_1,x_2)}\right) &\vee \\
\left((x_1 \oplus x_3) \wedge \frac{\partial f(x_1,x_2,x_3,x_4)}{\partial(x_1,x_3)}\right) &\vee \\
\left((x_1 \oplus x_4) \wedge \frac{\partial f(x_1,x_2,x_3,x_4)}{\partial(x_1,x_4)}\right) &= 0 \\
((x_1 \oplus x_2) \wedge ((x_1x_2 \oplus x_3x_4) \oplus (x_1 \oplus x_2)(x_3 \oplus x_4) \quad &\oplus \\
(\overline{x}_1\overline{x}_2 \oplus x_3x_4) \oplus (\overline{x}_1 \oplus \overline{x}_2)(x_3 \oplus x_4))) &\vee \\
((x_1 \oplus x_3) \wedge ((x_1x_2 \oplus x_3x_4) \oplus (x_1 \oplus x_2)(x_3 \oplus x_4) \quad &\oplus \\
(\overline{x}_1x_2 \oplus \overline{x}_3x_4) \oplus (\overline{x}_1 \oplus x_2)(\overline{x}_3 \oplus x_4))) &\vee \\
((x_1 \oplus x_4) \wedge ((x_1x_2 \oplus x_3x_4) \oplus (x_1 \oplus x_2)(x_3 \oplus x_4) \quad &\oplus \\
(\overline{x}_1x_2 \oplus x_3\overline{x}_4) \oplus (\overline{x}_1 \oplus x_2)(x_3 \oplus \overline{x}_4))) &= 0 \\
((x_1 \oplus x_2) \wedge (x_1x_2 \oplus \overline{x}_1\overline{x}_2)) &\vee \\
((x_1 \oplus x_3) \wedge (x_1x_3 \oplus \overline{x}_1\overline{x}_3)) &\vee \\
((x_1 \oplus x_4) \wedge (x_1x_4 \oplus \overline{x}_1\overline{x}_4)) &= 0 \\
(x_1 \oplus x_2)\overline{(x_1 \oplus x_2)} \vee (x_1 \oplus x_3)\overline{(x_1 \oplus x_3)} \vee (x_1 \oplus x_4)\overline{(x_1 \oplus x_4)} &= 0 \\
0 &= 0 \,.
\end{aligned}$$

Condition (2.9) is satisfied for the function (2.156); hence, this function is symmetric with regard to all pairs of variables.

(d) Using the distributivity (1.46) and the commutativity (1.36) we get for for the function (2.156):

$$\begin{aligned} f(x_1, x_2, x_3, x_4) &= (x_1x_2 \oplus x_3x_4) \oplus (x_1 \oplus x_2)(x_3 \oplus x_4) \\ &= (x_1x_2 \oplus x_3x_4) \oplus (x_1x_3 \oplus x_1x_4 \oplus x_2x_3 \oplus x_2x_4) \\ &= x_1x_2 \oplus x_1x_3 \oplus x_1x_4 \oplus x_2x_3 \oplus x_2x_4 \oplus x_3x_4 \; . \end{aligned}$$

All six pairs of variables appear in exactly one conjunction of this antivalence form and confirm that the function (2.156) is symmetric with regard to all pairs of variables. The transformation into the requested antivalence normal form is done within two swaps by extensions of the conjunctions by missing variables and possible simplifications:

$$\begin{aligned} f(x_1, x_2, x_3, x_4) &= x_1x_2 \oplus x_1x_3 \oplus x_1x_4 \oplus x_2x_3 \oplus x_2x_4 \oplus x_3x_4 \\ &= x_1x_2(x_3 \oplus \overline{x}_3) \oplus x_1x_3(x_2 \oplus \overline{x}_2) \oplus x_1x_4(x_2 \oplus \overline{x}_2) \oplus \\ &\quad\; x_2x_3(x_4 \oplus \overline{x}_4) \oplus x_2x_4(x_1 \oplus \overline{x}_1) \oplus x_3x_4(x_2 \oplus \overline{x}_2) \\ &= x_1x_2\overline{x}_3 \oplus x_1\overline{x}_2x_3 \oplus x_1\overline{x}_2x_4 \oplus x_2x_3\overline{x}_4 \oplus \overline{x}_1x_2x_4 \oplus \overline{x}_2x_3x_4 \\ &= x_1x_2\overline{x}_3(x_4 \oplus \overline{x}_4) \oplus x_1\overline{x}_2x_3(x_4 \oplus \overline{x}_4) \oplus x_1\overline{x}_2x_4(x_3 \oplus \overline{x}_3) \oplus \\ &\quad\; x_2x_3\overline{x}_4(x_1 \oplus \overline{x}_1) \oplus \overline{x}_1x_2x_4(x_3 \oplus \overline{x}_3) \oplus \overline{x}_2x_3x_4(x_1 \oplus \overline{x}_1) \\ &= x_1x_2\overline{x}_3x_4 \oplus x_1x_2\overline{x}_3\overline{x}_4 \oplus x_1\overline{x}_2x_3\overline{x}_4 \oplus x_1\overline{x}_2\overline{x}_3x_4 \oplus \\ &\quad\; x_1x_2x_3\overline{x}_4 \oplus \overline{x}_1x_2x_3\overline{x}_4 \oplus \overline{x}_1x_2x_3x_4 \oplus \overline{x}_1x_2\overline{x}_3x_4 \oplus \\ &\quad\; x_1\overline{x}_2x_3x_4 \oplus \overline{x}_1\overline{x}_2x_3x_4 \\ &= x_1x_2\overline{x}_3\overline{x}_4 \oplus x_1\overline{x}_2x_3\overline{x}_4 \oplus x_1\overline{x}_2\overline{x}_3x_4 \oplus \\ &\quad\; \overline{x}_1x_2\overline{x}_3x_4 \oplus \overline{x}_1x_2x_3\overline{x}_4 \oplus \overline{x}_1\overline{x}_2x_3x_4 \oplus \\ &\quad\; x_1x_2x_3\overline{x}_4 \oplus x_1x_2\overline{x}_3x_4 \oplus x_1\overline{x}_2x_3x_4 \oplus \overline{x}_1x_2x_3x_4 \\ &= S_4^2(\mathbf{x}) \oplus S_4^3(\mathbf{x}) = S_4^2(\mathbf{x}) \vee S_4^3(\mathbf{x}) \; . \end{aligned}$$

The function (2.156) is equal to 1 if either two or three of the four variables are equal to 1.

2.3 Using Definition 2.2 we get:

$$\begin{aligned} \frac{\partial f(\mathbf{x})}{\partial \mathbf{x}_0} \wedge \overline{\max_{\mathbf{x}_0} f(\mathbf{x})} &= \\ (f(\mathbf{x}_0, \mathbf{x}_1) \oplus f(\overline{\mathbf{x}}_0, \mathbf{x}_1)) \wedge \overline{(f(\mathbf{x}_0, \mathbf{x}_1) \vee f(\overline{\mathbf{x}}_0, \mathbf{x}_1))} &= \\ (f(\mathbf{x}_0, \mathbf{x}_1) \oplus f(\overline{\mathbf{x}}_0, \mathbf{x}_1)) \wedge \overline{f(\mathbf{x}_0, \mathbf{x}_1)} \wedge \overline{f(\overline{\mathbf{x}}_0, \mathbf{x}_1)} &= \\ f(\mathbf{x}_0, \mathbf{x}_1)\; \overline{f(\mathbf{x}_0, \mathbf{x}_1)}\; \overline{f(\overline{\mathbf{x}}_0, \mathbf{x}_1)} \oplus f(\overline{\mathbf{x}}_0, \mathbf{x}_1)\; \overline{f(\mathbf{x}_0, \mathbf{x}_1)}\; \overline{f(\overline{\mathbf{x}}_0, \mathbf{x}_1)} &= \\ 0 \oplus 0 = 0 \; . \end{aligned}$$

In each of the conjunctions a function and its negation occur; hence, both conjunctions are equal to 0, and the rule (2.18) is satisfied.

2.4 The Shannon decomposition is applied to the left-hand side of Equation (2.51) (vectorial minimum with regard to x_i) taken from Definition 2.2, and transformations using the rules to simplify Boolean expressions lead to the right-hand side taken from Definition 2.4 (single minimum with regard to x_i) as follows:

$$
\begin{aligned}
\min_{x_i} f(\mathbf{x}) &= f(x_i, \mathbf{x}_1) \wedge f(\overline{x}_i, \mathbf{x}_1) \\
&= \overline{x}_i(f(x_i = 0, \mathbf{x}_1) \wedge f(x_i = 1, \mathbf{x}_1)) \vee x_i(f(x_i = 1, \mathbf{x}_1) \wedge f(x_i = 0, \mathbf{x}_1)) \\
&= (\overline{x}_i \vee x_i)(f(x_i = 0, \mathbf{x}_1) \wedge f(x_i = 1, \mathbf{x}_1)) \\
&= f(x_i = 0, \mathbf{x}_1) \wedge f(x_i = 1, \mathbf{x}_1) \ .
\end{aligned}
$$

2.5 The five single derivatives with regard to one of the variables $x_1, \ldots, x_5$ have the following results:

$$
\begin{aligned}
\frac{\partial f(\mathbf{x})}{\partial x_1} &= (\overline{x}_2 x_3 \vee x_2 x_3 x_4 \vee x_2 \overline{x}_4 \vee \overline{x}_4 x_5 \vee \overline{x}_2 \overline{x}_3 \overline{x}_4) \oplus (\overline{x}_4 \overline{x}_5 \vee \overline{x}_4 x_5) \\
&= x_4(\overline{x}_2 x_3 \vee x_2 x_3) \oplus \overline{x}_4(\overline{x}_2 x_3 \vee x_2 \vee x_5 \vee \overline{x}_2 \overline{x}_3) \oplus \overline{x}_4 \\
&= x_3 x_4 \oplus \overline{x}_4 \oplus \overline{x}_4 \\
&= x_3 x_4 \neq 0 \ , \\
\frac{\partial f(\mathbf{x})}{\partial x_2} &= (\overline{x}_1 x_3 \vee x_1 \overline{x}_4 \overline{x}_5 \vee \overline{x}_4 x_5 \vee \overline{x}_1 \overline{x}_3 \overline{x}_4) \oplus (\overline{x}_1 x_3 x_4 \vee x_1 \overline{x}_4 \overline{x}_5 \vee \overline{x}_1 \overline{x}_4 \vee \overline{x}_4 x_5) \\
&= (\overline{x}_1 x_3 \vee x_1 \overline{x}_4 \overline{x}_5 \vee \overline{x}_4 x_5 \vee \overline{x}_1 \overline{x}_4) \oplus (\overline{x}_1 x_3 \vee x_1 \overline{x}_4 \overline{x}_5 \vee \overline{x}_1 \overline{x}_4 \vee \overline{x}_4 x_5) \\
&= 0 \ , \\
\frac{\partial f(\mathbf{x})}{\partial x_3} &= (x_1 \overline{x}_4 \overline{x}_5 \vee \overline{x}_1 x_2 \overline{x}_4 \vee \overline{x}_4 x_5 \vee \overline{x}_1 \overline{x}_2 \overline{x}_4) \oplus \\
&\quad (\overline{x}_1 \overline{x}_2 \vee \overline{x}_1 x_2 x_4 \vee x_1 \overline{x}_4 \overline{x}_5 \vee \overline{x}_1 x_2 \overline{x}_4 \vee \overline{x}_4 x_5) \\
&= \overline{x}_4(x_1 \vee \overline{x}_1 x_2 \vee x_5 \vee \overline{x}_1 \overline{x}_2) \oplus (\overline{x}_1 \overline{x}_2 \vee \overline{x}_1 x_2 \vee x_1 \overline{x}_4 \overline{x}_5 \vee \overline{x}_4 x_5) \\
&= \overline{x}_4 \oplus (\overline{x}_1 \vee \overline{x}_4) \\
&= \overline{x}_1 x_4 \neq 0 \ , \\
\frac{\partial f(\mathbf{x})}{\partial x_4} &= (\overline{x}_1 \overline{x}_2 x_3 \vee x_1 \overline{x}_5 \vee \overline{x}_1 x_2 \vee x_5 \vee \overline{x}_1 \overline{x}_2 \overline{x}_3) \oplus (\overline{x}_1 \overline{x}_2 x_3 \vee \overline{x}_1 x_2 x_3) \\
&= (\overline{x}_1 \overline{x}_2 \vee x_1 \vee x_2 \vee x_5) \oplus (\overline{x}_1 x_3) \\
&= 1 \oplus \overline{x}_1 x_3 \\
&= x_1 \vee \overline{x}_3 \neq 0 \ , \\
\frac{\partial f(\mathbf{x})}{\partial x_5} &= (\overline{x}_1 \overline{x}_2 x_3 \vee \overline{x}_1 x_2 x_3 x_4 \vee x_1 \overline{x}_4 \vee \overline{x}_1 x_2 \overline{x}_4 \vee \overline{x}_1 \overline{x}_2 \overline{x}_3 \overline{x}_4) \oplus \\
&\quad (\overline{x}_1 \overline{x}_2 x_3 \vee \overline{x}_1 x_2 x_3 x_4 \vee \overline{x}_1 x_2 \overline{x}_4 \vee \overline{x}_4 \vee \overline{x}_1 \overline{x}_2 \overline{x}_3 \overline{x}_4) \\
&= (\overline{x}_1 \overline{x}_2 x_3 \vee \overline{x}_1 x_2 x_3 x_4 \vee x_1 \overline{x}_4 \vee x_2 \overline{x}_4 \vee \overline{x}_1 \overline{x}_2 \overline{x}_4) \oplus (\overline{x}_1 x_3 \vee \overline{x}_4) \\
&= (\overline{x}_1 x_3 \vee \overline{x}_4) \oplus (\overline{x}_1 x_3 \vee \overline{x}_4) \\
&= 0 \ .
\end{aligned}
$$

The same results can be computed much faster using the XBOOLE-Monitor. The derivatives with regard to x_2 and x_5 are equal to 0; hence, the function (2.157) does not depend on these variables. A 2-fold maximum with regard to (x_2, x_5) generates the simplified function. Instead of the computation of two consecutive maxima with regard to one of these variables Equation (2.105) is used:

$$\begin{aligned} f(x_1, x_3, x_4) &= \max_{(x_2,x_5)}{}^2 f(x_1, x_2, x_3, x_4, x_5) = \bigvee_{(c_2,c_5)\in\mathbb{B}^2} f(x_1, x_2 = c_2, x_3, x_4, x_5 = c_5) \\ &= \overline{x}_1 x_3 \vee \overline{x}_1 x_3 x_4 \vee x_1 \overline{x}_4 \vee \overline{x}_1 \overline{x}_4 \vee \overline{x}_4 \vee \overline{x}_1 \overline{x}_3 \overline{x}_4 \\ &= \overline{x}_1 x_3 \vee \overline{x}_4 \,. \end{aligned}$$

The function (2.157) could be mapped from $\mathbb{B}^5$ to $\mathbb{B}^3$. The number of conjunctions in the disjunctive form is reduced form 6 to 2, and the number of literals even from 19 to 3.

2.6 The results of the derivatives with regard to x_1 are:

$$\begin{aligned} g_1(x_2, x_3) &= \frac{\partial f_1(x_1, x_2, x_3)}{\partial x_1} \\ &= (x_2 \oplus x_3) \oplus (1 \oplus x_2 \oplus x_3) \\ &= 1 \,, \\ g_2(x_2, x_3) &= \frac{\partial f_2(x_1, x_2, x_3)}{\partial x_1} \\ &= (x_2 \oplus x_3) \oplus (1 \oplus x_2 \oplus x_3 \oplus x_2) \\ &= \overline{x}_2 \,, \\ g_3(x_2, x_3) &= \frac{\partial f_3(x_1, x_2, x_3)}{\partial x_1} \\ &= (x_2 \oplus x_3) \oplus (1 \oplus x_2 \oplus x_3 \oplus x_2 x_3) \\ &= \overline{x}_2 \vee \overline{x}_3 \,. \end{aligned}$$

The related degrees of linearity are:

$$\begin{aligned} \mathbf{degree}_{x_1}^{lin} f_1(x_1, x_2, x_3) &= \frac{1}{2^{3-1}} * \left| \frac{\partial f_1(x_1, x_2, x_3)}{\partial x_1} \right| = \frac{4}{4} = 1 \,, \\ \mathbf{degree}_{x_1}^{lin} f_2(x_1, x_2, x_3) &= \frac{1}{2^{3-1}} * \left| \frac{\partial f_2(x_1, x_2, x_3)}{\partial x_1} \right| = \frac{2}{4} = 0.5 \,, \\ \mathbf{degree}_{x_1}^{lin} f_3(x_1, x_2, x_3) &= \frac{1}{2^{3-1}} * \left| \frac{\partial f_3(x_1, x_2, x_3)}{\partial x_1} \right| = \frac{3}{4} = 0.75 \,. \end{aligned}$$

2.7 This task can be solved by means of 2-fold derivative operations.

(a) The 2-fold derivative with regard to (x_4, x_5) is equal to 1 for the subspaces $(x_1, x_2, x_3) = (c_1, c_2, c_3)$ containing an odd number of function values 1. The following computation in one step uses (2.97) and Theorem 1.11 about the orthogonality of conjunctions:

$$\begin{aligned}
\frac{\partial^2 f(\mathbf{x})}{\partial x_4 \partial x_5} &= \bigoplus_{(c_4,c_5)\in\mathbb{B}^2} f(x_1, x_2, x_3, x_4 = c_4, x_5 = c_5) \\
&= (\overline{x}_2\overline{x}_3 \vee x_3) \oplus (x_1\overline{x}_2 \vee \overline{x}_1) \oplus (\overline{x}_2\overline{x}_3) \oplus (x_1\overline{x}_2 \vee \overline{x}_1\overline{x}_3) \\
&= \overline{x}_2\overline{x}_3 \oplus x_3 \oplus x_1\overline{x}_2 \oplus \overline{x}_1 \oplus \overline{x}_2\overline{x}_3 \oplus x_1\overline{x}_2 \oplus \overline{x}_1\overline{x}_3 \\
&= x_3 \oplus \overline{x}_1 \oplus \overline{x}_1\overline{x}_3 \\
&= x_1 x_3 \ .
\end{aligned}$$

(b) The 2-fold minimum with regard to (x_4, x_5) is equal to 1 for the subspaces $(x_1, x_2, x_3) = (c_1, c_2, c_3)$ containing only function values 1. The consecutive calculation using Definition (2.99) leads to:

$$\begin{aligned}
\min_{(x_4,x_5)}{}^2 f(\mathbf{x}) &= \min_{x_4}\left(\min_{x_5} f(\mathbf{x})\right) \ , \\
\min_{x_5} f(\mathbf{x}) &= (\overline{x}_2\overline{x}_3 \vee x_3\overline{x}_4) \wedge (x_1\overline{x}_2 \vee \overline{x}_1\overline{x}_3x_4 \vee \overline{x}_1\overline{x}_4) \\
&= x_1\overline{x}_2\overline{x}_3 \vee \overline{x}_1\overline{x}_2\overline{x}_3x_4 \vee \overline{x}_1\overline{x}_2\overline{x}_3\overline{x}_4 \vee x_1\overline{x}_2x_3\overline{x}_4 \vee \overline{x}_1x_3\overline{x}_4 \\
&= \overline{x}_2\overline{x}_3 \vee \overline{x}_2x_3\overline{x}_4 \vee \overline{x}_1x_3\overline{x}_4 \ , \\
\min_{(x_4,x_5)}{}^2 f(\mathbf{x}) &= (\overline{x}_2\overline{x}_3 \vee \overline{x}_2x_3 \vee \overline{x}_1x_3) \wedge (\overline{x}_2\overline{x}_3) \\
&= \overline{x}_2\overline{x}_3 \ .
\end{aligned}$$

(c) The 2-fold maximum with regard to (x_4, x_5) is equal to 1 for the subspaces $(x_1, x_2, x_3) = (c_1, c_2, c_3)$ containing at least one function value 1. Equation (2.105) is used for the computation within one step:

$$\begin{aligned}
\max_{(x_4,x_5)}{}^2 f(\mathbf{x}) &= \bigvee_{(c_4,c_5)\in\mathbb{B}^2} f(x_1, x_2, x_3, x_4 = c_4, x_5 = c_5) \\
&= (\overline{x}_2\overline{x}_3 \vee x_3) \vee (x_1\overline{x}_2 \vee \overline{x}_1) \vee (\overline{x}_2\overline{x}_3) \vee (x_1\overline{x}_2 \vee \overline{x}_1\overline{x}_3) \\
&= \overline{x}_1 \vee \overline{x}_2 \vee x_3 \ .
\end{aligned}$$

(d) The Δ-operation with regard to (x_4, x_5) is equal to 1 for the subspaces $(x_1, x_2, x_3) = (c_1, c_2, c_3)$ containing different function values 1. Using Definition (2.109) we get:

$$\begin{aligned}
\Delta_{(x_4,x_5)} f(\mathbf{x}) &= (\overline{x}_2\overline{x}_3) \oplus (\overline{x}_1 \vee \overline{x}_2 \vee x_3) \\
&= \overline{x}_2\overline{x}_3 \wedge x_1x_2\overline{x}_3 \vee (x_2 \vee x_3) \wedge (\overline{x}_1 \vee \overline{x}_2 \vee x_3) \\
&= \overline{x}_1x_2 \vee x_3 \ .
\end{aligned}$$

2.8 Instead of the $\oplus$-operation of (2.150) an $\wedge$-operation fitting to the m-fold minimum is desirable; such a transformation is possible using (1.56), (1.18), and (1.23).

$$
\begin{aligned}
\min_{\mathbf{x}_0}{}^m (g(\mathbf{x}_1) \oplus f(\mathbf{x}_0, \mathbf{x}_1)) &= \min_{\mathbf{x}_0}{}^m (\overline{g(\mathbf{x}_1)} f(\mathbf{x}_0, \mathbf{x}_1) \vee g(\mathbf{x}_1) \overline{f(\mathbf{x}_0, \mathbf{x}_1)}) \\
&= \min_{\mathbf{x}_0}{}^m \Big((\overline{g(\mathbf{x}_1)} \vee g(\mathbf{x}_1)) \wedge (\overline{g(\mathbf{x}_1)} \vee \overline{f(\mathbf{x}_0, \mathbf{x}_1)}) \wedge \\
&\qquad (f(\mathbf{x}_0, \mathbf{x}_1) \vee g(\mathbf{x}_1)) \wedge (f(\mathbf{x}_0, \mathbf{x}_1) \vee \overline{f(\mathbf{x}_0, \mathbf{x}_1)}) \Big) \\
&= \min_{\mathbf{x}_0}{}^m \Big((\overline{g(\mathbf{x}_1)} \vee \overline{f(\mathbf{x}_0, \mathbf{x}_1)}) \wedge (f(\mathbf{x}_0, \mathbf{x}_1) \vee g(\mathbf{x}_1)) \Big) .
\end{aligned}
$$

The right-hand side of this intermediate result can be transformed using (2.138) and (2.146):

$$
\begin{aligned}
\min_{\mathbf{x}_0}{}^m (g(\mathbf{x}_1) \oplus f(\mathbf{x}_0, \mathbf{x}_1)) &= \min_{\mathbf{x}_0}{}^m (\overline{g(\mathbf{x}_1)} \vee \overline{f(\mathbf{x}_0, \mathbf{x}_1)}) \wedge \min_{\mathbf{x}_0}{}^m (g(\mathbf{x}_1) \vee f(\mathbf{x}_0, \mathbf{x}_1)) \\
&= (\overline{g(\mathbf{x}_1)} \vee \min_{\mathbf{x}_0}{}^m \overline{f(\mathbf{x}_0, \mathbf{x}_1)}) \wedge (g(\mathbf{x}_1) \vee \min_{\mathbf{x}_0}{}^m f(\mathbf{x}_0, \mathbf{x}_1)) ,
\end{aligned}
$$

and the application of (1.19), (1.20), and (1.23) leads to the wanted result:

$$
\begin{aligned}
\min_{\mathbf{x}_0}{}^m (g(\mathbf{x}_1) \oplus f(\mathbf{x}_0, \mathbf{x}_1)) &= \overline{g(\mathbf{x}_1)} \wedge g(\mathbf{x}_1) \vee \overline{g(\mathbf{x}_1)} \wedge \min_{\mathbf{x}_0}{}^m f(\mathbf{x}_0, \mathbf{x}_1) \vee \\
&\qquad \min_{\mathbf{x}_0}{}^m \overline{f(\mathbf{x}_0, \mathbf{x}_1)} \wedge g(\mathbf{x}_1) \vee \min_{\mathbf{x}_0}{}^m \overline{f(\mathbf{x}_0, \mathbf{x}_1)} \wedge \min_{\mathbf{x}_0}{}^m f(\mathbf{x}_0, \mathbf{x}_1) \\
&= g(\mathbf{x}_1) \min_{\mathbf{x}_0}{}^m \overline{f(\mathbf{x}_0, \mathbf{x}_1)} \vee \overline{g(\mathbf{x}_1)} \min_{\mathbf{x}_0}{}^m f(\mathbf{x}_0, \mathbf{x}_1) .
\end{aligned}
$$

The conjunction $(\min_{\mathbf{x}_0}{}^m \overline{f(\mathbf{x}_0, \mathbf{x}_1)} \wedge \min_{\mathbf{x}_0}{}^m f(\mathbf{x}_0, \mathbf{x}_1))$ is equal to 0 due to (2.111) and because $(\overline{f(\mathbf{x}_0, \mathbf{x}_1)} \wedge f(\mathbf{x}_0, \mathbf{x}_1))$ is equal to 0.

2.9 (a) All conjunctions of the given expression (2.162) are orthogonal to each other.

(b) The TVLs of the given function $f(\mathbf{x})$ (2.162) and the calculated 3-fold derivative in ODA are:

$$
\text{ODA}\,(f(\mathbf{x}_0, \mathbf{x}_1)) = \begin{array}{ccccc} x_1 & x_2 & x_3 & x_4 & x_5 \\ \hline 1 & 0 & - & 1 & 0 \\ 0 & 0 & 1 & 0 & 1 \\ - & 1 & 0 & - & 1 \\ 0 & 1 & 0 & - & 0 \\ 1 & 0 & 1 & 0 & - \\ \hline \end{array} \longrightarrow \text{ODA}\,(g\,(\mathbf{x}_1)) = \begin{array}{cc} x_2 & x_4 \\ \hline 0 & 0 \\ 1 & - \\ \hline \end{array}
$$

The yellow rows are deleted due to the dash elements in the columns $\mathbf{x}_0 = (x_1, x_3, x_5)$ and, thereafter, these light blue columns are deleted.

(c) The 3-fold derivative is:

$$
\frac{\partial^3 f(x_1, x_2, x_3, x_4, x_5)}{\partial x_1 \partial x_3 \partial x_5} = x_2 \vee \overline{x}_4 .
$$

6.3 SOLUTIONS OF CHAPTER 3

3.1 (a) The application of the distributive law shows that the mark functions f_q (3.108) and f_r (3.109) satisfy the restriction (3.5) of a Boolean lattice:

$$\begin{aligned} f_q(\mathbf{x}) \wedge f_r(\mathbf{x}) &= (\overline{x}_1(x_4 \oplus x_5) \vee x_1 x_2(x_3 \oplus x_4) \vee \overline{x}_2(x_3 \oplus x_5)) \wedge \\ &\quad (x_1\overline{x}_2(\overline{x}_3 \oplus x_5) \vee \overline{x}_2(\overline{x}_3\overline{x}_4\overline{x}_5 \vee x_3x_4x_5)) \\ &= 0 \,. \end{aligned}$$

(b) The number of function values belonging to the don't-care function $f_\varphi(\mathbf{x})$ of this lattice of five variables is:

$$\left|f_\varphi(\mathbf{x})\right| = 2^5 - \left|f_q(\mathbf{x})\right| - \left|f_r(\mathbf{x})\right| = 32 - 18 - 6 = 8 \,.$$

Hence, this lattice defined by the lattice Equation (3.9) and the mark functions f_q (3.108) and f_r (3.109) contains

$$N_l = 2^8 = 256$$

functions.

(c) A lattice (3.9) can contain a function $f_i(\mathbf{x})$ that does not depend on the simultaneous change of the variables $\mathbf{x}_0$ only if

$$\left(\frac{\partial f_q(\mathbf{x})}{\partial \mathbf{x}_0} \vee \frac{\partial f_r(\mathbf{x})}{\partial \mathbf{x}_0}\right) \wedge \overline{\frac{\partial f_\varphi(\mathbf{x})}{\partial \mathbf{x}_0}} = 0 \,.$$

For the lattice with the mark functions f_q (3.108) and f_r (3.109) we get:

$$\begin{aligned} \frac{\partial f_q(\mathbf{x})}{\partial(x_3, x_4, x_5)} &= (\overline{x}_1(x_4 \oplus x_5) \vee x_1x_2(x_3 \oplus x_4) \vee \overline{x}_2(x_3 \oplus x_5)) \oplus \\ &\quad (\overline{x}_1(\overline{x}_4 \oplus \overline{x}_5) \vee x_1x_2(\overline{x}_3 \oplus \overline{x}_4) \vee \overline{x}_2(\overline{x}_3 \oplus \overline{x}_5)) \\ &= 0 \,, \\ \frac{\partial f_r(\mathbf{x})}{\partial(x_3, x_4, x_5)} &= (x_1\overline{x}_2(\overline{x}_3 \oplus x_5) \vee \overline{x}_2(\overline{x}_3\overline{x}_4\overline{x}_5 \vee x_3x_4x_5)) \oplus \\ &\quad (x_1\overline{x}_2(x_3 \oplus \overline{x}_5) \vee \overline{x}_2(x_3x_4x_5 \vee \overline{x}_3\overline{x}_4\overline{x}_5)) \\ &= 0 \,. \end{aligned}$$

Hence, the lattice defined by the lattice Equation (3.9) and the mark functions f_q (3.108) and f_r (3.109) contains at least one function that does not depend on the simultaneous change of the variables (x_3, x_4, x_5).

(d) The vectorial derivative in (3.110) requires that pairs of the eight function values 1 of $f_\varphi(\mathbf{x})$ must have the same function value. Hence, all $2^{8/2} = 2^4 = 16$ functions of the explored lattice do not depend on the simultaneous change of the variables (x_3, x_4, x_5).

3.2 (a) All vectorial derivatives

$$g_i(\mathbf{x}) = \frac{\partial f_i(\mathbf{x})}{\partial(x_1, x_3, x_4)}$$

belong to a lattice of Boolean functions specified by

$$f_q^{\partial \mathbf{x}_0}(\mathbf{x}_0, \mathbf{x}_1) \wedge \overline{g_i(\mathbf{x})} \vee g_i(\mathbf{x}) \wedge f_r^{\partial \mathbf{x}_0}(\mathbf{x}_0, \mathbf{x}_1) \vee \frac{\partial f(\mathbf{x})}{\partial(x_3, x_4, x_5)} \vee \frac{\partial f(\mathbf{x})}{\partial(x_1, x_3, x_4)} = 0 .$$

(b) Using (3.30) and (3.31) with $\mathbf{x}_0 = (x_1, x_3, x_4)$ we get:

$$\begin{aligned}
f_q^{\partial \mathbf{x}_0}(\mathbf{x}_0, \mathbf{x}_1) &= \max_{\mathbf{x}_0} f_q(\mathbf{x}_0, \mathbf{x}_1) \wedge \max_{\mathbf{x}_0} f_r(\mathbf{x}_0, \mathbf{x}_1) \\
&= ((\overline{x}_1(x_4 \oplus x_5) \vee x_1 x_2(x_3 \oplus x_4) \vee \overline{x}_2(x_3 \oplus x_5)) \vee \\
&\quad (x_1(\overline{x}_4 \oplus x_5) \vee \overline{x}_1 x_2(\overline{x}_3 \oplus \overline{x}_4) \vee \overline{x}_2(\overline{x}_3 \oplus x_5))) \wedge \\
&\quad ((x_1 \overline{x}_2(\overline{x}_3 \oplus x_5) \vee \overline{x}_2(\overline{x}_3 \overline{x}_4 \overline{x}_5 \vee x_3 x_4 x_5)) \vee \\
&\quad (\overline{x}_1 \overline{x}_2(x_3 \oplus x_5) \vee \overline{x}_2(x_3 x_4 \overline{x}_5 \vee \overline{x}_3 \overline{x}_4 x_5))) \\
&= \overline{x}_1 \overline{x}_2(\overline{x}_3 \overline{x}_4 x_5 \vee x_3 x_4 \overline{x}_5) \vee \overline{x}_1 \overline{x}_2(x_3 \oplus x_5) \vee \overline{x}_2(x_3 x_4 \overline{x}_5 \vee \overline{x}_3 \overline{x}_4 x_5) \vee \\
&\quad x_1 \overline{x}_2(\overline{x}_3 \overline{x}_4 \overline{x}_5 \vee x_3 x_4 x_5) \vee x_1 \overline{x}_2(\overline{x}_3 \oplus x_5) \vee \overline{x}_2(\overline{x}_3 \overline{x}_4 \overline{x}_5 \vee x_3 x_4 x_5) \\
&= \overline{x}_2(\overline{x}_1(x_3 \oplus x_5) \vee (\overline{x}_3 \oplus x_4) \vee x_1(\overline{x}_3 \oplus x_5)) ,
\end{aligned}$$

$$\begin{aligned}
f_r^{\partial \mathbf{x}_0}(\mathbf{x}_0, \mathbf{x}_1) &= \min_{\mathbf{x}_0} f_q(\mathbf{x}_0, \mathbf{x}_1) \vee \min_{\mathbf{x}_0} f_r(\mathbf{x}_0, \mathbf{x}_1) \\
&= ((\overline{x}_1(x_4 \oplus x_5) \vee x_1 x_2(x_3 \oplus x_4) \vee \overline{x}_2(x_3 \oplus x_5)) \wedge \\
&\quad (x_1(\overline{x}_4 \oplus x_5) \vee \overline{x}_1 x_2(\overline{x}_3 \oplus \overline{x}_4) \vee \overline{x}_2(\overline{x}_3 \oplus x_5))) \vee \\
&\quad ((x_1 \overline{x}_2(\overline{x}_3 \oplus x_5) \vee \overline{x}_2(\overline{x}_3 \overline{x}_4 \overline{x}_5 \vee x_3 x_4 x_5)) \wedge \\
&\quad (\overline{x}_1 \overline{x}_2(x_3 \oplus x_5) \vee \overline{x}_2(x_3 x_4 \overline{x}_5 \vee \overline{x}_3 \overline{x}_4 x_5))) \\
&= \overline{x}_1 x_2(\overline{x}_3 x_4 \overline{x}_5 \vee x_3 \overline{x}_4 x_5) \vee \overline{x}_1 \overline{x}_2(\overline{x}_3 x_4 \overline{x}_5 \vee x_3 \overline{x}_4 x_5) \vee \\
&\quad x_1 x_2(\overline{x}_3 x_4 x_5 \vee x_3 \overline{x}_4 \overline{x}_5) \vee x_1 \overline{x}_2(x_3 \overline{x}_4 \overline{x}_5 \vee \overline{x}_3 x_4 x_5) \\
&= \overline{x}_1(\overline{x}_3 x_4 \overline{x}_5 \vee x_3 \overline{x}_4 x_5) \vee x_1(\overline{x}_3 x_4 x_5 \vee x_3 \overline{x}_4 \overline{x}_5) .
\end{aligned}$$

(c) All functions of the given lattice depend on five variables. The number of function values 1 of the don't-care function of the lattice of the vectorial derivatives is:

$$\left| f_\varphi(\mathbf{x}) \right| = 2^5 - \left| f_q^{\partial \mathbf{x}_0}(\mathbf{x}_0, \mathbf{x}_1) \right| - \left| f_r^{\partial \mathbf{x}_0}(\mathbf{x}_0, \mathbf{x}_1) \right| = 32 - 12 - 8 = 12 .$$

All calculated vectorial derivatives are independent of the simultaneous change of both (x_3, x_4, x_5) and (x_1, x_3, x_4). Hence, all derivatives can be partitioned into subsets of 2 * 2 = 4 identical function values so that the lattice of the calculated vectorial derivatives contains

$$N_l = 2^{12/4} = 2^3 = 8$$

functions.

(d) The independence matrices of the given lattice $\mathrm{IDM}(f)$, the lattice of the calculated derivatives $\mathrm{IDM}(g)$ and the generated binary vectors are:

IDM(f) =

$i \backslash j$	1	2	3	4	5
1	0	0	0	0	0
2	0	0	0	0	0
3	0	0	1	1	1
4	0	0	0	0	0
5	0	0	0	0	0

IDM(g) =

$i \backslash j$	1	2	3	4	5
1	1	0	0	0	1
2	0	0	0	0	0
3	0	0	1	1	1
4	0	0	0	0	0
5	0	0	0	0	0

$$\mathbf{s}_0 = (1 \quad 0 \quad 1 \quad 1 \quad 0)$$
$$\mathbf{s}_{min} = (1 \quad 0 \quad 0 \quad 0 \quad 1).$$

Due to $\mathbf{s}_{min}$ all calculated vectorial derivative functions are also independent of the simultaneous change of (x_1, x_5).

(e) The number of values 1 in the main diagonal of the independence matrices determine the requested ranks as follows:

$$\mathbf{rank}(\mathrm{IDM}(f)) = 1 ,$$
$$\mathbf{rank}(\mathrm{IDM}(g)) = 2 .$$

6.4 SOLUTIONS OF CHAPTER 4

4.1 (a) The graph to engrave the character 2 is:

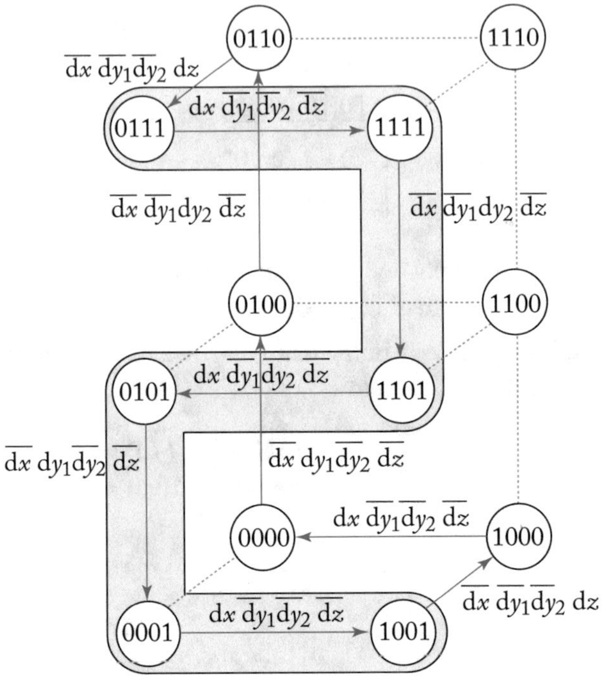

(b) The associated graph equation is:

$$\begin{aligned}
&\overline{x}\,\overline{y}_1\,\overline{y}_2\,\overline{z}\,\overline{\mathrm{d}x}\,\mathrm{d}y_1\,\overline{\mathrm{d}y_2}\,\overline{\mathrm{d}z} \vee \overline{x}\,y_1\,\overline{y}_2\,\overline{z}\,\overline{\mathrm{d}x}\,\overline{\mathrm{d}y_1}\,\mathrm{d}y_2\,\overline{\mathrm{d}z} \vee \\
&\overline{x}\,y_1\,y_2\,\overline{z}\,\overline{\mathrm{d}x}\,\overline{\mathrm{d}y_1}\,\overline{\mathrm{d}y_2}\,\mathrm{d}z \vee \overline{x}\,y_1\,y_2\,z\,\mathrm{d}x\,\overline{\mathrm{d}y_1}\,\overline{\mathrm{d}y_2}\,\overline{\mathrm{d}z} \vee \\
&x\,y_1\,y_2\,z\,\overline{\mathrm{d}x}\,\overline{\mathrm{d}y_1}\,\mathrm{d}y_2\,\overline{\mathrm{d}z} \vee x\,y_1\,\overline{y}_2\,z\,\mathrm{d}x\,\overline{\mathrm{d}y_1}\,\overline{\mathrm{d}y_2}\,\overline{\mathrm{d}z} \vee \\
&\overline{x}\,y_1\,\overline{y}_2\,z\,\overline{\mathrm{d}x}\,\mathrm{d}y_1\,\overline{\mathrm{d}y_2}\,\overline{\mathrm{d}z} \vee \overline{x}\,\overline{y}_1\,\overline{y}_2\,z\,\mathrm{d}x\,\overline{\mathrm{d}y_1}\,\overline{\mathrm{d}y_2}\,\overline{\mathrm{d}z} \vee \\
&x\,\overline{y}_1\,\overline{y}_2\,z\,\overline{\mathrm{d}x}\,\overline{\mathrm{d}y_1}\,\overline{\mathrm{d}y_2}\,\mathrm{d}z \vee x\,\overline{y}_1\,\overline{y}_2\,\overline{z}\,\mathrm{d}x\,\overline{\mathrm{d}y_1}\,\overline{\mathrm{d}y_2}\,\overline{\mathrm{d}z} \quad = 1\,.
\end{aligned}$$

4.2 Definition (4.39) of the 2-fold differential of $f(x_1, x_2)$ with regard to (x_1, x_2):

$$\mathrm{d}^2_{(x_1,x_2)} f(x_1, x_2) = \mathrm{d}_{x_2}(\mathrm{d}_{x_1} f(x_1, x_2))$$

can be expanded due to Definition (4.21) as:

$$\begin{aligned}
\mathrm{d}^2_{(x_1,x_2)} f(x_1, x_2) &= \mathrm{d}_{x_2}(f(x_1, x_2) \oplus f(x_1 \oplus \mathrm{d}x_1, x_2)) \\
&= f(x_1, x_2) \oplus f(x_1 \oplus \mathrm{d}x_1, x_2) \oplus \\
&\quad\; f(x_1, x_2 \oplus \mathrm{d}x_2) \oplus f(x_1 \oplus \mathrm{d}x_1, x_2 \oplus \mathrm{d}x_2)\,.
\end{aligned}$$

The application of the Shannon decomposition with regard to the differentials $\mathrm{d}x_1$ and $\mathrm{d}x_2$ leads to:

$$\begin{aligned}\mathrm{d}^2_{(x_1,x_2)} f(x_1,x_2) = {} & \overline{\mathrm{d}x_1}\,\overline{\mathrm{d}x_2}(f(x_1,x_2) \oplus f(x_1,x_2) \oplus f(x_1,x_2) \oplus f(x_1,x_2)) \oplus \\ & \overline{\mathrm{d}x_1}\mathrm{d}x_2(f(x_1,x_2) \oplus f(x_1,x_2) \oplus f(x_1,\overline{x}_2) \oplus f(x_1,\overline{x}_2)) \oplus \\ & \mathrm{d}x_1\overline{\mathrm{d}x_2}(f(x_1,x_2) \oplus f(\overline{x}_1,x_2) \oplus f(x_1,x_2) \oplus f(\overline{x}_1,x_2)) \oplus \\ & \mathrm{d}x_1\mathrm{d}x_2(f(x_1,x_2) \oplus f(\overline{x}_1,x_2) \oplus f(x_1,\overline{x}_2) \oplus f(\overline{x}_1,\overline{x}_2)) \; ,\end{aligned}$$

where the first three conjunctions are equal to 0, and the expression in parentheses of the last one can be transformed, using Definition (2.1), into the 2-fold derivative with regard to (x_1, x_2):

$$\begin{aligned}\mathrm{d}^2_{(x_1,x_2)} f(x_1,x_2) &= \left(\frac{\partial f(x_1,x_2)}{\partial x_1} \oplus \frac{\partial f(x_1,\overline{x}_2)}{\partial x_1}\right) \mathrm{d}x_1\mathrm{d}x_2 \\ &= \frac{\partial^2 f(x_1,x_2)}{\partial x_1 \partial x_2}\,\mathrm{d}x_1\mathrm{d}x_2 \; .\end{aligned}$$

6.5 SOLUTIONS OF CHAPTER 5

5.1 (a) The circuit is realizable because $F(\mathbf{x},\mathbf{y})$ (5.208) satisfies the condition

$$\max_{\mathbf{y}}{}^m F(\mathbf{x},\mathbf{y}) = 1 \ .$$

(b) The function $y_1 = f_1(\mathbf{x})$ is uniquely determined because $F(\mathbf{x},\mathbf{y})$ (5.208) satisfies the condition

$$\frac{\partial\left(\max_{y_2} F(\mathbf{x}, y_1, y_2)\right)}{\partial y_1} = 1 \ .$$

The function $y_2 = f_2(\mathbf{x})$ is not uniquely determined because

$$\frac{\partial\left(\max_{y_1} F(\mathbf{x}, y_1, y_2)\right)}{\partial y_2} = \overline{x}_1 x_3 \vee x_2 \neq 1 \ .$$

(c) The function $y_1 = f_1(\mathbf{x})$ can be calculated by

$$f_1(\mathbf{x}) = \max_{(y_1,y_2)}{}^2 \left(y_1 \wedge F(\mathbf{x},\mathbf{y})\right) = (x_1 \oplus x_2)x_3 \vee \overline{x}_2\overline{x}_3 \ .$$

The mark functions of the lattice of the functions $y_2 = f_2(\mathbf{x})$ are

$$f_{q2}(\mathbf{x}) = \overline{\max_{(y_1,y_2)}{}^2 \left(\overline{y}_2 \wedge F(\mathbf{x},\mathbf{y})\right)} = \overline{x}_1 x_3 \vee x_2\overline{x}_3 \ ,$$

$$f_{r2}(\mathbf{x}) = \overline{\max_{(y_1,y_2)}{}^2 \left(y_2 \wedge F(\mathbf{x},\mathbf{y})\right)} = x_1 x_2 x_3 \ .$$

A simple function of this lattice is

$$f_2(\mathbf{x}) = \overline{x}_1 \vee \overline{x}_3 \ .$$

5.2 Using (5.112) we get following test patterns.

For the Source g_4						For the Sink s_1						For the Sink s_2					
x_1	x_2	x_3	x_4	y	T	x_1	x_2	x_3	x_4	y	T	x_1	x_2	x_3	x_4	y	T
0	0	0	0	0	1	0	1	0	0	0	1	–	–	0	0	0	1
0	0	1	1	0	1	0	1	0	1	1	1	–	0	1	1	0	1
0	1	0	1	1	1	1	0	1	0	1	1	–	1	1	1	0	0
1	0	1	0	1	1	1	0	1	1	0	1						
1	–	0	1	1	1	1	–	0	1	1	1						
–	1	1	0	0	0	1	–	0	0	0	1						
						–	1	1	–	0	0						

The XBOOLE-Monitor is a very helpful tool for these calculations.

5.3 (a) The following strong bi-decompositions are possible.

OR-Bi-Decomposition	AND-Bi-Decomposition	EXOR-Bi-Decomposition
$x_1 - x_3$	$x_1 - x_2$	$x_1 - x_4$
$x_1 - x_4$	$x_1 - x_4$	$x_2 - x_4$
	$x_2 - x_3$	
	$x_3 - x_4$	

(b) Due to the results of the previous subtask the OR-bi-decomposition $x_1 - (x_3, x_4)$ must be verified; however, these dedicated sets of variables do not satisfy the condition (5.130).

The condition (5.131) of the AND-bi-decomposition is satisfied for the dedicated sets $\mathbf{x}_a = (x_1, x_3)$ and $\mathbf{x}_b = (x_2, x_4)$; hence, the given lattice with f_q (5.209) and f_r (5.210) contains a function that is disjoint AND-bi-decomposable.

Based on the results of the previous subtask the EXOR-bi-decomposition $(x_1, x_2) - x_4$ must be verified; however, these dedicated sets of variables do not satisfy the condition (5.132).

(c) Using (5.137) and (5.138) we get the mark functions $g_q(x_1, x_3)$ and $g_r(x_1, x_3)$ for the AND-bi-decomposition of f_q (5.209) and f_r (5.210) with regard to the dedicated sets $\mathbf{x}_a = (x_1, x_3)$ and $\mathbf{x}_b = (x_2, x_4)$ as

$$g_q(x_1, x_3) = x_1 \vee \overline{x}_3 \ ,$$
$$g_r(x_1, x_3) = \overline{x}_1 x_3 \ .$$

Hence, the decomposition function $g(x_1, x_3) = x_1 \vee \overline{x}_3$ is uniquely specified.

Using (5.139) and (5.140) we get the associated mark functions $h_q(x_2, x_4)$ and $h_r(x_2, x_4)$ of this AND-bi-decomposition as

$$h_q(x_2, x_4) = \overline{x}_2 \oplus x_4 \ ,$$
$$h_r(x_2, x_4) = x_2 \oplus x_4 \ .$$

Hence, the decomposition function $h(x_2, x_4) = \overline{x}_2 \oplus x_4$ is also uniquely specified, and we get the circuit structure of Figure 6.1.

5.4 (a) A function $f(x_i, \mathbf{x}_1)$ can be represented in the OR-separated form

$$f(x_i, \mathbf{x}_1) = x_i \vee h(\mathbf{x}_1)$$

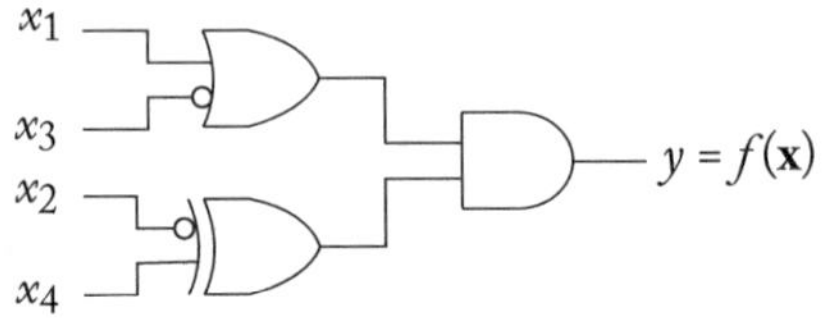

Figure 6.1: Disjoint AND-bi-decomposition.

if the condition

$$\min_{x_i} (\overline{x}_i \vee f(x_i, \mathbf{x}_1)) = 1$$

is satisfied. The substitution of the OR-separated form of $f(x_i, \mathbf{x}_1)$ into this condition confirms its correctness:

$$\begin{aligned}
\min_{x_i} (\overline{x}_i \vee f(x_i, \mathbf{x}_1)) &= 1 \\
\min_{x_i} (\overline{x}_i \vee x_i \vee h(\mathbf{x}_1)) &= 1 \\
\min_{x_i} (1 \vee h(\mathbf{x}_1)) &= 1 \\
\min_{x_i} (1) &= 1 \\
1 &= 1 \,.
\end{aligned}$$

(b) The OR-separation function $h(\mathbf{x}_1)$ can be calculated by

$$h(\mathbf{x}_1) = \max_{x_i} (\overline{x}_i \wedge f(x_i, \mathbf{x}_1)) \ .$$

The substitution of the OR-separated form of $f(x_i, \mathbf{x}_1)$ into this formula followed by equivalent transformations also confirms this formula:

$$\begin{aligned}
h(\mathbf{x}_1) &= \max_{x_i} (\overline{x}_i \wedge f(x_i, \mathbf{x}_1)) \\
h(\mathbf{x}_1) &= \max_{x_i} (\overline{x}_i \wedge (x_i \vee h(\mathbf{x}_1))) \\
h(\mathbf{x}_1) &= \max_{x_i} (0 \vee \overline{x}_i \wedge h(\mathbf{x}_1)) \\
h(\mathbf{x}_1) &= \max_{x_i} (\overline{x}_i \wedge h(\mathbf{x}_1)) \\
h(\mathbf{x}_1) &= h(\mathbf{x}_1) \wedge \max_{x_i} (\overline{x}_i) \\
h(\mathbf{x}_1) &= h(\mathbf{x}_1) \wedge 1 \\
h(\mathbf{x}_1) &= h(\mathbf{x}_1) \,.
\end{aligned}$$

Bibliography

J. S. B. Akers. On a theory of Boolean functions. *Journal of the Society for Industrial and Applied Mathematics (SIAM)*, 7(4):487–498, 1959. DOI: 10.1137/0107041.

Dieter Bochmann. *Binary Systems—A BOOLEAN Book*. TUDpress, Verlag der Wissenschaft GmbH, Dresden, Germany, 2008.

Dieter Bochmann and Christian Posthoff. *Binäre Dynamische Systeme*. Berlin, Akademie-Verlag, Oldenbourg, Germany, 1981.

George Boole. *An Investigation of the Laws of Thought*. Watchmaker Publishing, London, 1854, Reprint: 2010. DOI: 10.1017/cbo9780511693090.

Marco Davio, Jean-Pierre Deschamps, and André Thayse. *Discrete and Switching Functions*. McGraw-Hill International, 1978.

David A. Huffman. Solvability criterion for simultaneous logical equations. *Quarterly Progress Report*, 1(56):87–88, 1958.

Trung Quoc Le. *Testability of Combinational Circuits—Theory and Design*. Ph.D. thesis, University of technology Karl-Marx-Stadt, 1989. Original German title: Testbarkeit kombinatorischer Schaltungen—Theorie und Entwurf.

Alan Mishchenko, Bernd Steinbach, and Marek Perkowski. An algorithm for bi-decomposition of logic functions. In *Proc. of the 38th Design Automation Conference*, pages 103–108, Las Vegas, Nevada, 2001. DOI: 10.1145/378239.378353.

Christian Posthoff and Bernd Steinbach. *Logic Functions and Equations—Binary Models for Computer Science*. Springer, Dordrecht, The Netherlands, 2004.

I. S. Reed. A class of multiple-error-correcting codes and the decoding scheme. *Transactions of the IRE Professional Group on Information Theory*, 4(4):38–49, 1954. DOI: 10.1109/TIT.1954.1057465.

Tsutomu Sasao and Jon T. Butler. Progress in applications of Boolean functions, *Synthesis Lecturers on Digital Circuits and Systems*. Morgan & Claypool Publishers, San Rafael, CA, 2010.

Bernd Steinbach. *Solution of Boolean Differential Equations and their Application to Binary Systems*. Ph.D. thesis, TH Karl-Marx-Stadt, 1981. Original German title: Lösung binärer Differentialgleichungen und ihre Anwendung auf binäre Systeme.

Bernd Steinbach. XBOOLE—a toolbox for modelling, simulation, and analysis of large digital systems. *Systems Analysis and Modelling Simulation*, 9(4):297–312, 1992.

Bernd Steinbach. Generalized lattices of Boolean functions utilized for derivative operations. *Materialy konferencyjne KNWS'13*, pages 1–17, 2013. DOI: 10.13140/2.1.1874.3680.

Bernd Steinbach. Vectorial bi-decompositions of logic functions. In *Proc. Reed-Muller Workshop*, RM, pages 1–10, Waterloo, Canada, 2015.

Bernd Steinbach. Relationships between vectorial bi-decompositions and strong exor-bi-decompositions. In *Proc. of the 25th International Workshop on Post-Binary ULSI Systems*, page 44, Sapporo, Hokkaido, Japan, 2016.

Bernd Steinbach and Christian Posthoff. *Logic Functions and Equations—Examples and Exercises*. Springer Science + Business Media B.V., 2009. DOI: 10.1007/978-1-4020-9595-5_3.

Bernd Steinbach and Christian Posthoff. Boolean differential calculus—theory and applications. *Journal of Computational and Theoretical Nanoscience*, 7(6):933–981, 2010a. ISSN 1546-1955. DOI: 10.1166/jctn.2010.1441.

Bernd Steinbach and Christian Posthoff. Boolean differential calculus. In Tsutomu Sasao and Jon T. Butler, Eds., *Progress in Applications of Boolean Functions*, pages 55–78, 121–126. Morgan & Claypool Publishers, San Rafael, CA, 2010b. DOI: 10.1007/978-1-4020-2938-7_4.

Bernd Steinbach and Christian Posthoff. *Boolean Differential Equations*. Morgan & Claypool Publishers, San Rafael, CA, 2013a. DOI: 10.2200/S00511ED1V01Y201305DCS042.

Bernd Steinbach and Christian Posthoff. Derivative operations for lattices of Boolean functions. In *Proc. Reed-Muller Workshop*, pages 110–119, Toyama, Japan, 2013b. DOI: 10.13140/2.1.2398.6568.

Bernd Steinbach and Christian Posthoff. Boolean differential equations—a common model for classes, lattices, and arbitrary sets of Boolean functions. *Facta Universitatis Niš*, 28:51–76, 2015. DOI: 10.2298/FUEE1501051S.

Bernd Steinbach and Christian Posthoff. Vectorial bi-decompositions for lattices of Boolean functions. In Bernd Steinbach, Ed., *Proc. of the 12th International Workshops on Boolean Problems*, IWSBP, pages 93–104, Freiberg, Germany, 2016.

Bernd Steinbach and Mathias Stöckert. Design of fully testable circuits by functional decomposition and implicit test pattern generation. In *Proc. of the 12th IEEE VLSI Test Symposium*, pages 22–27, 1994. DOI: 10.1109/VTEST.1994.292339.

André Thayse. *Boolean Calculus of Differences*. Springer, Berlin, Germany, 1981. DOI: 10.1007/3-540-10286-8.

Svetlana N. Yanushkevich. *Logic Differential Calculus in Multi-Valued Logic Design*. Ph.D. thesis, Tech. University of Szczecin, Szczecin, Poland, 1998.

Arkadij D. Zakrevskij. *Logic Equations*. Наука и Техника, Минск, 1975. In Russian (Логические Уравнения).

Authors' Biographies

BERND STEINBACH

From 1973–1977, Bernd Steinbach studied Information Technology at the University of Technology in Chemnitz (Germany) and graduated with an M.Sc. in 1977. He graduated with a Ph.D. and with a Dr. sc. techn. (Doctor scientiae technicarum) for his second doctoral thesis from the Faculty of Electrical Engineering of the Chemnitz University of Technology in 1981 and 1984, respectively. In 1991, Steinbach obtained the habilitation (Dr.-Ing. habil.) from the same faculty. Topics of his theses involved Boolean equations, Boolean differential equations, and their application in the field of circuit design using efficient algorithms and data structures on computers.

Steinbach worked in industry as an electrician, where he had tested professional controlling systems at the Niles Company. After his studies he taught as Assistant Lecturer at the Department of Information Technology of the Chemnitz University of Technology. As a research engineer he developed programs for test pattern generation for computer circuits at the company Robotron. He later returned to the Department of Information Technology of the Chemnitz University of Technology as Associate Professor for design automation in logic design.

Since 1992 he has worked as a Full Professor of Computer Science/Software Engineering and Programming at the Freiberg University of Mining and Technology, Department of Computer Science. He has served as Head of the Department of Computer Science and Vice-Dean of the Faculty of Mathematics and Computer Science.

His research areas include logic functions and equations and their application in many fields, such as artificial intelligence, UML-based testing of software, and UML-based hardware/software co-design. He is the head of a group that developed the XBOOLE software system.

He published three books about logic synthesis. The first one (together with D. Bochmann) covers *Logic Design using XBOOLE* (in German), Technik 1991. The following two, co-authored by Christian Posthoff, are *Logic Functions and Equations—Binary Models for Computer Science* and *Logic Functions and Equations—Examples and Exercises*, Springer 2004, and 2009, respectively. As one application of the Boolean Differential Calculus, he co-authored another book with Christian Posthoff, *Boolean Differential Equations*, Morgan & Claypool Publishers 2013. He is the editor and co-author of several sections of the books *Recent Problems in the Boolean Domain* and *Problems and New Solutions in the Boolean Domain*, both of which were published by Cambridge Scholars Publishing in 2014 and 2016, respectively. Again co-authored by Christian Posthoff, he published three textbooks in German: *Logic Functions—Boolean Models*, *Efficient Calculations Using XBOOLE*, and *Java Programming for Beginners* EAGLE 2014, 2015, and 2016. He published more than 250 chapters in books, complete issues of journals, and papers in journals and proceedings.

He has served as Program Chairman for the IEEE International Symposium on Multiple-Valued Logic (ISMVL), and as guest editor of the *Journal of Multiple-Valued Logic and Soft Computing*. He is the initiator and general chair of a biennial series of International Workshops on Boolean Problems (IWSBP) which started in 1994, now with 12 workshops.

He received the Barkhausen Award from the University of Technology Dresden in 1983.

CHRISTIAN POSTHOFF

From 1963–1968, Christian Posthoff studied Mathematics at the University of Leipzig. His thesis was titled: "Axiomatic Description of a Finite Class Calculus" (Prof. Dr. Klaua). From 1968-1972, he worked as a programmer and in the field of Operations Research; simultaneously, he did his Ph.D. in 1975 with the thesis "Application of Mathematical Methods in Communicative Psychotherapy."

In 1972, he joined the Department of Information Technology at the Chemnitz University of Technology; until 1983, his research activities concentrated on logic design, particularly on numerical methods for Boolean problems. His cooperation with B. Steinbach goes back to these days. Important results have been algorithms and programs for solving Boolean equations with a high number of variables and the Boolean Differential Calculus for the analytical treatment of different problems in the field of logic design.These results have been collected in a monograph *Binary Dynamic Systems* published simultaneously in Akademie-Verlag Berlin, Oldenbourg-Verlag Munich-Vienna and in the Soviet Union, and allowed the habilitation (Dr.-Ing. habil.) at the Faculty of Electrical Engineering in 1979 and the promotion to Associate Professor. He wrote two textbooks at this time, aimed at a higher level in the theoretical and mathematical training of graduate engineers of information technology. About 1976, he started research activities in Artificial Intelligence.

In 1983, Posthoff started as Full Professor of Computer Science in the Department of Computer Science at the same university, with the aim of starting the program in Computer Science in 1984. In 1984, he became Head of the Institute of Theoretical Computer Science and Artificial Intelligence and Research Director of the Department of Computer Science. An independent direction of research activities within AI, investigations of computer chess and other strategic games, arose from his love for chess. His research activities concentrated on the application of fuzzy logic for the modeling of human-like "thinking" methods, the learning from examples, the construction of intelligent tutoring systems, the application of inference mechanisms, the construction of systems for diagnosis, and configuration. In cooperation with colleagues from different areas of mechanical engineering and medicine, he has been supervising the construction of several expert systems. He received the Scientific Award of the Chemnitz University of Technology four times.

In 1994, he moved to the Chair of Computer Science at the University of The West Indies, St. Augustine, Trinidad & Tobago. From 1996–2002 he was Head of the Department of Mathematics & Computer Science. His main focus was the development of Computer Science education at the undergraduate and graduate level to attain international standard. In 2001, he received the Vice-Chancellor's Award of Excellence.

He is the author or co-author of 19 books and many publications in journals and conference proceedings.

Index

Printed in the United States
by Baker & Taylor Publisher Services